# Biomechanics of Tendons and Ligaments

Woodhead Publishing Series in Biomaterials

# Biomechanics of Tendons and Ligaments

## Tissue Reconstruction and Regeneration

*Johanna Buschmann*
*and*
*Gabriella Meier Bürgisser*

University Hospital Zurich, Zurich, Switzerland

ELSEVIER

WP
WOODHEAD PUBLISHING
An imprint of Elsevier • elsevier.com

Woodhead Publishing is an imprint of Elsevier
The Officers' Mess Business Centre, Royston Road, Duxford, CB22 4QH, United Kingdom
50 Hampshire Street, 5th Floor, Cambridge, MA 02139, United States
The Boulevard, Langford Lane, Kidlington, OX5 1GB, United Kingdom



**Library of Congress Cataloging-in-Publication Data**
A catalog record for this book is available from the Library of Congress

**British Library Cataloguing-in-Publication Data**
A catalogue record for this book is available from the British Library

ISBN: 978-0-08-100489-0 (print)
ISBN: 978-0-08-100492-0 (online)

For information on all Woodhead publications visit our website at https://www.elsevier.com/



*Publisher:* Matthew Deans
*Acquisition Editor:* Laura Overend
*Editorial Project Manager:* Lucy Beg
*Production Project Manager:* Poulouse Joseph
*Cover Designer:* Mark Rogers

Typeset by SPi Global, India

To Lea and Livio.

# Contents

# Part One

# Fundamentals and biomechanics of tendons and ligaments

# Structure and function of tendon and ligament tissues

**1**

## Abstract

Tendons and ligaments are connective tissues that serve as the force transmitting entities and enable musculoskeletal motion. Typical features of normal tendon tissue are parallel-aligned collagen I fibers and tenocytes. Moreover, the extracellular matrix (ECM) is composed of proteoglycans, glycoproteins, and elastin. The tissue has almost no vessels and the nutrition as well as oxygen are supplied at the vascularized myotendinous and osteotendinous junctions. Growth factors such as transforming growth factor beta are important for tendon development, homeostasis, and regeneration. Structural changes upon aging and tendinopathy include the extent of vascularization (aging causes less tendinopathy and more vascularization), the ECM (age-related lower collagen content and in tendinopathy collagen disorganization), and the proteoglycan content (older tendons having less, tendinopathic tendons more proteoglycans), which will be addressed in detail in this chapter.

**Keywords:** Tendon cells, Collagen, Extracellular matrix, Elastin, Growth factors.

## Abbreviations

| | |
|---|---|
| **ACL** | anterior cruciate ligament (tendons and ligaments) |
| **ADAM** | a disintegrin and metalloproteinase (enzyme in living cells) |
| **ADAMTS** | a disintegrin and metalloproteinase with thromospondin motifs (enzyme in living cells) |
| **AGEs** | advanced glycation end products (substances in degenerative diseases) |
| **AT** | Achilles tendon (tendons and ligaments) |
| **C** | calcaneus (elements of the foot ankle) |
| **CD34** | hematopoietic progenitor cell antigen (antigen protein) |
| **CDET** | common digital extensor tendon (tendon and ligaments) |
| **COMP** | collagen oligomeric matrix protein (protein in soft tissues) |
| **COX2** | cyclo-oxigenase2 (enzyme in living cells) |
| **CT** | connective tissue (elements of the body) |
| **CTGF** | connective tissue growth factor (protein in living cells) |
| **ECM** | extra cellular matrix (matrix in living cells) |
| **EH** | extensor hood (elements of the body) |
| **ET** | extensor tendons (tendons and ligaments) |
| **FACIT** | fibril-associated collagen with interrupted triple-helix (protein in soft tissues) |
| **FDL** | flexor *digitorum longus* (tendon and ligaments) |
| **FDP** | flexor *digitorum profundus* (tendons and ligaments) |
| **FDS** | flexor *digitorum superficialis* (tendon and ligaments) |
| **FHL** | flexor *hallucis longus* (tendons and ligaments) |

**Biomechanics of Tendons and Ligaments. http://dx.doi.org/10.1016/B978-0-08-100489-0.00001-6**

| | |
|---|---|
| **FR** | flexor *retinaculum* (tendons and ligaments) |
| **FT** | flexor tendons (tendons and ligaments) |
| **FT (also FDP)** | flexor *digitorum profundus* tendon (tendons and ligaments) |
| **GAGs** | glycosaminoglycans (called also mucopolysaccharide) (polysaccharides in living cells) |
| **GDF (-*n*)** | growth/differentiation factor *n* (protein in living cells) |
| **HGF** | hepatocyte growth factor (protein in living cells) |
| **IFM** | interfascicular matrix (called also endotenon) (element in tendons and ligaments) |
| **IGF (-*n*)** | insulin-like growth factor *n* (protein in living cells) |
| **IPJ** | interphalangeal joints (elements of the foot) |
| **JT** | *juncturae tendinum* (elements of the body) |
| **KP** | Kager's fat pad (also known as the precalcaneal fat pad or pre-Achilles fat pad) (fat pad in the food ankle) |
| **L** | lumbricals: intrinsic muscles of the hand (also present in the foot) (elements of the body) |
| **MRI** | magnetic resonance imaging (imaging technique) |
| **mRNA** | messenger ribonucleic acid (element of living cells) |
| **MSC** | mesenchymal stem cell (element of the body) |
| **MTJ** | metatarsophalangeal joint (elements of the foot) |
| **nm** | nano meter (measure of length) |
| **NMDAR1** | *N*-methyl-D-aspartate receptor (receptor protein) |
| **PDGF** | platelet-derived growth factor (protein in living cells) |
| **PDGF-R** | platelet-derived growth factor receptors (receptor for growth factor) |
| **PLCL** | poly(L-lactide-*co*-ε-caprolactone) (synthetic polymer for tissue engineering) |
| **ROCK** | rho-associated protein kinase (enzyme in living cells) |
| **Scx** | scleraxis (transcription factor protein) |
| **SDFTs** | superficial digital flexor tendons (tendon and ligaments) |
| **SLCs** | synovial lining cells (cell type of the body) |
| **SLRP** | small leucine rich proteoglycans (glycosylated proteins in living cells) |
| **ST** | superior tuberosity (part of the calcaneus) |
| **TGF (-β*n*)** | transforming growth factor beta *n* (protein in living cells) |
| **TIMPs** | tissue inhibitors of metalloproteinase (enzyme in living cells) |
| **US** | ultrasound/ultrasonography (imaging technique) |
| **VEGF** | vascular endothelial growth factor (protein in living cells) |

## 1.1 Introduction

The main function of tendons and ligaments is to transfer force from muscle to bone (in tendons) or from bone to bone (in ligaments) in order to provoke motion. In the hand and foot, tendon networks occur associated by the links between each other; the concept of "supertendons" was proposed to describe the fact that such networks exhibit a functional range exceeding that of its individual members (Benjamin, 2010).

In this chapter, tendon and ligament structures and functions are presented for three states: the normal healthy state, the aging state, and the tendino-pathological state.

Nourissat et al. (2013) have nicely put the main structural features of these three states together (Fig. 1.1). As such, the main differences are found in terms of collagen fiber organization and morphology, vascularization, and cell density as well as cell morphology. Moreover, also the proteins in the extracellular matrix (ECM) do change from normal to aging and degenerative state of tendon and ligament tissue.

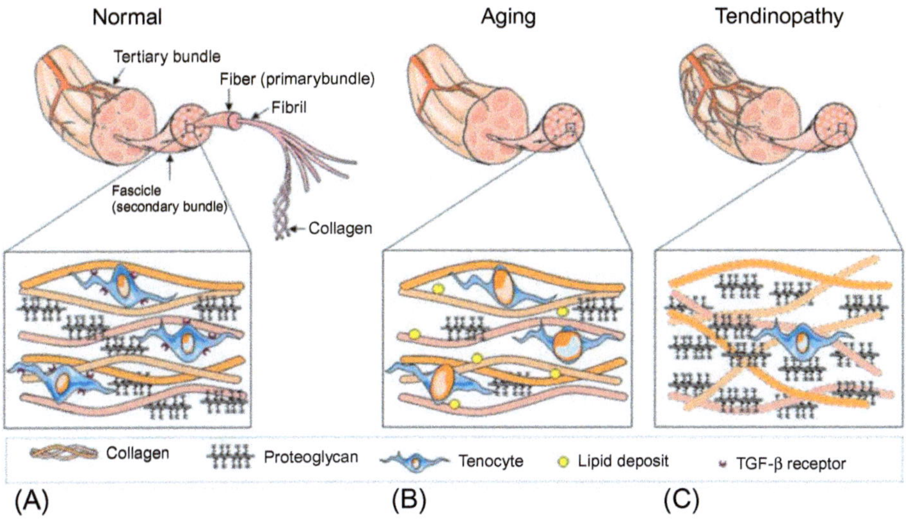


**Fig. 1.1** Structural changes of tendons from normal (A) through aging (B) and chronic tendinopathy (C). Normal tendon is a well-organized network of collagen fibrils. Collagen is arranged in hierarchical levels of increasing complexity, beginning with tropocollagen, a triple-helix polypeptide chain, which unites into fibrils, fibers (primary bundles), fascicles (secondary bundles), tertiary bundles, and the tendon itself. The extra cellular matrix (ECM) is dense, with a fibrillar network of predominantly parallel-aligned collagen fibers, principally consisting of 95% type I collagen. In addition, the ECM is composed of proteoglycans, glycosaminoglycans, and glycoproteins. The synthesis of the tendon ECM is under the control of two main growth factors: TGF-β (transforming growth factor beta) and FGF (fibroblast growth factor). Blood supply is necessary for nutrition and healing of the tendon and come from two places. The first source of blood supply is the extremity of the tendon; the second source is the peripheral zone of the tendon, called the peritendon and synovial sheath. With aging, tenocytes decrease in volume, becoming longer, slender, with an increased nucleus-to-cytoplasm ratio, and produce less ECM, but with increase in type I collagen volume density (mostly by less degradation). Deposit of lipid is routinely seen in aging tendon. TGF-β receptors disappear from tenocytes membranes. Tendon blood flow decreases with increasing age. In chronic tendinopathy, histological examination shows intra-tendinous collagen degeneration with fiber disorientation and glycosaminoglycans accumulation in between thinning fibrils without inflammatory cells or inflammatory signs. Tenocytes look normal but decrease in number. Hypervascularization is frequently found.
Based on Nourissat, G., Houard, X., Sellam, J., Duprez, D., Berenbaum, F., 2013. Use of autologous growth factors in aging tendon and chronic tendinopathy. Front. Biosci. E5, 911–921, reprinted by permission from Frontiers in Bioscience, copyright 2013.

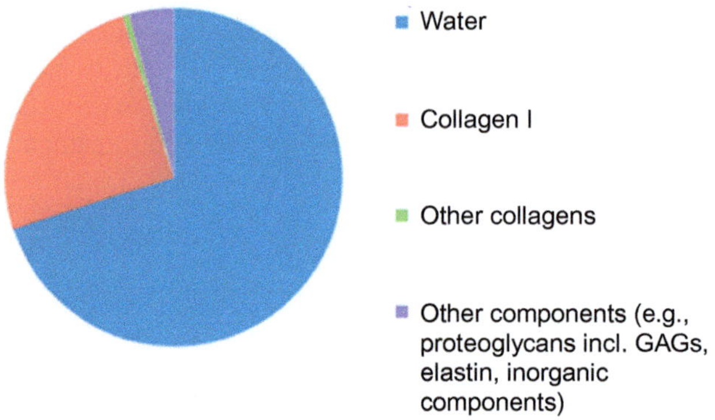


**Fig. 1.2** Composition of tendon tissue. Note: *GAG*, glycosaminoglucans. Data taken from literature mentioned in the text.

Briefly, healthy tendon tissue is a connective tissue composed of collagen fibers densely packed in an amorphous ground substance, including water (60–80% of the total wet weight), collagen (65–86% of the dry weight, mostly type I collagen 95–98%), proteoglycans (1–5%), elastin (1–2%), and 0.2% inorganic components (Fig. 1.2; Kannus, 2000; Riley, 2005; Lin et al., 2004; Jozsa and Kannus, 1997; Woo et al., 2000). Also tendon cells occur in parallel arranged lines (Kahn et al., 2013). Aging tendon tissue differs from healthy tissue in terms of tendon cell morphology, turning towards more slender tenocytes with larger nuclei at older ages. As for the vascularization, it is diminished and there are also fat deposits in the connective tissue. The tendinopathic tendon finally is much more vascularized than the normal tendon with disorganized collagen fibers, and the ECM is enriched with proteoglycans.

## 1.2 Anatomy

Tendon and ligament anatomy has been reviewed and reported frequently. For example, tendons in the hand include extensor and flexor tendons (FTs). While extensor tendons are flattened and have a slightly aponeurotic character, FTs are rather roundish or oval (Fig. 1.3; Benjamin et al., 2008). The longest tendons are those of the hand and the feet. In the foot, the tendons not only serve as force transmitters of muscle contraction to the bones, but also as modulators of the speed at which the distal limbs can be moved. This is achieved by their attachment to sites that are nearer or further away from the center of rotation—meaning the point through which the axis of movement passes (Benjamin et al., 2008).

Typically, bony surfaces are acting as pulleys at the attachment sites of many tendons. As an example for a pulley, the Achilles tendon's (AT) attachment to the calcaneus with the superior tuberosity acting as a pulley during dorsiflexion is shown in Fig. 1.4.

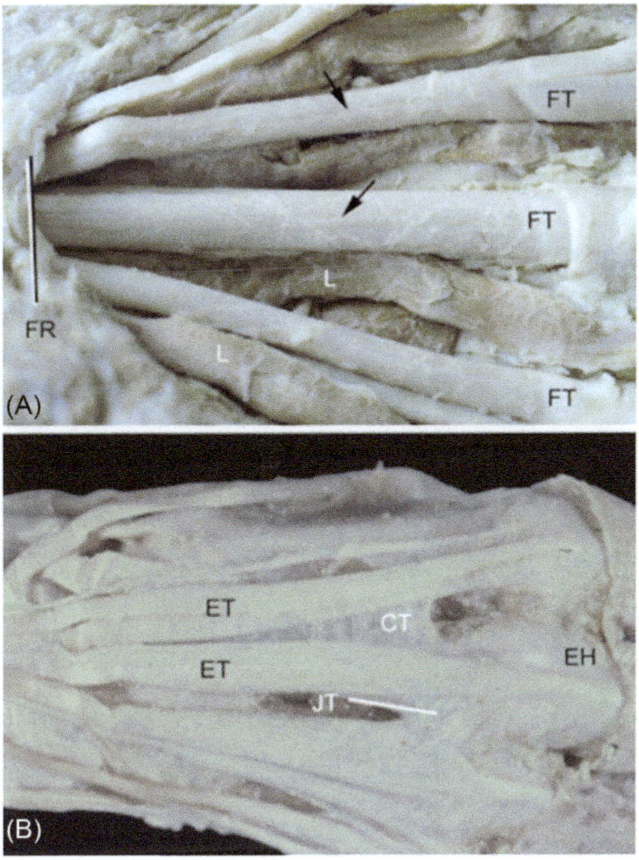


**Fig. 1.3** The gross anatomy of tendons in the hand. (A) The *flexor digitorum superficialis tendons* (FT) emerging from beneath the *flexor retinaculum* (FR) to enter the palm of the hand. Note their rounded character and the shallow grooves that are occasionally evident on their surface *(arrows)*. *L*, Lumbricals (intrinsic muscles of the hand, also found in the foot). (B) The web of extensor tendons (ET) on the dorsum of the hand collectively form a "supertendon" complex in which the individual components are interconnected by films of connective tissue (CT) and obliquely-orientated *juncturae tendinum* (JT). Note the extensor hood (EH) over the metacarpophalangeal joints.
Based on Benjamin, M., Kaiser, E., Milz, S., 2008. Structure-function relationships in tendons: a review. J Anat. 212, 211–228, © by Journal of Anatomy with permission from Wiley.

Some anatomical studies also compare human and animal anatomy. Doherty et al. reported in a magnetic resonance imaging (MRI) and US based investigation the similarities and differences of human and rabbit ATs. The rabbit flexor *digitorum superficialis* (FDS) tendon which corresponds to the human flexor *digitorum longus* (FDL) and is comparable in size with the gastrocnemius tendons, was located anterior to the medial gastrocnemius tendon proximally and rotated

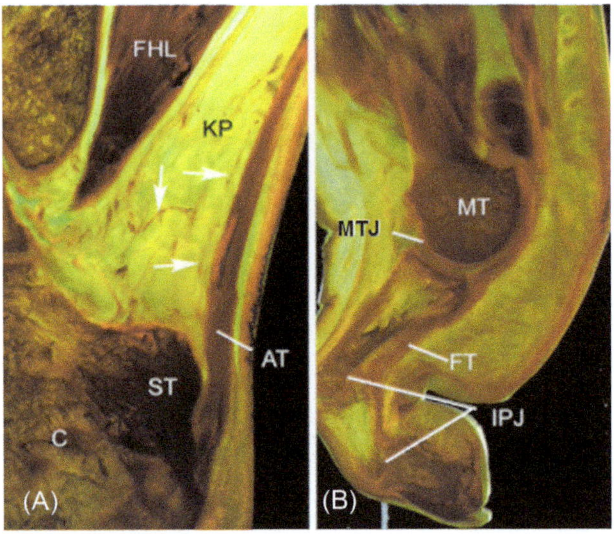


**Fig. 1.4** Foot and foot ankle-images performed with MRI (magnetic resonance imaging). (A) A sagittal section of the attachment of the human Achilles tendon (AT) to the calcaneus (C), showing its relation to the superior tuberosity (ST) that acts as a tendon pulley during dorsiflexion. Note also the presence of Kager's fat pad (KP) filling the space between the AT and flexor hallucis longus (FHL). It contains numerous blood vessels (arrows), some of which enter the deep surface of the AT to supply it. (B) A sagittal section of a toe that is hyperextended at the metatarsophalangeal joint (MTJ) and flexed at both interphalangeal joints (IPJ). Note how the head of the metatarsal acts as a pulley not only for the plantar fascia in maintaining the medial longitudinal arch of the foot, but also for the flexor tendons (FT) when the phalanges are dorsiflexed at the MTJ.
Based on Benjamin, M., Kaiser, E., Milz, S., 2008. Structure-function relationships in tendons: a review. J Anat. 212, 211–228, © by Journal of Anatomy with permission from Wiley.

dorsally and laterally to run posterior to the insertion site of AT at the calcaneus. In humans, the FDL tracks posteriorly to the medial *malleolus*. In the rabbit, the soleus muscle and tendon are negligible. Therefore, the rabbit AT shows distinctive gross anatomical and MRI features that must be considered when using the rabbit as a research model, especially for biomechanical testing, or when generalizing results from rabbits to humans (Doherty et al., 2006). Sometimes, other animal tendons may serve as better translational models for human tendons such as shown for the rabbit AT acting as a model for human FTs in terms of biomechanics (Meier Buergisser and Buschmann, 2014). Further anatomical tendon and ligament studies deal with the human shoulder (Allen, 2008; Tagliafico et al., 2010), the human knee (Steckel et al., 2007; Kaz et al., 2007), the equine superficial digital flexor tendons (SDFTs; Denoix and Busoni, 1999), and murine supraspinatus (Bell et al., 2015), among others.

## 1.3 The structure of tendons and ligaments

### 1.3.1 Healthy tendons and ligaments

Tendons and ligaments are well-organized hierarchical structures. From bottom-up, the smallest and main entity is the collagen I molecule which is a peptide having glycine as each third amino acid. Triple-helix rods are formed by three collagen molecules together and this secondary structure is named tropocollagen. Furthermore, five tropocollagen entities constitute a microfibril and they are connected to each other by covalent bonds leading to fibrils which are themselves grouped to the fibers. The fibers are also denoted as primary bundles which are assembled in the next higher hierarchical structure, the fascicle (= secondary bundle). The primary and secondary bundles are separated from each other by endotenons, cell-rich layers between the highly collagen-rich ECM and facilitating the sliding between fibers and/or fascicles. Finally, fascicles are bundled together to give the tertiary bundles surrounded by the epitenon and building up the whole tendon. In addition, some tendons have a paratenon, which is a sheath (but not the tendon sheath) that is quite distinct from the tendon itself, the best example being the paratenon around the AT. The paratenon is known as the false tendon sheath but may be of interest during the proper choice of an animal model where the paratenon plays an important role in adhesion formation (Buschmann et al., 2013). Many schemes are available to depict this structural hierarchy and organization (Sharma and Maffulli, 2005, 2006; Silver et al., 2003; Killian et al., 2012; Screen et al., 2004), exemplified with one from Screen and coworkers (without paratenon) (Fig. 1.5).

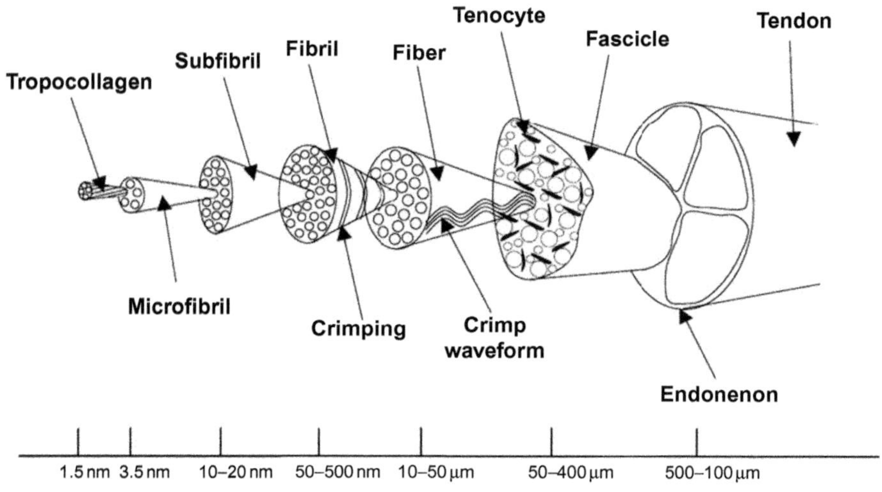


**Fig. 1.5** The hierarchical structure of a typical tendon.
From Screen, H.R.C., Bader, D.L., Lee, D.A., Shelton, J.C., 2004. Local strain measurement within tendon. Strain 40, 157–163, © by Strain with permission from Wiley.

### 1.3.2 The components of the ECM

Healthy native tendon tissue consists of a dense ECM with parallel-aligned collagen I fibers that are slightly crimped (Buschmann et al., 2014; Fig. 1.6). Besides collagen I, other types of collagen are also present, however, in much fewer amounts compared to collagen I; while collagen I represents around 95% of the tendon collagen dry weight, other collagens such as II, III, V, VI, IX, and XI amount for around 3% (Riley, 2005). In addition, the FACIT (fibril-associated collagen with interrupted triple-helix) collagens (types IX, XII, and XIV) were determined in low amounts in tendon tissue (Riley, 2005).

Mechanical loading leads to an upregulation of procollagen mRNA and to an increased collagen synthesis. Collagen synthesis is enhanced from 1% at rest to 2–3% after exercise and this increased synthesis remains for 2–3 days after acute exercise (Heinemeier et al., 2003). The collagen fibrils are bio-generated in activated tenocytes and have precursor structures such as procollagen filaments with a length of 30–60 nm that are excreted to the ECM as has been shown in the rabbit AT (Santander et al., 1999). Lateral fusion of fibril intermediates result in the linear as well as the lateral growth of fibrils (Silver et al., 2003). Fibrillogenesis is regulated by the interaction of quantitatively minor fibrillar collagens and proteoglycans (Zhang et al., 2005). Although all these processes may occur in the absence of collagen cross-linking, the development of a strong tendon or ligament tissue requires the conversion of collagen macromolecule assemblies which are viscous into solid components that are strong and energy storing. Therefore, the collagen molecules must be cross-linked end-to-end within a fibril (Silver et al., 2003). Incomplete collagen processing was reported to be responsible for smaller diameter fibrils ending up in weaker tendons (Sluss et al., 2001).

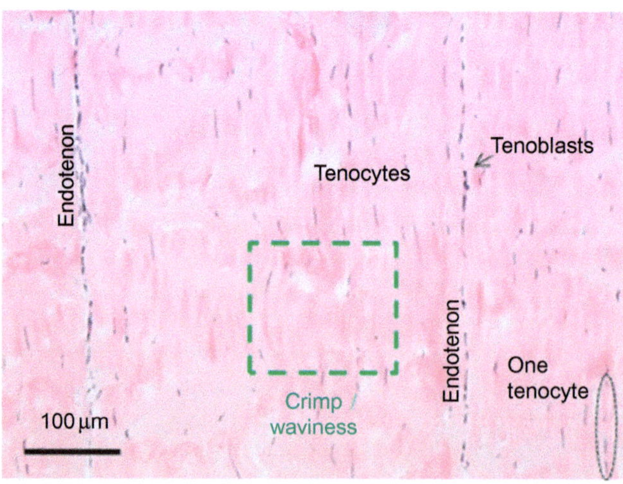


**Fig. 1.6** Typical histological section (100 × magnification) of rabbit AT depicting cellular rich endotenon, tenoblasts, tenocytes, and the waviness (crimping) of the collagen fibers (H&E staining; for further sections see Buschmann et al., 2014).

As for the crimp pattern that is observed in fibers, Morgan et al. reported that the in situ tension of unloaded digital FTs in rabbits is high enough for the majority of the fibers to be completely straightened and that the crimp pattern is only characteristic for excised tendon tissue without any tension (Morgan et al., 2006). On the next higher hierarchical level, the fibril, crimps are also observed and the fibrillar crimp is claimed to be acting as a shock absorber (Franchi et al., 2007). However, this is only true for the fibrils of the mid-substance tendon tissue, not for the fibrils of the tendon sheath. In contrast, the tendon sheath was reported to display small and uniform fibrils parallel with a wavy course, however, without any ultrastructural aspect of crimps—as such, the sheath is suggested to be not involved as a shock absorber (Franchi et al., 2007). Furthermore, the angles of these crimps are different in healthy tendon tissue compared to lacerated tissue and during the healing process, where the angles are significantly higher, the collagen fibers are highly disorganized (Buschmann et al., 2014). On the other hand, collagen organization under load is changed towards lower crimp angles associated with an increase in fiber alignment, which has been nicely shown by Riggin et al. (2014) in a murine AT model where they analyzed the histograms of the localized fiber directions through crossed polarizer imaging analysis. At the molecular level, however, no significant structural changes upon tendon loading could be observed (Masic et al., 2011). There was no change in the collagen backbone during tension; the deformation was redistributed through higher levels in the hierarchical structures, i.e., the collagen fibers and fibrils.

A typical feature of tendon collagen type I fibrils is their banding periodicity of 67 nm (Venturoni et al., 2003). Moreover, as reported by Venturoni et al. who used rat tail tendon collagen fibrils for their analysis, there are additional periodicities at 23 nm (1/3 of 67 nm) and at 210 nm ($3 \times 67$ nm). Other collagen types exhibit other periodicities like example shown for collagen IV where the periodicity is 100 nm, which was found for fibrils in rat tail tendons (Fig. 1.7). This tendon tissue had numerous 100-nm periodic fibrils that were situated adjacent to the plasma membrane of tenocytes and among the bundles of type I collagen fibrils. In contrast to the collagen I fibrils, the collagen IV fibrils were not distributed uniformly throughout the tendon tissue (Bruns et al., 1986).

The collagen fibers are interconnected via cross-linking, playing an important role during stretching processes and being an important biomarker for aging (see below). Mature tendon collagen fibers have a diameter 1–500 μm (Ushiki, 2002; Fratzl, 2003) and an elastic modulus of around 300 MPa measured in rabbit AT (Kahn et al., 2013). Collagen cross-linking is achieved by trivalent intermolecular pyridinoline cross-links that stabilize the fibrillar structure of collagen—as such contributing to the specific biomechanical properties of the tendon (Avery and Bailey, 2005; Barnard et al., 1987). Enzymatically derived covalent immature bonds form such cross-links, which later undergo a spontaneous conversion into more mature trivalent cross-links during collagen maturation (Avery and Bailey, 2005). Further cross-linking occurs via the non-enzymatic reactions of glucose with the lysine and arginine amino acid residues of the collagen triple-helix (= Maillard reaction) (Maillard, 1912a,b). This process results in the accumulation of advanced glycation end products (AGEs) in tendon tissue, among them being pentosidine the most widely studied AGE. In a study by Couppé et al. it was demonstrated that pentosidine was significantly increased in the human patellar tendons

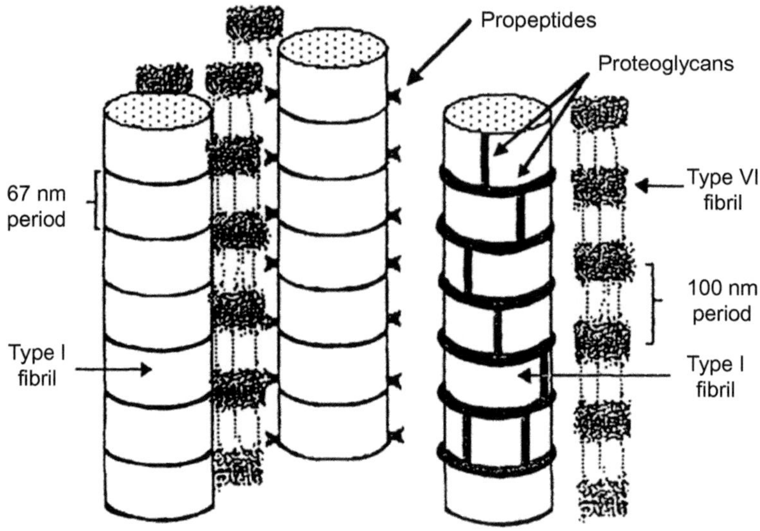


**Fig. 1.7** Summarizing diagram showing a concept of the major fibrous components in the extracellular compartment of adult rat tail tendon: collagen fibrils (from Bruns, R.R., Press, W., Engvall, E., Timpl, R., Gross, J., 1986. Type VI collagen in extracellular, 100-nm periodic filaments and fibrils: identification by immunoelectron microscopy. J. Cell Biol. 103, 393–404, http://dx.doi.org/10.1083/jcb.103.2.393 ©1986). Band structure periodicity of collagens: 64–67 nm, depending on tissue and organ (Ushiki, 2002), diameter of a fibril: 10–800 nm depending on tensile strength (Birk et al., 1990; Christiansen et al., 2000; Morgan et al., 2006; Fratzl, 2003).

(PTs) of old people compared to young people (Couppé et al., 2009). Therefore, pentosidine was considered to be a biomarker for tendon aging.

Although there are differences in the noncollagenous matrix composition among different tendon types such as purely positional versus energy storing tendons (Thorpe et al., 2013a), all tendons and ligaments have a noncollagenous part of the ECM that is (in various amounts) composed of proteoglycans, glycosaminoglycans (GAGs), and glycoproteins. Proteoglycans are heavily glycosylated proteins; a proteoglycan consists of a core protein and GAG side chains. The chains are long, linear carbohydrate polymers that are negatively charged under physiological conditions due to the occurrence of sulfate and uronic acid groups. There are different proteoglycans and their categorization is based on their size and the nature of the side chain. Large proteoglycans are aggrecan, lubricin, and versican; small proteoglycans are decorin, biglycan, fibromodulin, and lumican. Moreover, there are small leucine rich proteoglycans (SLRP), including decorin, biglycan, fibromodulin, and lumican. Among the tendon ECM GAGs, there are typically chondroitin sulfate, dermatan sulfate, heparan sulfate, and keratan sulfate.

The proteoglycans are responsible for the tissue hydration and for the viscoelastic behavior of the tendon—especially decorin is responsible for hydration and to a lesser extent fibromodulin and lumican. Decorin and fibromodulin are also important in binding to FACIT collagen XII during fibrillogenesis (Font et al., 1996). Also mechanical properties of tendons are dependent on fibromodulin and lumican expression as shown in a study where tendon stiffness was correlated to body weights of four different murine genotypes lacking either fibromodulin or lumican, both, or none (Jepsen et al., 2002). The authors clearly showed that fibromodulin was a key regulator and lumican a modulator of tendon strength. A disproportionate increase in small diameter collagen fibrils and a lack of progression to mature, large diameter fibrils in the absence of fibromodulin were speculated to constitute one of the underlying causes of tendon weakness.

Other components in the tendon ECM are fibronectin, a glycoprotein and thrombospondin, a multifunctional protein (inhibiting angiogenesis) which are both responsible and involved in tendon repair. Moreover, there is tenascin-C which is sensitive towards mechanical stimulation and important in terms of 3D organization and elasticity. Tenascin-C is not specific for tendons alone, it is also found in bone and smooth muscle (Chiquet-Ehrismann and Tucker, 2004). Another component makes up 2% of the ECM; elastin, which is very important—as its name implies—for the elasticity of the tendon. While collagen provides the qualities of strength and stiffness, elastin has a low stiffness and a high tolerance for deformation. Besides tendon and ligament tissues, elastin is also found in other tissues that have a high potential for elastic recovery, including skin and blood vessels (Korol et al., 2007). Tropoelastin is deposited onto preformed microfibril bundles during elastogenesis and is stabilized through cross-links (Baldwin et al., 2013). Mature elastic fibers have a diameter of 200–800 nm (Lorber, 1989) and an elastic modulus of 300–600 kPa (Mithieux and Weiss, 2005). Elastic fibers in tendon tissue are organized with a high density around the tenocytes and a lower density between the cells (Fig. 1.8; Grant et al., 2013).

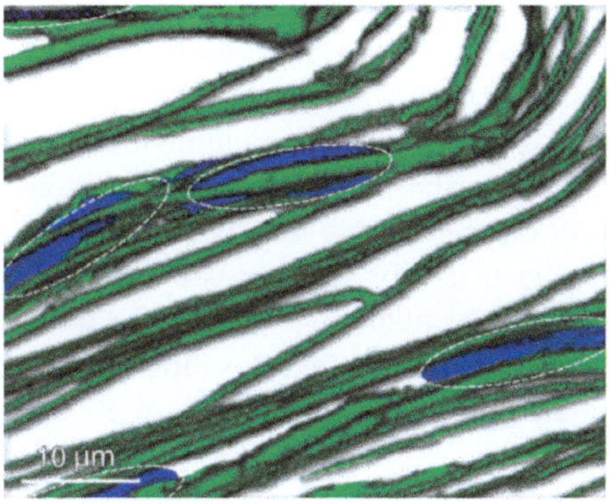


**Fig. 1.8** Three-dimensional reconstruction of elastic fiber organization (*green*) with outline of cell nuclei (*blue*) highlighted by dotted lines. Multiple elastic fibers surround groups of cells and have a branching structure. Fibers are also present between cells, but in fewer numbers. From Grant, T.M., Thompson, M.S., Urban, J., Yu, J., 2013. Elastic fibers are broadly distributed in tendon and highly localized around tenocytes. J. Anat. 222, 573–579, © by Journal of Anatomy with permission from Wiley.

Furthermore, elaunin is also a component of the elastic fibers of tendons. It is formed after the deposition of elastin. Interestingly, it is found in pressure-withstanding zones of flexor *digitorum profundus* (FDP) tendons (exemplified through the analysis of tendons of rabbits and dogs) in the form of microfibrils with sizes of around 12 nm (Carvalho et al., 1994).

The tendon and ligament tissues also contain water, which makes up a non-negligible part with around 60–80 wt% (Jozsa and Kannus, 1997; Woo et al., 2000). Water influences the viscoelastic behavior of those tissues. It was reported that by varying the water content of ligaments through osmosis before mechanical testing, different results were obtained in terms of viscoelastic response during cyclic loading (Chimich et al., 1992). Moreover, an MRI based study examined the water diffusion coefficient while the water was redistributed from the core to the rim of the tendon under loading in a rabbit AT model (Wellen et al., 2005).

And finally, 0.2% of inorganic components such as copper, manganese, and calcium are also found in the ECM. While calcium is present at the highest concentration of the inorganic metal species and plays a key role in the development of the osteotendinous junction, copper is a trace element acting in the formation of collagen cross-linking, and manganese is required for several enzymatic reactions during synthesis of ECM molecules (Kannus, 2000).

### 1.3.3 The cells

Tenocytes are tendon cells that secrete and build up the ECM with its components. These cells are longish and slender in their morphology and similar to fibroblasts. They are "specialized" fibroblasts (see Fig. 1.6 for the morphology of tenocytes and tenoblasts). Tenocytes are arranged in rows one behind another in the longitudinal direction of the tendon. The intertenocyte communication is allowed by thin cytoplasmatic projections like sheets via gap junctions (see also Chapter 3 for gap junction permeability under physical load).

Then, there are tenoblasts which are the immature tenocytes or the precursors of the tenocytes. Tenoblasts are primarily situated in the endotenons and the epitenon. They are spindle-shaped and compared to the tenocytes, they are rather roundish (lower length/width ratio). Tenoblasts exhibit numerous cytoplasmatic organelles reflecting their high metabolic activity. The nucleus-to-cytoplasm ratio is similar for young tenoblasts and tenocytes, however, upon aging, this ratio is increased; in other words, the nucleus almost completely occupies the cytoplasm in tenocytes of aged species (Nourissat et al., 2013).

The morphology of tenocytes is not only altered with aging, but also due to mechanical loading (Abraham et al., 2011). Rat ATs subjected to mechanical loading demonstrated micro-regions of abnormal tenocyte morphology, with the cells having a high nucleus-to-cytoplasm ratio and rounded morphology (similar to aged tenocytes). Furthermore, if loading is constantly increased as in the experiments performed by Miyazaki et al., biomechanical properties such as the failure load and the elongation at break of single rabbit PT tenocytes could be assessed. The maximum load and elongation at failure were 0.9±0.2 μN and 86±24 μm, respectively (Miyazaki et al., 2000).

Typical surface markers for tenocytes include tenomodulin, a transmembrane glycoprotein which regulates tenocyte proliferation and plays a pivotal role in the maturation of collagen fibrils. For example gene targeted mice with a loss of tenomodulin exhibited a reduced tenocyte density and the collagen fibrils in the tendon tissue had significantly increased maximum diameters (Docheva et al., 2005). The expression of tenomodulin is induced by scleraxis (Scx). This transcription factor has been used to identify tendon cells at all stages of tendon development and is important in terms of collagen synthesis (Schweitzer et al., 2001; Murchison et al., 2007; for more information about Scx, see Chapter 3).

Moreover, there are peripheral cells which are vascular in nature and which bring blood to the tendon cells. Compared to muscles, tendons have a vascular supply that is much lower, which is associated to the significantly higher metabolic activity of muscles compared to tendons. Therefore, excised tendons are "white" and "shiny" in their appearance, whereas muscles are reddish. Also, the osteotendinous insertion site is more vascularized than the tendon itself (Fig. 1.9).

Nevertheless, tendons are slightly vascularized—and the surrounding tissues of the tendons such as the tendon sheath or tendon-associated adipose tissue provide typically more blood supply. While the vessels within the tendons are small and thin-walled, the vessels in the tendon sheath are normal in terms of size and morphology.

Finally, synovial cells in the tendon sheath and synovial lining cells (SLCs) are further cell types found in the tendon tissue. As for the SLCs, two major cell types were determined by Steinberg and Hodde (1990); the so called A-cells producing hyaluronic acid acting as a lubricant and responsible for phagocytosis, and the B-cells that synthesize proteins and are capable of phagocytosis, too. Morphologically, A-cells are similar to macrophages with cauliflower-like appearance and many cytoplasmatic protrusions, while B-cells and their processes are slender and tend to orient in the length-axis of the synovium. A- and B-cells are found in different ratios in different synovia; in the knee joint, for example, where there is a large mechanical load on the anterior cruciate ligament (ACL) and where the joint is subjected to wear and tear, the removal of the wear by-products should be adequate and it is not surprising that A-cells have been reported to be dominant in this kind of ligament.

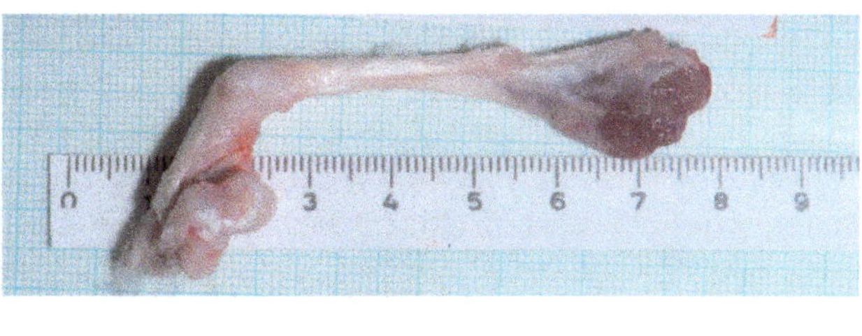

**Fig. 1.9** Excised rabbit AT including the calcaneus and the muscle; note the white and shiny surface of the practically avascular tendon (scale in cm).

The existence of a population of cells with differentiation potential has been reported for human tendons (De Mos et al., 2007). These tendon stem/progenitor cells reside in the ECM of the tendon and biglycan and fibromodulin are reported to be critical in terms of forming the corresponding stem cell niche (Bi et al., 2007). Tendon stem cells have not only been isolated and characterized from humans (Bi et al., 2007), but also from rabbits (Zhang and Wang, 2010), rats (Yin et al., 2009), and mice (Rui et al., 2010). Tendon stem cells share the general characteristics of other adult stem cells. They form colonies, are able to self-renew and can differentiate into multiple cell types (Yin et al., 2009). Upon mild mechanical stretching of up to 4%, they differentiate into tenocytes, while stretching to 8% induces differentiation towards chondrocytes, osteocytes, and/or adipocytes (Zhang and Wang, 2010). However, tendon stem/progenitor cells are able to express more tenogenic differentiation-related mRNA than what other mesenchymal stem cells (MSCs) do from the same species under the same conditions. For example, tendon stem cells express more tenomodulin, Scx, collagen type I, decorin, and biglycan than MSCs (Tan et al., 2012). Hence, it is self-evident, that tendon stem cells are and were often applied in the field of tissue engineering of tendons and ligaments. For example, PT stem cells were successfully isolated and applied to seed poly(L-lactide-*co*-ε-caprolactone) (PLCL)/collagen constructs in vitro and were shown to promote tendon healing in a rabbit PT injury model in vivo (Xu et al., 2014). Moreover, Mifune et al. (2012) were able to show that human ACL-derived stem cells were useful for ACL reconstruction. In their study, they sorted the ACL-derived stem cells and obtained a subpopulation that was $CD34^+$. CD34 is known as a hematopoietic stem/endothelial progenitor cell marker. In contrast to $CD34^-$ and unsorted ACL-derived stem cells, the $CD34^+$ ACL-derived subpopulation had a higher proliferation rate in vitro and a better in vivo outcome (injection into the joint capsules of reconstructed ACLs). This has been demonstrated by the better functional healing between bone and tendon graft which was assessed through tensile testing 8 weeks postreconstruction.

### 1.3.4 Growth factors

Growth factors play a pivotal role in tendon homeostasis as well as in tendon healing (Docheva et al., 2015). Transforming growth factor beta (TGF-β) is one of the most important growth factors in tendons as it influences the proliferation of tenocytes and the matrix synthesis in tendon tissue. Important to note is the fact that TGF-β was reported to interact with other growth factors such as vascular endothelial growth factor (VEGF) and hepatocyte growth factor (HGF) and that their presence or absence directed the actions of TGF-β (Anitua et al., 2007). Furthermore, also the interaction with platelet-derived growth factor (PDGF) was determined; the production of TGF-β in tenocytes was significantly reduced in the presence of PDGF (Duffy et al., 1995). The TGF-β family is involved in the wound healing and in inducing collagen production via smads and Scx (see Chapter 3). During exercise, TGF-β1 is upregulated preceding the rise in collagen expression. It is similarly upregulated like the mRNA for the connective tissue growth factor (CTGF; Heinemeier et al., 2007).

Another important growth factor in tendons is IGF-1 (insulin-like growth factor 1; Dahlgren et al., 2005). As shown by Abrahamsson, protein synthesis in tendons was accelerated when IGF-1 was administered to young rabbit FTs in vitro (Abrahamsson, 1997). Furthermore it has been shown that short-term training induced the expression of IGF-1 (and also TGF-β1) in rat ATs (Heinemeier et al., 2007).

PDGF is also produced by tenocytes, especially during wound healing (Banes et al., 1995; Chan et al., 2006). One of its isoforms, PDGF-β isoform, is involved in cell division and in matrix synthesis (Chan et al., 2006) and is reported to be the first trigger of the healing process in lacerated tendons. The interaction of PDGF with IGF-1 is essential; as without this interaction, the efficiency of PDGF is significantly reduced. Therefore, PDGF is suggested to accelerate IGF-1 synthesis and to upregulate IGF receptors (Lynch et al., 1989). On the other hand, the presence of TGF-β has been shown to decrease the effects of PDGF in terms of matrix synthesis (Hildebrand et al., 1998).

The angiogenic VEGF is worthwhile to mention; not primarily for the healthy tendon tissue which is practically avascular, however, during the healing of a tendon, VEGF is of high importance. For example, VEGF concentration is increased in human AT rupture in contrast to healthy tendons (Pufe et al., 2001). Moreover, also in chronic degenerative tendons, there is a high expression of VEGF as shown by Pufe et al. (2005).

Also growth/differentiation factor-5 (GDF-5) is important for tendon homeostasis, as it has been shown in a murine AT model, that GDF-5 deficient mice exhibited 40% less collagen in the ECM of their ATs (Mikic et al., 2001). GDF-5 deficient mice served as models for the rare disease called acromesomelic chondrodysplasia (Hunter–Thompson and Grebe diseases) which is associated with dramatic dislocations of the ankles and the knees caused by tendon and ligament laxities.

### 1.3.5 Aging tendons and ligaments

Due to the accumulation of micro-damages in the ECM of tendon and ligament tissues while individuals are getting older, a decrease in mechanical strength is found for aging people (Riley, 2008). However, not only tissue damages occur when people are getting older; as also the tissue as such is subjected to alterations (Figs. 1.1 and 1.10). It has been found, for example, that human PTs of old and young men differed in their collagen concentrations; with lower collagen in the old group than in the young group ($0.49 \pm 0.27$ versus $0.73 \pm 0.14$ mg/mg dry weight; Couppé et al., 2009). After analyzing biopsies, the authors found that in the old group there was significantly higher hydroxylysyl pyridinoline than in younger people. The trivalent intermolecular pyridinoline cross-links (primarily hydroxylysyl pyridinoline and lysyl pyridinoline) are reported to be stabilizing the fibrillar structure of collagen (Avery and Bailey, 2005).

Age-related tendinopathic changes of the tendon tissue include also a stiffening of the interfascicular matrix (IFM) (called also endotenon) that is mainly composed of collagen type III and proteoglycans, binding the fascicles together (Thorpe et al., 2013b). This has been shown by comparing two functionally distinct equine tendons,

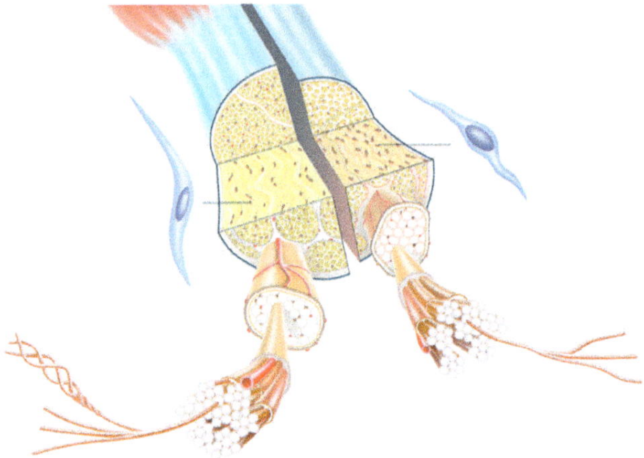

**Fig. 1.10** Schematic comparison of a normal tendon (left) and aged tendon (right). Aged tendons have lower thickness and vascularisation, and cell nuclei are bigger than in normal tendons.
Courtesy of Stefan Schwyter.

the SDFT, a mechanically and functionally well characterized tendon (Dowling and Dart, 2005) and the common digital extensor tendon (CDET). These two tendons have significantly different gross failure strains, although demonstrating no differences in the mechanical characteristics of the tendon fascicles; in other words the fascicles had very similar failure loads for both tendon types (Thorpe et al., 2012). The structural difference leading to different strains was found in a different sliding behavior of the fascicles; SDFT had significantly higher and better sliding than CDET at physiological loads—and this sliding behavior was attributed to the IFM. As such, the interfascicular sliding was proposed to be critical for energy storing tendons (such as the SDFT). And as age evokes changes in the IFM like a general decrease of the area occupied by the IFM (Thorpe et al., 2013b), particularly the energy storing tendons may be subjected to age-related tendinopathy.

Changes in the noncollagenous part of the ECM caused by aging include a decrease in GAGs as shown for human *supraspinati* (Riley et al., 1994). Moreover, also collagen oligomeric matrix protein (COMP) levels are decreased with aging—however, regional and tendon specifically. COMP plays a pivotal role in tendon development taking over an organizational role in collagenous matrix formation and is otherwise a noncollagenous ECM protein primarily found in cartilage. It has been found that upon aging COMP decreased in the tension-subjected parts of the SDFT, however, no change of COMP levels were found for the compression-subjected parts of the SDFT, and none for the CDET in all regions (Smith et al., 2002).

There is an increase in protein cross-linking during aging. The collagen cross-link levels are important in terms of biomechanics as has been shown in an equine model.

Enzyme-derived intermolecular cross-linking of collagen correlated with mechanical properties of the SDFT (Thorpe et al., 2010). To prevent excessive collagen cross-linking during aging, green tea has been shown to be effective in terms of suppression of collagen cross-linking (Rutter et al., 2013). Moreover, there is considerable evidence that a Maillard reaction based mechanism (nonenzymatic addition of reducing sugar and dicarbonyl moieties) is responsible for the formation of AGEs; Baynes, 1994; Naresh and Brodsky, 1992). During aging, there is an accumulation of AGEs. This accumulation of AGEs has been studied in terms of biomechanics—with a significantly reduced fiber-fiber sliding in the presence of higher amounts of AGEs (Li et al., 2013), resulting in a loss of tissue viscoelasticity during aging.

Furthermore, the elastin levels were reported to increase during aging in a rat tail tendon model, going along with lower ultimate load, ultimate strain (under increasing load), elastic modulus, and failure stress (Vogel, 1980). As the authors compared these biomechanical readouts to changes upon maturation (meaning from young rats to middle-aged rats), they were able to show that in contrast maturation effects the biomechanics by an increase in all the above mentioned parameters, while aging (from middle-aged to old rats) was significantly decreasing these biomechanical characteristics. This was not only related to increasing elastin content in the ECM upon aging, but also to a decreasing total (including soluble and insoluble) collagen fraction in the tendon tissue. Besides, aging effects were also studied for other tissues such as skin, bone, cartilage, and aorta (Vogel, 1980).

Changes at the cell level include morphological changes such as an increase in the nucleus-to-cytoplasm ratio of the tenocytes that overall get smaller compared to young tenocytes, more slender and do produce less ECM. Young and old tenocytes do also show a different behavior towards antibacterial agents such as fluoroquinolones, which was shown in an in vitro study of young and old rabbit Achilles primary tenocytes exposed to quinolones in different concentrations (Pouzaud et al., 2006). Old tenocytes reacted much more sensitively than young tenocytes. Age-related teno-toxicity induced by quinolones was explained by the fact that young tenocytes' mitochondrial activity was unaltered, while it was altered in mature tenocytes.

Also the tendon stem/progenitor cells are affected by aging (Kohler et al., 2013). The capacity for self-renewal and clonogenicity as well as migration were reported to be significantly reduced for old human AT stem/progenitors compared to young ones, although the capacity to differentiate towards several cell lines remained the same. Specifically, an altered ECM production, integrin expression, and rho-associated protein kinase (ROCK) activity were reported; collagen I and fibronectin, which are essential for cell adhesion and migration were downregulated at the gene-level; the collagen I-binding integrins $\alpha 1\beta 1$, $\alpha 2\beta 1$, and $\alpha 11\beta 1$, and the fibronectin-binding $\alpha 5\beta 1$, $\alpha\nu\beta 3$, and $\alpha\nu\beta 5$ were affected upon aging (Docheva et al., 2007) as well as the ROCK1 and 2 which downstream signaling regulates actin stress fiber formation via myosin light chain (Schmitz et al., 2000). Finally also wound closure capacity was found to be reduced upon senescence of tendon stem/progenitor cells (Kohler et al., 2013).

### 1.3.6 Tendinopathy

Tendinopathy as the degenerative state of the tendon or ligament tissue includes tendinitis and tendinosis. While tendinitis is associated with an acute event accompanied by inflammation (end: *-itis*) and pain, tendinosis is rather a chronic tendon degeneration without inflammation (end: *-osis*), however, also going along with pain. Why is tendinopathy painful? After van Sterkenburg and van Dijk, AT tendinopathy is primarily painful because of innervation of the surrounding tissue. Chronic painful tendons have an ingrowth of sensory and sympathetic nerves from the paratenon—and only denervation of chronic painful tendons leads to a relief of pain in the majority of patients (Van Sterkenburg and Van Dijk, 2011).

Tendinopathies are reported to be a result of a variety of pathological processes leading to the loss of tissue integrity as well as to partial or full tissue ruptures (Riley, 2005). Extrinsic factors that are likely to be involved in the pathogenesis of tendon disorders are high mechanical loading as achieved through sport activities, drug abuse, or environmental conditions like walking surface or the footwear—leading to local hyperthermia, hypoxia, oxidative stress, and/or apoptosis (Rees et al., 2006). Also the complete absence of load has been reported to change the tendon matrix similar as found for overload. Stress-shielded tendons (completely unloaded) exhibit decreases in the mechanical integrity of tendon tissue and typical changes in terms of cells and ECM are histologically found (Ohno et al., 1993). On the other hand, intrinsic factors such as age, gender, body weight, anatomical variants, systemic diseases (Magnan et al., 2014), genetics (Mokone et al., 2006; September et al., 2006), and blood group (Jozsa et al., 1989) also play a nonnegligible role in the generation of tendinopathies.

Tendinopathological concepts include not only the degenerative tendinopathy with hypoxic, hyaline or mucoid degeneration of the tissue, but also angio-fibroblastic hyperplasia, which refers to a transition state of injured tendon tissues during the healing process—and in the case of chronic tendinopathy, is termed "failed healing" (Clancy, 1989). Furthermore, a continuum of tendon pathology was proposed and presented by Cook and Purdam (2009). In this model, addition or removal of load is proposed to be the first stimulus driving the tendon forth or back along a continuum of pathological stages—either to a stage entitled "stress shielded" (unloaded tendons) or to a stage called "reactive tendinopathy" (excessive load and individual factors).

The "reactive tendinopathy" is characterized by higher cell proliferation in the matrix, however, without inflammation. This occurs as a response to an acute tensile or compressive overload and results in a short-term thickening of the tendon at a specific location in order to reduce stress at that site. Compared to normal tendon adaptation to tensile load, where the tendon stiffens without thickening, the short-term thickening as a reaction of acute overload is reported to be reversible. This quick adaptation goes along with the cells turning toward a more chondroid morphology and an increased production of large proteoglycans such as aggrecan, versican, and glycoproteins such as hyaluronan (Scott et al., 2007). Typically, upregulation of these molecules occurs within minutes up to a few days—in contrast to the upregulation of small proteoglycans found primarily in healthy tendons (Samiric et al., 2004).

A further stage in the continuum model is reached after excessive overloading for a longer time period (chronic) and is entitled "tendon disrepair." This stage is reported to go along with an increase in cell density, mostly chondrocytes and some myofibroblasts, a marked increase in proteins such as proteoglycans and collagen and a clear disorganization of the ECM. Finally, the last stage is "degenerative tendinopathy" with areas of apoptosis and trauma. The ECM is completely disordered and filled with vessels, there is little collagen, but a lot of matrix breakdown products. This stage is reported to be irreversible.

At the molecular level, chronic tendinopathy includes an increase in mRNA of collagen types I and III, fibronectin (Tillander et al., 2002), tenascin-C, aggrecan, and biglycan (Corps et al., 2006). While the upregulation of the collagen mRNAs and the fibronectin as well as the tenascin-C (Riley et al., 1996) are associated with a healing response (Riley, 2005), an increased expression of aggrecan and biglycan are consistent with a mechanobiological response to compression and shear (Corps et al., 2006). Furthermore there is an increase of proteoglycans and sulfated GAGs (Fu et al., 2007), both being responses to mechanical stimulation. GAGs have been reported to protect tendon fascicles from fatigue effects, possibly by enabling sample hydration, with GAG depletion leading to tendon fascicle's higher stress relaxation and lower failure stress as shown with bovine extensor tendon fascicles (Legerlotz et al., 2013).

Metalloproteinase expressions and activities are also affected in tendinopathy (Riley, 2008; Xu and Murrell, 2008); there is an increase in MMP-1 (collagenase), MMP-2 (gelatinase), and MMP-23, while there is a decrease in MMP-3 (stromelysin), MMP-10, MMP-12, and MMP-27. Also the levels of tissue inhibitors of metalloproteinase (TIMPs) are consistent with the proteolytic activity going on in tendinopathic tissue. In addition, increased expression of a disintegrin and metalloproteinase (ADAM) and ADAMTS (ADAM with thromospondin motifs), have been associated with painful tendons: increased ADAM12, ADAMTS2, and ADAMTS3—where ADAM12 has been reported to be involved in the change of cells type (myogenesis and lipidogenesis; Guo et al., 2005).

In terms of growth factors, cytokines and signaling factors, tendinopathy evokes an increase in TGF-β, IGF-1, and PDGF-R (platelet-derived growth factor receptor), cyclo-oxigenase-2 (COX-2), and neurotransmitters such as glutamate and substance P as well as glutamate receptors *N*-methyl-D-aspartate receptor (NMDAR1; Alfredson et al., 2001). A typical feature of tendinopathic tendons and ligaments is also hypervascularization, with the growth factor VEGF stimulating the vessel ingrowth into the degenerative tissue (Fig. 1.11 on the right side).

VEGF is significantly increased in tendinopathy (Pufe et al., 2005). Typically, structural changes with neovessels penetrating the connective tissue are clinically envisaged by color Doppler ultrasound (US), as for example reported in studies of patients with jumper's knee (patellar tendinopathy) where sclerosing treatment with polidocanol was investigated (Hoksrud et al., 2008, Willberg et al., 2011). Moreover, the higher the blood volume of a degenerative tendon, the weaker the tendon in terms of biomechanics (Stange et al., 2014). Therefore, it is not surprising that VEGF is also increased by mechanical loading (which leads to tendinopathy in the case it is excessive), which has been shown in a tendinopathic model of rabbit flexor *digitorum profundi* (Nakama et al., 2006).

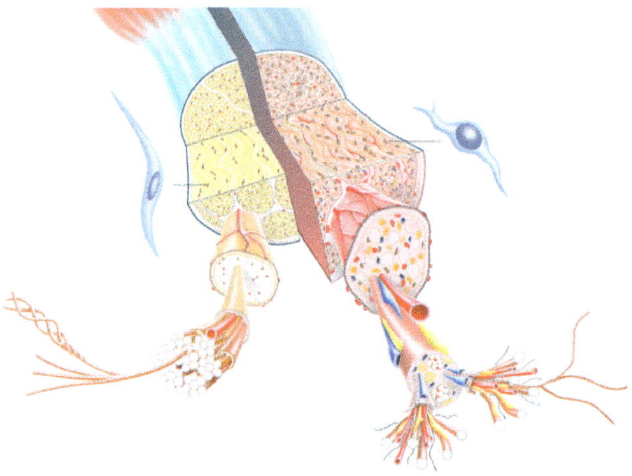

**Fig. 1.11** Schematic comparison of a normal tendon (left) and tendinopatic tendon (right). In chronic tendinopathy, histological examination shows intra-tendinous collagen degeneration with fiber disorientation and glycosaminoglycans accumulation in between thinning fibrils without inflammatory cells or inflammatory signs. Tenocytes look normal but increase in number. Hypervascularization is frequently found (see Section 1.1).
Courtesy of Stefan Schwyter.

## 1.4 Summary

Tendons and ligaments connect muscle to bone and bone to bone, respectively. Their main function relies in enabling motion of the musculoskeletal system by the transmittance of force between the respective entities. In this chapter, we presented the hierarchical structure of tendons and ligaments, focusing on the ECM components and the cells. By comparison of the healthy state with the degenerative state and the aging situation, a clear picture was drawn in terms of tendon and ligament homeostasis as well as alterations caused by senescence or accidents. The specific anatomical structures and their functions were also discussed briefly with further references to other more in depth articles or reviews.

## References

Abraham, T., Fong, G., Scott, A., 2011. Second harmonic generation analysis of early Achilles tendinosis in response to in vivo mechanical loading. BMC Musculoskelet. Disord. 12, 26.

Abrahamsson, S.O., 1997. Similar effects of recombinant human insulin-like growth factor-I and II on cellular activities in flexor tendons of young rabbits: experimental studies in vitro. J. Orthop. Res. 15, 256–262.

Alfredson, H., Forsgren, S., Thorsen, K., Lorentzon, R., 2001. In vivo microdialysis and immunohistochemical analyses of tendon tissue demonstrated high amounts of free glutamate and glutamate NMDAR1 receptors, but no signs of inflammation, in Jumper's knee. J. Orthop. Res. 19, 881–886.

Allen, G.M., 2008. Shoulder ultrasound imaging-integrating anatomy, biomechanics and disease processes. Eur. J. Radiol. 68, 137–146.

Anitua, E., Sanchez, M., Nurden, A.T., Zalduendo, M., De La Fuente, M., Azofra, J., Andia, I., 2007. Reciprocal actions of platelet-secreted TGF-beta 1 on the production of VEGF and HGF by human tendon cells. Plast. Reconstr. Surg. 119, 950–959.

Avery, N.C., Bailey, A.J., 2005. Enzymic and non-enzymic cross-linking mechanisms in relation to turnover of collagen: relevance to aging and exercise. Scand. J. Med. Sci. Sports 15, 231–240.

Baldwin, A.K., Simpson, A., Steer, R., Cain, S.A., Kielty, C.M., 2013. Elastic fibres in health and disease. Expert Rev. Mol. Med 15, 1–30.

Banes, A.J., Tsuzaki, M., Hu, P.Q., Brigman, B., Brown, T., Almekinders, L., Lawrence, W.T., Fischer, T., 1995. PDGF-BB, IGF-I and mechanical load stimulate DNA synthesis in avian tendon fibroblasts in vitro. J. Biomech. 28, 1505–1513.

Barnard, K., Light, N.D., Sims, T.J., Bailey, A.J., 1987. Chemistry of the collagen cross-links—origin and partial characterization of a putative mature cross-link of collagen. Biochem. J. 244, 303–309.

Baynes, J.W., 1994. Role of metal-catalyzed autoxidation in maillard reaction damage to proteins in-vivo. Redox Rep. 1, 31–34.

Bell, R., Taub, P., Cagle, P., Flatow, E.L., Andarawis-Puri, N., 2015. Development of a mouse model of supraspinatus tendon insertion site healing. J. Orthop. Res. 33, 25–32.

Benjamin, R., 2010. The structure of tendons and ligaments. In: Archer, C., Ralphs, J. (Eds.), Regenerative Medicine and Biomaterials for the Repair of Connective Tissues. Woodhead Publishing Limited, Oxford, Cambridge, New Delhi.

Benjamin, M., Kaiser, E., Milz, S., 2008. Structure-function relationships in tendons: a review. J. Anat. 212, 211–228.

Bi, Y., Ehirchiou, D., Kilts, T.M., Inkson, C.A., Embree, M.C., Sonoyama, W., Li, L., Leet, A.I., Seo, B.-M., Zhang, L., Shi, S., Young, M.F., 2007. Identification of tendon stem/progenitor cells and the role of the extracellular matrix in their niche. Nat. Med. 13, 1219–1227.

Birk, D.E., Fitch, J.M., Babiarz, J.P., Doane, K.J., Linsenmayer, T.F., 1990. Collagen fibrillogenesis in vitro: interaction of types I and V collagen regulates fibril diameter. J. Cell Sci. 95, 649–657.

Bruns, R.R., Press, W., Engvall, E., Timpl, R., Gross, J., 1986. Type VI collagen in extracellular, 100-nm periodic filaments and fibrils: identification by immunoelectron microscopy. J. Cell Biol. 103, 393–404.

Buschmann, J., Meier Buergisser, G., Bonavoglia, E., Neuenschwander, P., Milleret, V., Giovanoli, P., Calcagni, M., 2013. Cellular response of healing tissue to DegraPol tube implantation in rabbit Achilles tendon rupture repair: an in vivo histomorphometric study. J. Tissue Eng. Regen. Med. 7, 413–420.

Buschmann, J., Puippe, G., Meier Buergisser, G., Bonavoglia, E., Giovanoli, P., Calcagni, M., 2014. Correspondence of high-frequency ultrasound and histomorphometry of healing rabbit Achilles tendon tissue. Connect. Tissue Res. 55, 123–131.

Carvalho, H.F.D., Lino Neto, J., Taboga, S.R., 1994. Microfibrils: neglected components of pressure-bearing tendons. Ann. Anat. 176, 155–159.

Chan, B.P., Fu, S.C., Qin, L., Rolf, C., Chan, K.M., 2006. Supplementation-time dependence of growth factors in promoting tendon healing. Clin. Orthop. Relat. Res., 448, 240–247.

Chimich, D., Shrive, N., Frank, C., Marchuk, L., Bray, R., 1992. Water-content alters viscoelastic behavior of the normal adolescent rabbit medial collateral ligament. J. Biomech. 25, 831–837.

Chiquet-Ehrismann, R., Tucker, R.P., 2004. Connective tissues: signalling by tenascins. Int. J. Biochem. Cell Biol. 36, 1085–1089.

Christiansen, D.L., Huang, E.K., Silver, F.H., 2000. Assembly of type I collagen: fusion of fibril subunits and the influence of fibril diameter on mechanical properties. Matrix Biol. 19, 409–420.

Clancy, W., 1989. Failed healing responses. In: Leadbetter, W., Buckwater, J., Gordon, G. (Eds.), Sports-Induced Inflammation: Clinical and Basic Concepts. American Orthopedic Society for Sports Medicine, Park Ridge, IL.

Cook, J.L., Purdam, C.R., 2009. Is tendon pathology a continuum? A pathology model to explain the clinical presentation of load-induced tendinopathy. Br. J. Sports Med. 43, 409–416.

Corps, A.N., Robinson, A.H.N., Movin, T., Costa, M.L., Hazleman, B.L., Riley, G.P., 2006. Increased expression of aggrecan and biglycan mRNA in Achilles tendinopathy. Rheumatology 45, 291–294.

Couppé, C., Hansen, P., Kongsgaard, M., Kovanen, V., Suetta, C., Aagaard, P., Kjær, M., Magnusson, S.P., 2009. Mechanical properties and collagen cross-linking of the patellar tendon in old and young men. J. Appl. Physiol. 107 (3), 880–886.

Dahlgren, L.A., Mohammed, H.O., Nixon, A.J., 2005. Temporal expression of growth factors and matrix molecules in healing tendon lesions. J. Orthop. Res. 23, 84–92.

De Mos, M., Koevoet, W.J.L.M., Jahr, H., Verstegen, M.M.A., Heijboer, M.P., Kops, N., Van Leeuwen, J.P.T.M., Weinans, H., Verhaar, J.A.N., Van Osch, G.J.V.M., 2007. Intrinsic differentiation potential of adolescent human tendon tissue: an in-vitro cell differentiation study. BMC Musculoskelet. Disord. 8.

Denoix, J.M., Busoni, V., 1999. Ultrasonographic anatomy of the accessory ligament of the superficial digital flexor tendon in horses. Equine Vet. J. 31, 186–191.

Docheva, D., Hunziker, E.B., Fassler, R., Brandau, O., 2005. Tenomodulin is necessary for tenocyte proliferation and tendon maturation. Mol. Cell. Biol. 25, 699–705.

Docheva, D., Popov, C., Mutschler, W., Schieker, M., 2007. Human mesenchymal stem cells in contact with their environment: surface characteristics and the integrin system. J. Cell. Mol. Med. 11, 21–38.

Docheva, D., Mueller, S.A., Majewski, M., Evans, C.H., 2015. Biologics for tendon repair. Adv. Drug Deliv. Rev. 84, 222–239.

Doherty, G.P., Koike, Y., Uhthoff, H.K., Lecompte, M., Trudel, G., 2006. Comparative anatomy of rabbit and human achilles tendons with magnetic resonance and ultrasound imaging. Comp. Med. 56, 68–74.

Dowling, B.A., Dart, A.J., 2005. Mechanical and functional properties of the equine superficial digital flexor tendon. Vet. J. 170, 184–192.

Duffy Jr., F.J., Seiler, J.G., Gelberman, R.H., Hergrueter, C.A., 1995. Growth factors and canine flexor tendon healing: initial studies in uninjured and repair models. J. Hand Surg. 20, 645–649.

Font, B., Eichenberger, D., Rosenberg, L.M., Vanderrest, M., 1996. Characterization of the interactions of type XII collagen with two small proteoglycans from fetal bovine tendon, decorin and fibromodulin. Matrix Biol. 15, 341–348.

Franchi, M., Fini, M., Quaranta, M., De Pasquale, V., Raspanti, M., Giavaresi, G., Ottani, V., Ruggeri, A., 2007. Crimp morphology in relaxed and stretched rat Achilles tendon. J. Anat. 210, 1–7.

Fratzl, P., 2003. Cellulose and collagen: from fibres to tissues. Curr. Opin. Colloid Interface Sci. 8, 32–39.

Fu, S.-C., Chan, K.-M., Rolf, C.G., 2007. Increased deposition of sulfated glycosaminoglycans in human patellar tendinopathy. Clin. J. Sport Med. 17, 129–134.

Grant, T.M., Thompson, M.S., Urban, J., Yu, J., 2013. Elastic fibres are broadly distributed in tendon and highly localized around tenocytes. J. Anat. 222, 573–579.

Guo, L.T., Shelton, G.D., Wewer, U.M., Engvall, E., 2005. ADAM 12 overexpression does not improve outcome in mice with laminin alpha 2-deficient muscular dystrophy. Neuromuscul. Disord. 15, 786–789.

Heinemeier, K., Langberg, H., Olesen, J.L., Kjaer, M., 2003. Role of TGF-beta(1) in relation to exercise-induced type I collagen synthesis in human tendinous tissue. J. Appl. Physiol. 95, 2390–2397.

Heinemeier, K.M., Olesen, J.L., Haddad, F., Langberg, H., Kjaer, M., Baldwin, K.M., Schjerling, P., 2007. Expression of collagen and related growth factors in rat tendon and skeletal muscle in response to specific contraction types. J. Physiol. Lond. 582, 1303–1316.

Hildebrand, K.A., Woo, S.L.Y., Smith, D.W., Allen, C.R., Deie, M., Taylor, B.J., Schmidt, C.C., 1998. The effects of platelet-derived growth factor-BB on healing of the rabbit medial collateral ligament—an in vivo study. Am. J. Sports Med. 26, 549–554.

Hoksrud, A., Ohberg, L., Alfredson, H., Bahr, R., 2008. Color Doppler ultrasound findings in patellar tendinopathy (jumper's knee). Am. J. Sports Med. 36, 1813–1820.

Jepsen, K.J., Wu, F., Peragall, O,.J.H., Paul, J., Roberts, L., Ezura, Y., Oldberg, A., Birk, D.E., Chakravarti, S., 2002. A syndrome of joint laxity and impaired tendon integrity in lumican- and fibromodulin-deficient mice. J. Biol. Chem. 277, 35532–35540.

Jozsa, L., Kannus, P., 1997. Human Tendons: Anatomy, Physiology, and Pathology. Human Kinetics, Champaign, IL, pp. 1–576.

Jozsa, L., Balint, J.B., Kannus, P., Reffy, A., Barzo, M., 1989. Distribution of blood-groups in patients with tendon-rupture—an analysis of 832 cases. J. Bone Joint Surg. 71, 272–274.

Kahn, C.J.F., Dumas, D., Arab-Tehrany, E., Marie, V., Tran, N., Wang, X., Cleymand, F., 2013. Structural and mechanical multi-scale characterization of white New-Zealand rabbit Achilles tendon. J. Mech. Behav. Biomed. Mater. 26, 81–89.

Kannus, P., 2000. Structure of the tendon connective tissue. Scand. J. Med. Sci. Sports 10, 312–320.

Kaz, R., Starman, J.S., Fu, F.H., 2007. Anatomic double-bundle anterior cruciate ligament reconstruction revision surgery. Arthroscopy 23, 3.

Killian, M.L., Cavinatto, L., Galatz, L.M., Thomopoulos, S., 2012. The role of mechanobiology in tendon healing. J. Shoulder Elb. Surg. 21, 228–237.

Kohler, J., Popov, C., Klotz, B., Alberton, P., Prall, W.C., Haasters, F., Mueller-Deubert, S., Ebert, R., Klein-Hitpass, L., Jakob, F., Schieker, M., Docheva, D., 2013. Uncovering the cellular and molecular changes in tendon stem/progenitor cells attributed to tendon aging and degeneration. Aging Cell 12, 988–999.

Korol, R.M., Finlay, H.M., Josseau, M.J., Lucas, A.R., Canham, P.B., 2007. Fluorescence spectroscopy and birefringence of molecular changes in maturing rat tail tendon. J. Biomed. Opt. 12, 024011.

Legerlotz, K., Riley, G.P., Screen, H.R.C., 2013. GAG depletion increases the stress-relaxation response of tendon fascicles, but does not influence recovery. Acta Biomater. 9, 6860–6866.

Li, Y., Fessel, G., Georgiadis, M., Snedeker, J.G., 2013. Advanced glycation end-products diminish tendon collagen fiber sliding. Matrix Biol. 32, 169–177.

Lin, T.W., Cardenas, L., Soslowsky, L.J., 2004. Biomechanics of tendon injury and repair. J. Biomech. 37, 865–877.

Lorber, M., 1989. Elastic fibers in the rat exorbital lacrimal gland duct system. Invest. Ophthalmol. Vis. Sci. 30, 2002–2011.

Lynch, S.E., Colvin, R.B., Antoniades, H.N., 1989. Growth-factors in wound-healing—single and synergistic effects on partial thickness porcine skin wounds. J. Clin. Investig. 84, 640–646.

Magnan, B., Bondi, M., Pierantoni, S., Samaila, E., 2014. The pathogenesis of Achilles tendinopathy: a systematic review. Foot Ankle Surg. 20, 154–159.

Maillard, L.C., 1912a. The action of amino acids on sugar; the formation of melanoidin by a methodic route. Comptes Rendus Hebd. Seances Acad. Sci. 154, 66–68.

Maillard, L.C., 1912b. General reaction of amino acids on sugars: its biological consequences. Comptes Rendus Seances Soc. Biol. 72, 599–601.

Masic, A., Bertinetti, L., Schuetz, R., Galvis, L., Timofeeva, N., Dunlop, J.W., Seto, J., Hartmann, M.A., Fratzl, P., 2011. Observations of multiscale, stress-induced changes of collagen orientation in tendon by polarized Raman spectroscopy. Biomacromolecules 12, 3989–3996.

Meier Buergisser, G., Buschmann, J., 2014. History and performance of implant materials applied as peritendinous antiadhesives. J. Biomed. Mater. Res. B Appl. Biomater. 103, 212–228.

Mifune, Y., Matsumoto, T., Ota, S., Nishimori, M., Usas, A., Kopf, S., Kuroda, R., Kurosaka, M., Fu, F.H., Huard, J., 2012. Therapeutic potential of anterior cruciate ligament-derived stem cells for anterior cruciate ligament reconstruction. Cell Transplant. 21, 1651–1665.

Mikic, B., Schalet, B.J., Clark, R.T., Gaschen, V., Hunziker, E.B., 2001. GDF-5 deficiency in mice alters the ultrastructure, mechanical properties and composition of the Achilles tendon. J. Orthop. Res. 19, 365–371.

Mithieux, S.M., Weiss, A.S., 2005. Elastin. In: Parry, D.A.D., Squire, J.M. (Eds.), Fibrous Proteins: Coiled-Coils, Collagen and Elastomers. Advances in Protein Chemistry, vol. 70. Elsevier Academic Press, San Diego, CA.

Miyazaki, H., Hasegawa, Y., Hayashi, K., 2000. A newly designed tensile tester for cells and its application to fibroblasts. J. Biomech. 33, 97–104.

Mokone, G.G., Schwellnus, M.P., Noakes, T.D., Collins, M., 2006. The COL5A1 gene and Achilles tendon pathology. Scand. J. Med. Sci. Sports 16, 19–26.

Morgan, M., Kostyuk, O., Brown, R.A., Mudera, V., 2006. In situ monitoring of tendon structural changes by elastic scattering spectroscopy: correlation with changes in collagen fibril diameter and crimp. Tissue Eng. 12, 1821–1831.

Murchison, N.D., Price, B.A., Conner, D.A., Keene, D.R., Olson, E.N., Tabin, C.J., Schweitzer, R., 2007. Regulation of tendon differentiation by scleraxis distinguishes force-transmitting tendons from muscle-anchoring tendons. Development 134, 2697–2708.

Nakama, L.H., King, K.B., Abrahamsson, S., Rempel, D.M., 2006. VEGF, VEGFR-1, and CTGF cell densities in tendon are increased-with cyclical loading: an in vivo tendinopathy model. J. Orthop. Res. 24, 393–400.

Naresh, M.D., Brodsky, B., 1992. X-ray diffraction studies on human tendon show age-related changes in collagen packing. Biochim. Biophys. Acta 1122, 161–166.

Nourissat, G., Houard, X., Sellam, J., Duprez, D., Berenbaum, F., 2013. Use of autologous growth factors in aging tendon and chronic tendinopathy. Front. Biosci. E5, 911–921.

Ohno, K., Yasuda, K., Yamamoto, N., Kaneda, K., Hayashi, K., 1993. Effects of complete stress-shielding on the mechanical-properties and histology of in-situ frozen patellar tendon. J. Orthop. Res. 11, 592–602.

Pouzaud, F., Dutot, M., Martin, C., Debray, M., Warnet, J.M., Rat, P., 2006. Age-dependent effects on redox status, oxidative stress, mitochondrial activity and toxicity induced by fluoroquinolones on primary cultures of rabbit tendon cells. Comp. Biochem. Physiol. C: Toxicol. Pharmacol. 143, 232–241.

Pufe, T., Petersen, W., Tillmann, B., Mentlein, R., 2001. The angiogenic peptide vascular endothelial growth factor is expressed in foetal and ruptured tendons. Virchows Arch. 439, 579–585.

Pufe, T., Petersen, W.J., Mentlein, R., Tillmann, B.N., 2005. The role of vasculature and angiogenesis for the pathogenesis of degenerative tendons disease. Scand. J. Med. Sci. Sports 15, 211–222.

Rees, J.D., Wilson, A.M., Wolman, R.L., 2006. Current concepts in the management of tendon disorders. Rheumatology 45, 508–521.

Riggin, C.N., Sarver, J.J., Freedman, B.R., Thomas, S.J., Soslowsky, L.J., 2014. Analysis of collagen organization in mouse achilles tendon using high-frequency ultrasound imaging. J. Biomech. Eng. 136, 021029.

Riley, G., 2005. Chronic tendon pathology: molecular basis and therapeutic implications. Expert Rev. Mol. Med. 7, 1–25.

Riley, G., 2008. Tendinopathy—from basic science to treatment. Nat. Clin. Pract. Rheumatol. 4, 82–89.

Riley, G.P., Harrall, R.L., Constant, C.R., Chard, M.D., Cawston, T.E., Hazleman, B.L., 1994. Glycosaminoglycans of human rotator cuff tendons—changes with age and in chronic rotator cuff tendinitiS. Ann. Rheum. Dis. 53, 367–376.

Riley, G.P., Harrall, R.L., Cawston, T.E., Hazleman, B.L., Mackie, E.J., 1996. Tenascin-C and human tendon degeneration. Am. J. Pathol. 149, 933–943.

Rui, Y.-F., Lui, P.P.Y., Li, G., Fu, S.C., Lee, Y.W., Chan, K.M., 2010. Isolation and characterization of multipotent rat tendon-derived stem cells. Tissue Eng. A 16, 1549–1558.

Rutter, K., Sell, D.R., Fraser, N., Obrenovich, M., Zito, M., Starke-Reed, P., Monnier, V.M., 2013. Green tea extract suppresses the age-related increase in collagen crosslinking and fluorescent products in C57BL/6 mice. Int. J. Vitam. Nutr. Res. 73, 453–460.

Samiric, T., Ilic, M.Z., Handley, C.J., 2004. Characterisation of proteoglycans and their catabolic products in tendon and explant cultures of tendon. Matrix Biol. 23, 127–140.

Santander, R.G., Arriba, M.A.P., Cuadrado, G.M., Alonso, A.L., Martinez, M.G.S., Alonso, F.J.M., Monteagudo, M., Lobo, M.V.T., 1999. Intracellular biogenesis of collagen fibrils in 'activated fibroblasts' of tendo Achillis—an ultrastructural study in the New Zealand rabbit. J. Bone Joint Surg. 81B, 522–530.

Schmitz, A.A.P., Govek, E.E., Bottner, B., Van Aelst, L., 2000. Rho GTPases: signaling, migration, and invasion. Exp. Cell Res. 261, 1–12.

Schweitzer, R., Chyung, J.H., Murtaugh, L.C., Brent, A.E., Rosen, V., Olson, E.N., Lassar, A., Tabin, C.J., 2001. Analysis of the tendon cell fate using Scleraxis, a specific marker for tendons and ligaments. Development 128, 3855–3866.

Scott, A., Cook, J.L., Hart, D.A., Walker, D.C., Duronio, V., Khan, K.M., 2007. Tenocyte responses to mechanical loading in vivo: a role for local insulin-like growth factor 1 signaling in early tendinosis in rats. Arthritis Rheum. 56, 871–881.

Screen, H.R.C., Bader, D.L., Lee, D.A., Shelton, J.C., 2004. Local strain measurement within tendon. Strain 40, 157–163.

September, A.V., Mokone, G.G., Schwellnus, M.P., Collins, M., 2006. Genetic risk factors for Achilles tendon injuries. Int. Sportmed J. 7, 201–215.

Sharma, P., Maffulli, N., 2005. Tendon injury and tendinopathy: healing and repair. J. Bone Joint Surg. Am. 87, 187–202.

Sharma, P., Maffulli, N., 2006. Biology of tendon injury: healing, modeling and remodeling. J. Musculoskelet. Neuronal Interact. 6, 181–190.

Silver, F.H., Freeman, J.W., Seehra, G.P., 2003. Collagen self-assembly and the development of tendon mechanical properties. J. Biomech. 36, 1529–1553.

Sluss, J.R., Liberti, J.P., Jiranek, W.A., Wayne, J.S., Zuelzer, W.A., 2001. pN collagen type III within tendon grafts used for anterior cruciate ligament reconstruction. J. Orthop. Res. 19, 852–857.

Smith, R.K.W., Gerard, M., Dowling, B., Dart, A.J., Birch, H.L., Goodship, A.E., 2002. Correlation of cartilage oligomeric matrix protein (COMP) levels in equine tendon with mechanical properties: a proposed role for COMP in determining function-specific mechanical characteristics of locomotor tendons. Equine Vet. J. Suppl. 34, 241–244.

Stange, R., Sahin, H., Wieskotter, B., Persigehl, T., Ring, J., Bremer, C., Raschke, M.J., Vieth, V., 2014. In vivo monitoring of angiogenesis during tendon repair: a novel MRI-based technique in a rat patellar tendon model. Knee Surg. Sports Traumatol. Arthrosc. 23, 2433–2439.

Steckel, H., Starman, J.S., Baums, M.H., Klinger, H.M., Schultz, W., Fu, F.H., 2007. The double-bundle technique for anterior cruciate ligament reconstruction: a systematic overview. Scand. J. Med. Sci. Sports 17, 99–108.

Steinberg, P.J., Hodde, K.C., 1990. The morphology of synovial lining of various structures in several species as observed with scanning electron microscopy. Scanning Microsc. 4, 987–1019 (discussion 1019–20).

Tagliafico, A., Michaud, J., Capaccio, E., Derchi, L., Martinoli, C., 2010. Ultrasound demonstration of distal biceps tendon bifurcation: normal and abnormal findings. Eur. Radiol. 20, 202–208.

Tan, Q., Lui, P.P.Y., Rui, Y.F., Wong, Y.M., 2012. Comparison of potentials of stem cells isolated from tendon and bone marrow for musculoskeletal tissue engineering. Tissue Eng. A 18, 840–851.

Thorpe, C.T., Stark, R.J.F., Goodship, A.E., Birch, H.L., 2010. Mechanical properties of the equine superficial digital flexor tendon relate to specific collagen cross-link levels. Equine Vet. J. 42, 538–543.

Thorpe, C.T., Udeze, C.P., Birch, H.L., Clegg, P.D., Screen, H.R.C., 2012. Specialization of tendon mechanical properties results from interfascicular differences. J. R. Soc. Interface 9, 3108–3117.

Thorpe, C.T., Birch, H.L., Clegg, P.D., Screen, H.R.C., 2013a. The role of the non-collagenous matrix in tendon function. Int. J. Exp. Pathol. 94, 248–259.

Thorpe, C.T., Udeze, C.P., Birch, H.L., Clegg, P.D., Screen, H.R.C., 2013b. Capacity for sliding between tendon fascicles decreases with ageing in injury prone equine tendons: a possible mechanism for age-related tendinopathy? Eur. Cell. Mater. 25, 48–60.

Tillander, B., Franzen, L., Norlin, R., 2002. Fibronectin, MMP-1 and histologic changes in rotator cuff disease. J. Orthop. Res. 20, 1358–1364.

Ushiki, T., 2002. Collagen fibers, reticular fibers and elastic fibers. A comprehensive understanding from a morphological viewpoint. Arch. Histol. Cytol. 65, 109–126.

Van Sterkenburg, M.N., Van Dijk, C.N., 2011. Mid-portion Achilles tendinopathy: why painful? An evidence-based philosophy. Knee Surg. Sports Traumatol. Arthrosc. 19, 1367–1375.

Venturoni, M., Gutsmann, T., Fantner, G.E., Kindt, J.H., Hansma, P.K., 2003. Investigations into the polymorphism of rat tail tendon fibrils using atomic force microscopy. Biochem. Biophys. Res. Commun. 303, 508–513.

Vogel, H.G., 1980. Influence of maturation and aging on mechanical and biochemical-properties of connective-tissue in rats. Mech. Ageing Dev. 14, 283–292.

Wellen, J., Helmer, K.G., Grigg, P., Sotak, C.H., 2005. Spatial characterization of T1 and T2 relaxation times and the water apparent diffusion coefficient in rabbit Achilles tendon subjected to tensile loading. Magn. Reson. Med. 53, 535–544.

Willberg, L., Sunding, K., Forssblad, M., Fahlström, M., Alfredson, H., 2011. Sclerosing polidocanol injections or arthroscopic shaving to treat patellar tendinopathy/jumper's knee? A randomised controlled study. Br. J. Sports Med. 45, 411–415.

Woo, S.L., An, K., Frank, C.B., Livesay, G., Ma, A., Zeminski, J., Wayne, J., Myers, B., 2000. Anatomy, biology, and biomechanics of tendon and ligament. In: Buckwalter, J., Einhorn, T., Simon, S. (Eds.), Orthopaedic Basic Science: Biology and Biomechanics of the Musculoskeletal System, second ed. American Academy of Orthopaedic Surgeons, Rosemont, IL.

Xu, Y., Murrell, G.A.C., 2008. The basic science of tendinopathy. Clin. Orthop. Relat. Res. 466, 1528–1538.

Xu, Y., Dong, S., Zhou, Q., Mo, X., Song, L., Hou, T., Wu, J., Li, S., Li, Y., Li, P., Gan, Y., Xu, J., 2014. The effect of mechanical stimulation on the maturation of TDSCs-poly(L-lactide-co-e-caprolactone)/collagen scaffold constructs for tendon tissue engineering. Biomaterials 35, 2760–2772.

Yin, Z., Chen, X., Chen, J.L., Shen, W.L., Nguyen, T.M.H., Gao, L., Ouyang, H.W., 2009. The regulation of tendon stem cell differentiation by the alignment of nanofibers. Biomaterials 31, 2163–2175.

Zhang, J., Wang, J.H.C., 2010. Mechanobiological response of tendon stem cells: implications of tendon homeostasis and pathogenesis of tendinopathy. J. Orthop. Res. 28, 639–643.

Zhang, G., Young, B.B., Ezura, Y., Favata, M., Soslowsky, L.J., Chakravarti, S., Birk, D.E., 2005. Development of tendon structure and function: regulation of collagen fibrillogenesis. J. Musculoskelet. Neuronal. Interact. 5, 5–21.

# Biomechanical properties of tendons and ligaments in humans and animals

2

## Abstract

Although biomechanical tendon and ligament tissue properties are being improved during regeneration after rupture, laceration, and tendinopathies (tendinosis and tendinitis), the biomechanical characteristics of healthy native tendons and ligaments should be known in order to compare surgical and rehabilitation outcomes to a baseline. Such baseline values are presented in this chapter for humans as well as for animals. As preclinical studies often use animal models to investigate novel surgical techniques as well as implantation of tissue engineered constructs, there is an inevitable need to choose an appropriate animal model in terms of biomechanics. An overview of rabbit, sheep, mouse, rat, dog, pig, monkey, and horse tendon and ligament models is given and the models are not only discussed in terms of suitability to represent the human situation, they are also compared to each other and with respect to their practicability.

**Keywords:** Gender, Age, Exercise, Small and large animal models.

## Abbreviations

| | |
|---|---|
| **ACL** | anterior cruciate ligament (tendons and ligaments) |
| **AT** | Achilles tendon (tendons and ligaments) |
| **ATT** | anterior *tibialis* tendon (tendons and ligaments) |
| **CSA** | cross-sectional area (feature of tendons and ligaments) |
| **FDP** | *flexor digitorum profundus* (tendons and ligaments) |
| **FHL** | *flexis hallucis longus* (tendons and ligaments) |
| **FITB** | fan-folded iliotibial band (tendons and ligaments) |
| **ISp** | *infraspinatus* (tendons and ligaments) |
| **MCL** | medial collateral ligament (tendons and ligaments) |
| **MMPs** | matrix metalloproteinases (enzymes in living cells) |
| **MSC** | mesenchymal stem cells (cells for seeding) |
| **Ø** | mean (element of statistics) |
| **PB** | *peroneus brevis* tendon (tendons and ligaments) |
| **PCL** | posterior cruciate ligament (tendons and ligaments) |
| **PLLA** | poly(L-lactid acid) (synthetic scaffold material) |
| **PT** | patellar tendon (tendons and ligaments) |
| **QPRPT** | quadriceps–patellar retinaculum–patellar tendon (tendons and ligaments) |
| **RC** | rotator cuff (tendons and ligaments) |
| **SD** | standard deviation (element of statistics) |
| **SDFT** | superficial digital flexor tendon (tendons and ligaments) |

**Biomechanics of Tendons and Ligaments. http://dx.doi.org/10.1016/B978-0-08-100489-0.00002-8**

| | |
|---|---|
| **SEM** | standard error of the mean (element of statistics) |
| **SSp** | *supraspinatus* (tendons and ligaments) |
| **STT** | *semitendinosus* tendon (tendons and ligaments) |
| **TGF (-β*n*)** | transforming growth factor beta ($n = 1, 2, 3$) (protein in living cells) |
| **TM** | *teres minor* (tendons and ligaments) |

## 2.1 Introduction

Tissue engineered tendons and ligaments include whole organ replacements as well as reinforcements by augmenting the damaged tissue. They should have a mechanical behavior similar to undamaged tissues—or at least allow early active mobilization during the rehabilitation including the biomechanical strength necessary for that. Therefore, in order to be able to compare biomechanical properties of materials aimed at tendon tissue engineering, the first step is to consider the natural biomechanical characteristics of tendons and ligaments—including peak in vivo values as well as cadaver ultimate strength values (failure values). Such biomechanical properties can then be considered as an ideal baseline serving as evaluation milestones (Shearn et al., 2011), which can be used to tune fabrication processes of scaffold materials and optimize the final products aimed at clinical use. In this chapter, biomechanical properties of native tendons and ligaments for humans as well as animals are presented and listed in Tables 2.1 and 2.2, respectively. Furthermore, elastic moduli were correlated to ultimate stress data in order to (i) give an overview of the wide range covered by the biomechanical properties and (ii) discuss divergences from the linear correlation lines that have quite high correlation coefficients (Figs. 2.1 and 2.2 for humans and animals, respectively).

## 2.2 Regional differences of biomechanical properties and impact of size

Human tendons and ligaments have been studied extensively in terms of their biomechanical properties and the factors influencing them. For example, the human cadaver *supraspinatus* tendon of the rotator cuff (RC) has been divided into three stripes (anterior, middle, and posterior) and ultimate stress, elastic modulus as well as ultimate load have been determined (Itoi et al., 1995). Clearly, the anterior portion of the *supraspinatus* has been shown to be mechanically stronger compared to the middle and posterior portion and has therefore be assigned to perform the main functional role of this tendon. The anterior part had ultimate stresses and ultimate loads of twofold–fourfold compared to the other parts and an elastic modulus of at least twofold (Table 2.1 and Fig. 2.1).

A subsequent study performed in the same group dealt with the regional variations in the structural properties of the *subscapularis* tendon in two joint positions (angles were 0 degree for the first position (hanging arm) and 60 degrees for the second arm

**Table 2.1 Human native tendons and ligaments: Biomechanical and structural properties**

| Articulation | Tendon or ligament | Specifications | Ultimate stress (MPa) | Elastic modulus (MPa) | Ultimate load (N) | Stiffness (N/mm) | Ref. |
|---|---|---|---|---|---|---|---|
| Hand | FDP | Native FDP | 55 ± 19 | 520 ± 200 | | | Pridgen et al. (2011) |
| Elbow | Ulnar MCL | Native ulnar MCL | | | | 99 ± 67 | Prud'homme et al. (2008) |
| Shoulder | RC | Native RC (suprasp. + infrasp.) | | | 935 ± *5* | 127 ± *1* | Van Der Meijden et al. (2013) |
| | Supraspinatus | Anterior | 17 ± *7* | 140 ± *18* | 411 ± *159* | | Itoi et al. (1995) |
| | | Middle portion | 6 ± *3* | 68 ± *15* | 153 ± *87* | | |
| | | Posterior | 4 ± *1* | 58 ± *10* | 88 ± *32* | | |
| | Subscapularis | Inferior portion (0 degree)* | | | 75 ± 54 | 27 ± 18 | Halder et al. (2000b) |
| | | Inferior portion (60 degrees)** | | | 30 ± 13 | 10 ± 6 | |
| | | Superior (0 degree) | | | 623 ± 199 | 147 ± 32 | |
| | | Superior (60 degrees) | | | 478 ± 207 | 209 ± 61 | |
| | | Mid-superior (0 degree) | | | 706 ± 165 | | |
| | | Mid-superior (60 degrees) | | | 598 ± 268 | | |
| | Infraspinatus | Superior*** | 16 ± 7 | 120 ± 55 | 463 ± 237 | 134 ± 45 | Halder et al. (2000a) |
| | | Mid-superior | 27 ± 9 | 156 ± 60 | 677 ± 231 | 171 ± 60 | |
| | | Mid-inferior | 16 ± 7 | 111 ± 50 | 315 ± 182 | 97 ± 40 | |
| | | Inferior | 22 ± 14 | 140 ± 55 | 550 ± 285 | 149 ± 50 | |
| | | Teres minor | 2 ± 1 | 14 ± 9 | 67 ± 31 | 23 ± 14 | |
| Knee | PT | Young donors | 65 ± 15 | 660 ± 266 | | | Johnson et al. (1994) |
| | | Old donors | 54 ± 10 | 504 ± 222 | | | |
| | | PT as graft | | 225 ± 129 | 1784 ± 580 | 210 ± 66 | Wilson et al. (1999) |
| | | Native PT | 57–58 | | 2734–2900 | 162–208 | Noyes et al. (1984) |

*Continued*

**Table 2.1 Continued**

| Articulation | Tendon or ligament | Specifications | Ultimate stress (MPa) | Elastic modulus (MPa) | Ultimate load (N) | Stiffness (N/mm) | Ref. |
|---|---|---|---|---|---|---|---|
| | MPFL | Native MPFL | | | 209 ± 55 | | Burks et al. (1998) |
| | | | | | 208 ± 90 | | Mountney et al. (2005) |
| | | Femur–MPFL–patella | | | 145 ± 58 | 43 ± 10 | Criscenti et al. (2016) |
| | QT (QPRPT) | Native QT | 10–16 | | 249–371 | 17–24 | Noyes et al. (1984) |
| | ACL | Native younger humans | 38 ± 9 | 111 ± 26 | 1730 ± 660 | 182 ± 56 | Noyes and Grood (1976) |
| | | Native older humans | 13 ± 5 | 65 ± 24 | 734 ± 266 | 129 ± 39 | |
| | | Native ACL | | | 2160 ± 157 | 242 ± 28 | Woo et al. (1991) |
| | | | | | 886 ± 254 | | Delcroix et al. (2013) |
| | | Male | 26 ± 10 | 128 ± 35 | 1818 ± 699 | 308 ± 89 | Chandrashekar et al. (2006) |
| | | Female | 23 ± 9 | 99 ± 50 | 1266 ± 527 | 199 ± 88 | |
| | | Combined | 24 ± 9 | 113 ± 45 | 1526 ± 658 | 250 ± 102 | |
| | MCL | Native MCL | 39 ± 5 | ‡ 332 ± 58 | | | Quapp and Weiss (1998) |
| | ATT | Native ATT | | $1.2 \pm 0.2 \times 10^3$ | 530 ± *59* | 161 ± *26* | Maganaris and Paul (1999) |
| | | | 50 ± 12 | | 3357 ± 111 | | Delcroix et al. (2013) |

| | | | | | | | |
|---|---|---|---|---|---|---|---|
| | STT | Native STT | 89±*5* | | 1216±*50* | | Noyes et al. (1984) |
| | FITB | Native FITB | 48±11 | | 3459±939 | | Delcroix et al. (2013) |
| | Gracilis tendon | Native gracilis | 112±*4* | | 838±*30* | | Noyes et al. (1984) |
| | HT (STT +GT) | STT+gracilis as graft | | 145±58 | 2422±538 | 238±71 | Wilson et al. (1999) |
| | ALL | Native ALL | 33±4 | | 175±65 | 20±8 | Kennedy et al. (2015), Zens et al. (2015) |
| | | ALL as ACL-graft | | | 50±15 | | |
| Foot/Foot ankle | AT | Long graft 75 mm | | | | 88±36 | Defrate et al. (2004) |
| | | Medium graft 48 mm | | | | 116±37 | |
| | | Short graft 34 mm | | | | 154±42 | |
| | | Native AT | | | 1724±514 | 175±95 | Datta et al. (2006) |
| | | | | | 843–5579 | | Giza et al. (2011) |
| | | Donors aged 36–50 year | 73±8 | ǂ 459±54 | | 794±298 | Lewis and Shaw (1997) |
| | | Donors aged 52–67 year | 81±14 | 545±43 | | 920±120 | |
| | | Donors aged 79–100 year | 48±16 | 333±109 | | 634±300 | |
| | | All donors | 59±18 | 375±102 | | 685±262 | |
| | | Native AT, var. load rates | 79±22 | 819±208 | 5098±1199 | | Wren et al. (2001) |
| | PB | Native PB | | | 333±137 | 44±19 | Datta et al. (2006) |
| | FHL | Native FHL | | | 511±164 | 43±14 | |
| Others | Fascia lata | Native fascia lata | 79±*5* | | 628±*35* | 39±*2* | Noyes et al. (1984) |
| | | | | 389±189 | | 85±15 | Derwin et al. (2008) |

Data: Ø±SD (mean±standard deviation) or Ø±*SEM* (mean±*standard error of the mean*).

*Key*: Tendons and ligaments: *ACL*, anterior cruciate ligament; *AT*, Achilles tendon; *ATT*, anterior tibial tendon; *FDP*, flexor digitorum profundus; *FHL*, flexis hallucis longus; *FITB*, fan-folded iliotibial band; *MCL*, medial collateral ligament; *MPFL*, medial patellofemoral ligament; *PB*, peroneus brevis; *PT*, patellar tendon; *QT*, quadriceps tendon; *QPRPT*, quadriceps–patellar retinaculum–patellar tendon; *RC*, rotator cuff; *STT*, semitendinosus tendon.

Remarks: *Hanging-arm position (0 degree), **angulated arm position (60 degrees), ***combination of 0 and 60 degrees position.

Mechanical parameters: ǂ tangent modulus.

**Table 2.2 Animal native tendons and ligaments: Biomechanical and structural properties**

| Model | Articulation | Tendon or ligament | Specifications | Ultimate stress (MPa) | Elastic modulus (MPa) | Ultimate load (N) | Stiffness (N/mm) | Ref. |
|---|---|---|---|---|---|---|---|---|
| Mouse | Shoulder | Suprasp. | Native suprasp. | | | 1.2±0.5 | 2.4±1.6 | Bell et al. (2015) |
| | Foot ankle | AT | Native AT | | | 8±1 | | Mikic et al. (2001) |
| | | | | | | 8±1 | 6±1 | Palmes et al. (2002) |
| | Knee | PT | Native PT | 18±3 | 140±20 | 5±1 | 12±2 | Gilday et al. (2014) |
| Inbred mice | Knee | PT | Nat. PT C57BL/6 (1) | 16±3 | 105±20 | 4±1 | 10±1 | Lalley et al. (2015) |
| | | | Nat. PT MRL/MpJ (2) | 18±7 | 131±40 | 4±2 | 9±2 | |
| Rat | Foot ankle | AT | Native AT | | | 60–80 | 30–86 | Majewski et al. (2008, 2012) |
| | | | | 16±6 | | 37±12 | • 107±41 | Chamberlain et al. (2013) |
| | | | | 28±8 | | 53±8 | | Pietschmann et al. (2013) |
| | Tail | Tail T | Native fascicles | 54±7 | ‡ 892±82 | | | Li and Snedeker (2013) |
| | Knee | PT | Native PT | 30±4 | 324±56 | | | Sahin et al. (2012) |

| | | | | | | | | |
|---|---|---|---|---|---|---|---|---|
| | | ACL | Native ACL | | | 57±*2* | 40±*7* | Leong et al. (2015) |
| | Shoulder | Suprasp. | Native suprasp. | | | 29–42 | 24–66 | Zalavras et al. (2006) |
| | | | | 25±9 | 433±251 | | | Carpenter et al. (1998) |
| | | | | | | 18±3 | | Ide et al. (2009) |
| Rabbit | Paw | FDP | Native FDP, hind | 73±15 | 1166±281 | | | Saber et al. (2010) |
| | | | | 64±12 | 980±150 | | | Angelidis et al. (2010) |
| | | Flexors | Native, fore+hind | 59–66 | 628–630 | | | Chong et al. (2009) |
| | | | Native flexors | | | 93±18 | 38±5 | Henn et al. (2010) |
| | Foot ankle | AT | AT fibers | | 298±35 | | | Kahn et al. (2013) |
| | | | Interfibrils | | 223 | | | |
| | | | Fibrils | | 3000 | | | |
| | | | Native AT | 116–228 | ‡ 59–87 | | | Hosey et al. (1991) |
| | | | | | | | 32±9 | Ouyang et al. (2003) |
| | | | | | | 208±35 | | Ilhami et al. (2004) |
| | | | | 33±*4* | 180±*13* | 390±*50* | 93±*9* | Juncosa-Melvin et al. (2006a) |
| | | | | 10±2 | 19±*3* | | | Ni et al. (2012) |
| | | | | 34±7 | | 293±16 | | Buschmann et al. (2011) |

*Continued*

**Table 2.2** **Continued**

| Model | Articulation | Tendon or ligament | Specifications | Ultimate stress (MPa) | Elastic modulus (MPa) | Ultimate load (N) | Stiffness (N/mm) | Ref. |
|---|---|---|---|---|---|---|---|---|
| | | | | 20–34 | 1.5–3.3 | 212–359 | 57–215 | Meimandi-Parizi et al. (2013), Oryan et al. (2013), Moshiri et al. (2013), Oryan et al. (2015) |
| | | | | 32 ± 15 | 95 ± 12 | 330 ± 120 | 28 ± 5 | Meier Buergisser et al. (2014) |
| | | | Native/trained AT | | | 377/388 | 157/177 | Viidik (1969) |
| | Knee | PT | Native PT | 101 ± *16* | 861 ± *106* | 471 ± *67* | 160 ± *11* | Juncosa-Melvin et al. (2006b) |
| | | Time 0w | Acellular fascicles | | ‡ 190 ± 43 | | | Ikema et al. (2007) |
| | | | Ac fasc. + Fb | | 175 ± 40 | | | |
| | | Time 3w | Ac fasc. | | 180 ± 40 | | | |
| | | | Ac fasc. + Fb | | 95 ± 25 | | | |
| | | Time 6w | Ac fasc. | | 110 ± 30 | | | |
| | | | Ac fasc. + Fb | | 60 ± 35 | | | |

| | | | | | | | | |
|---|---|---|---|---|---|---|---|---|
| | | Time 0w | Acellular bundles | | 760 ± 100 | | | |
| | | | Ac bundl. + Fb | | 740 ± 100 | | | |
| | | Time 3w | Ac bundl. | | 220 ± 50 | | | |
| | | | Ac bundl. + Fb | | 230 ± 100 | | | |
| | | Time 6w | Ac bundl. | | 140 ± 50 | | | |
| | | | Ac bundl. + Fb | | 200 ± 50 | | | |
| | | ACL | Native ACL | 76 ± 10 | ‡ 464 ± 66 | | | Sakai et al. (2002) |
| | | | | 98 ± 18 | ‡ 680 ± 170 | | | Nagumo et al. (2005) |
| | | | | | | 291 ± 73 | | Tischer et al. (2010) |
| | | MCL | Native MCL | 76 ± 14 | ‡ 936 ± 284 | 332 ± 51 | 90 ± 15 | Musahl et al. (2004) |
| | | | Sham-OP MCL | 84 ± 19 | 1142 ± 412 | | | Liang et al. (2006) |
| | | | Native MCL | | | 106 ± 28 | 93 ± 27 | Irie et al. (2011) |
| | | STT | Native STT | | | 134 ± *13* | 59 ± *7* | Tischer et al. (2007) |
| | Shoulder | RC | Native infrasp. | 54 ± *13* | ‡ 583 ± *93* | | | Funakoshi et al. (2005) |
| Chicken | Foot | FDP | Native FDP | | | 110 ± 9 | | Cao et al. (2002) |
| Dog | Shoulder | RC | Native infrasp. | | | 187 ± 31 | | Dejardin et al. (2001) |
| | | | | | * 405 ± 86 | | | Derwin et al. (2006) |

*Continued*

**Table 2.2 Continued**

| Model | Articulation | Tendon or ligament | Specifications | Ultimate stress (MPa) | Elastic modulus (MPa) | Ultimate load (N) | Stiffness (N/mm) | Ref. |
|---|---|---|---|---|---|---|---|---|
| | | | | | | 1595 ± 285 | 515 ± 141 | Derwin et al. (2009) |
| | | | | | | | 100 ± 30 | Smith et al. (2012) |
| Goat | Knee | ACL | Native ACL | | | 3151 ± 407 | 480 ± 42 | Badylak et al. (1999) |
| Sheep | Foot ankle | AT | Native AT | | 140 ± 47 | 705 ± 86 | | Huri et al. (2013) |
| | Shoulder | RC | Native infrasp. | | | 5233 ± 736 | 545 ± 47 | Schlegel et al. (2006) |
| | | | | | | 3516 ± 839 | 244 ± 71 | Santoni et al. (2010) |
| | Knee | ACL | Native ACL | | | 780 ± 60 | 210 ± 60 | Yoshikawa et al. (2006) |
| | | | | | | 1490 ± 500 | 270 ± 120 | Kondo et al. (2012) |
| | | | | | | 1400 ± 100 | 230 ± 30 | Seitz et al. (2013) |
| Pig | Foot | FDS | Native FDS | 36–42 | | | | Liu et al. (2006) |
| | Knee | ACL | Native ACL | | | ° 713 ± 118 | 154 ± 20 | Joshi et al. (2009) |

| | | | | | | | | |
|---|---|---|---|---|---|---|---|---|
| Rhesus monkey | Knee | ACL | Native ACL | 66±8 | 186±26 | 830±110 | 194±28 | Noyes and Grood (1976) |
| Horse | Hoof | FDS | Foal (5 mt) native T | 99–119 | | $7–9 \times 10^3$ | | Dowling and Dart (2005) |
| | | | Foal (11 mt) | 100–116 | | $11–12 \times 10^3$ | | |
| | | | Adult native T | 109±*8* | 1189±*63* | $12 \pm \mathit{1} \times 10^3$ | | Dowling et al. (2002), |
| | | | | 65±*4* | | $8 \pm \mathit{1} \times 10^3$ | 1076±*93* | Crevier et al. (1996) |

Data: Ø±SD (mean±standard deviation) or *Ø±SEM* (mean±*standard error of the mean*).
Key: *Sham-OP*, operation without a special treatment.
Tendons and ligaments: *AT*, Achilles tendon; *ACL*, anterior cruciate ligament; *FDP*, flexor digitorum profundus; *FDS*, flexor digitorum superficialis; *Infrasp.*, infraspinatus; *MCL*, medial collateral ligament; *PT*, patellar tendon; *RC*, rotator cuff; Suprasp., supraspinatus; *STT*, semitendinosus tendon.
Mechanical parameters: ‡ tangent modulus, *measurement at 8% strain, °yield Load, and •normalized stiffness (MPa).
Key for inbred mice: (1) C57BL/6 = common inbred strain of laboratory mouse, (2) MRL/MpJ = Murphy Roths Large mice provide unique insights into wound repair and regeneration.

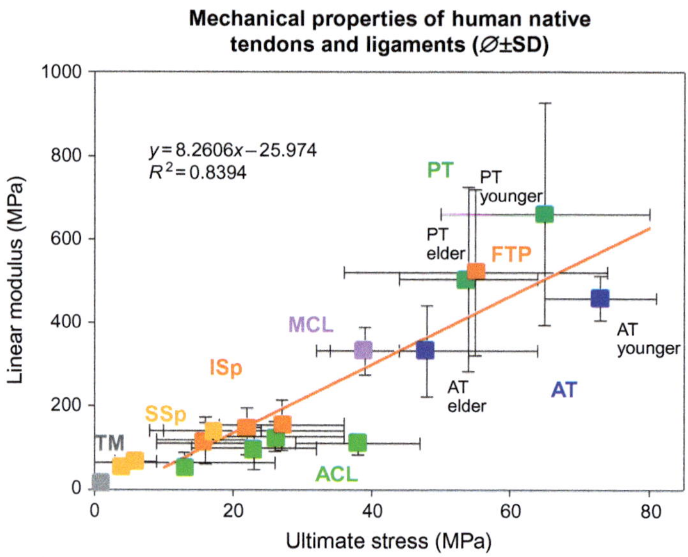


**Fig. 2.1** Human native tendons and ligaments: ex vivo biomechanical correlations with data from literature. Keys: *Tendons*: *ACL*, anterior cruciate ligament; *AT*, Achilles tendon; *FDP*, flexor digitorum profundus; *ISp*, infraspinatus; *MCL*, medial collateral ligament; *PT*, patellar tendon; *SSp*, supraspinatus; *TM*, teres minor. Data: *Ø* ± *SD*, mean ± standard deviation.
Data from Pridgen, B.C., Woon, C.Y.L., Kim, M., Thorfinn, J., Lindsey, D., Pham, H., et al., 2011. Flexor tendon tissue engineering: acellularization of human flexor tendons with preservation of biomechanical properties and biocompatibility. Tissue Eng. Part C Methods 17, 819–828; Itoi, E., Berglund, L.J., Grabowski, J.J., Schultz, F.M., Growney, E.S., Morrey, B.F., et al., 1995. Tensile properties of the supraspinatus tendon. J. Orthop. Res. 13, 578–584; Halder, A., Zobitz, M.E., Schultz, F., An, K.N. 2000a. Mechanical properties of the posterior rotator cuff. Clin. Biomech. (Bristol, Avon) 15, 456–462; Johnson, G.A., Tramaglini, D.M., Levine, R.E., Ohno, K., Choi, N.Y., Woo, S.L. 1994. Tensile and viscoelastic properties of human patellar tendon. J. Orthop. Res. 12, 796–803; Noyes, F.R., Grood, E.S. 1976. The strength of the anterior cruciate ligament in humans and Rhesus monkeys. J. Bone Joint Surg. Am. 58, 1074–1082; Chandrashekar, N., Mansouri, H., Slauterbeck, J., Hashemi, J. 2006. Sex-based differences in the tensile properties of the human anterior cruciate ligament. J. Biomech. 39, 2943–2950; Quapp, K.M., Weiss, J.A. 1998. Material characterization of human medial collateral ligament. J. Biomech. Eng. 120, 757–763; Lewis, G., Shaw, K.M. 1997. Tensile properties of human tendo achillis: effect of donor age and strain rate. J. Foot Ankle Surg. 36, 435–445.

position). For all three portions considered (inferior, superior, and mid-superior), the ultimate loads were significantly higher for the hanging arm position compared to the 60 degree angle (Table 2.1). Moreover, the superior portion and the mid-superior portion which had similar biomechanical properties, were found to have around 10 times higher loads at failure compared to the inferior portions, which led the authors to conclude that repair of the inferior portions of the *subscapularis* tendon should be considered in surgery for *glenohumeral* instability because significantly lower ultimate load and stiffness in the inferior tendon region could facilitate anterior dislocation of the humeral head (Halder et al., 2000b).

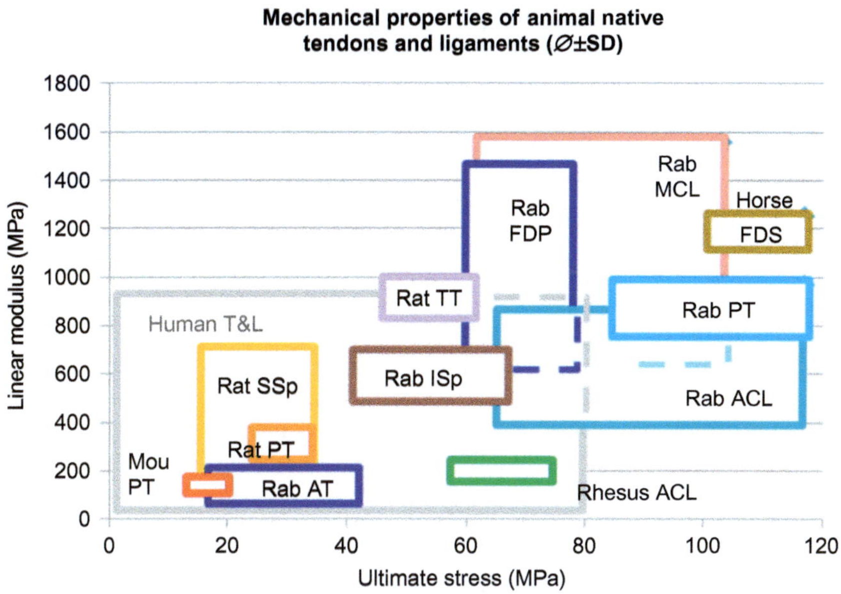


**Fig. 2.2** Animal native tendons and ligaments: ex vivo biomechanical correlations. Keys: *Tendons: ACL*, anterior cruciate ligament; *AT*, Achilles tendon; *FDP*, flexor digitorum profundus; *FDS*, flexor digitorum superficialis; *ISp*, infraspinatus; *MCL*, medial collateral ligament; *PT*, patellar tendon; *SSp*, supraspinatus; *T&L*, tendons and ligaments; *TT*, tail tendon. Data: $\varnothing \pm SD$, mean ± standard deviation.
Data from Gilday, S.D., Chris Casstevens, E., Kenter, K., Shearn, J.T., Butler, D.L. 2014. Murine patellar tendon biomechanical properties and regional strain patterns during natural tendon-to-bone healing after acute injury. J. Biomech. 47, 2035–2042; Carpenter, J.E., Thomopoulos, S., Flanagan, C.L., Debano, C.M., Soslowsky, L.J. 1998. Rotator cuff defect healing: a biomechanical and histologic analysis in an animal model. J. Shoulder Elb. Surg. 7, 599–605; Chong, A.K., Chang, J., Go, J.C.H. 2009. Mesenchymal stem cells and tendon healing. Front. Biosci. 14, 4598–4605; Saber, S., Zhang, A.Y., Ki, S.H., Lindsey, D.P., Smith, R.L., Riboh, J., et al., 2010. Flexor tendon tissue engineering: bioreactor cyclic strain increases construct strength. Tissue Eng. Part A 16, 2085–2090; Angelidis, I.K., Thorfinn, J., Connolly, I.D., Lindsey, D., Pham, H.M., Chang, J. 2010. Tissue engineering of flexor tendons: the effect of a tissue bioreactor on adipoderived stem cell-seeded and fibroblast-seeded tendon constructs. J. Hand Surg. Am. 35, 1466–1472; Buschmann, J., Müller, A., Feldman, K., Tervoort, T.A., Fessel, G., Snedeker, J.G., et al., 2011. Small hook thread (Quill) and soft felt internal splint to increase the primary repair strength of lacerated rabbit Achilles tendons: biomechanical analysis and considerations for hand surgery. Clin. Biomech. (Bristol, Avon) 26, 626–631; Meier Buergisser, G., Calcagni, M., Muller, A., Bonavoglia, E., Fessel, G., Snedeker, J.G., et al., 2014. Prevention of peritendinous adhesions using an electrospun DegraPol polymer tube: a histological, ultrasonographic, and biomechanical study in rabbits. Biomed Res. Int. 2014, 656240–656240; Juncosa-Melvin, N., Boivin, G.P., Galloway, M.T., Gooch, C., West, J.R., Butler, D. L. 2006a. Effects of cell-to-collagen ratio in stem cell-seeded constructs for Achilles tendon repair. Tissue Eng. 12, 681–689; Juncosa-Melvin, N., Boivin, G.P., Gooch, C., Galloway, M.T., West, J.R., Dunn, M.G., et al., 2006b. The effect of autologous mesenchymal stem cells on the biomechanics and histology of gel-collagen sponge constructs used for rabbit

A tendon that is frequently involved in RC tears is the *infraspinatus* which is an active and passive stabilizer of the glenohumeral joint. Halder et al. (2000a) investigated positional changes in mechanical properties of this tendon (Table 2.1) and found that although there were considerable changes of ultimate load and stiffness, these different properties do not play a role in the patho-mechanism of posterior shoulder dislocations.

Furthermore, the length of a tendon has an impact on its biomechanics. A study by Defrate et al. (2004) using different lengths of Achilles tendon (AT) grafts for posterior cruciate ligament (PCL) reconstruction, clearly indicated that short tendon grafts are significantly stiffer compared to medium or long grafts (Table 2.1); in order to mimic mid tunnel fixation of the PCL, inlay fixation and fixation near the ligament, 75, 48, and 34 mm long AT grafts were used, respectively, leading to 175% higher stiffness for the short graft and to 132% for the medium graft compared to the long tendon graft. Such information is useful not only in planning the dimensions of tendons aimed at transplantation, but also very important to be considered in artificial tendon graft design as for example collagen sponges seeded with mesenchymal stem cells (MSCs) aimed at tendon reconstruction (Nirmalanandhan et al., 2007). Going along with these findings, it was suggested not to take longer tendon grafts instead of a *palmaris longus* graft in medial collateral ligament (MCL) reconstruction when

**Fig. 2.2 continued.**
patellar tendon repair. Tissue Eng. 12, 369–379; Sakai, T., Yasuda, K., Tohyama, H., Azuma, H., Nagumo, A., Majima, T., et al., 2002. Effects of combined administration of transforming growth factor-beta 1 and epidermal growth factor on properties of the in situ frozen anterior cruciate ligament in rabbits. J. Orthop. Res. 20, 1345–1351; Nagumo, A., Yasuda, K., Numazaki, H., Azuma, H., Tanabe, Y., Kikuchi, S.-I., et al., 2005. Effects of separate application of three growth factors (TGF-beta1, EGF, and PDGF-BB) on mechanical properties of the in situ frozen-thawed anterior cruciate ligament. Clin. Biomech. (Bristol, Avon) 20, 283–290; Musahl, V., Abramowitch, S.D., Gilbert, T.W., Tsuda, E., Wang, J.H.C., Badylak, S.F., et al., 2004. The use of porcine small intestinal submucosa to enhance the healing of the medial collateral ligament-a functional tissue engineering study in rabbits. J. Orthop. Res. 22, 214–220; Liang, R., Woo, S.L.Y., Takakura, Y., Moon, D.K., Jia, F., Abramowitch, S.D. 2006. Long-term effects of porcine small intestine submucosa on the healing of medial collateral ligament: a functional tissue engineering study. J. Orthop. Res. 24, 811–819; Funakoshi, T., Majima, T., Iwasaki, N., Suenaga, N., Sawaguchi, N., Shimode, K., et al., 2005. Application of tissue engineering techniques for rotator cuff regeneration using a chitosan-based hyaluronan hybrid fiber scaffold. Am. J. Sports Med. 33, 1193–1201; Noyes, F.R., Grood, E.S. 1976. The strength of the anterior cruciate ligament in humans and Rhesus monkeys. J. Bone Joint Surg. Am. 58, 1074–1082; Crevier, N., Pourcelot, P., Denoix, J.M., Geiger, D., Bortolussi, C., Ribot, X., et al., 1996. Segmental variations of in vitro mechanical properties in equine superficial digital flexor tendons. Am. J. Vet. Res. 57, 1111–1117; Sahin, H., Tholema, N., Petersen, W., Raschke, M.J., Stange, R. 2012. Impaired biomechanical properties correlate with neoangiogenesis as well as VEGF and MMP-3 expression during rat patellar tendon healing. J. Orthop. Res. 30, 1952–1957; Li, X., Snedeker, J.G. 2013. Wired silk architectures provide a biomimetic ACL tissue engineering scaffold. J. Mech. Behav. Biomed. Mater. 22, 30–40.

the *palmaris longus* graft properties were compared to the properties of *gracilis* tendon, *semitendinosus* tendon (STT), and *patellar* tendon (PT; Prud'homme et al., 2008).

Fig. 2.1 shows the wide range of human tendon tensile properties; from 2 up to 80 MPa with respect to ultimate stress and from 3 to 680 MPa on the modulus axis, the modulus-stress pairs for different human tendons are given. Although there is quite a good positive linear correlation ($r^2=0.84$), some tendons have their modulus-stress pairs considerably below the linear correlation line and some of them above. For example, anterior cruciate ligament (ACL) as well as AT are lying below the correlation line; although completely different in their daily function (AT being a positional tendon, ACL an energy storing ligament), both of them exhibit higher ultimate stresses than would be predicted by the linear correlation equation ($y=8.3x-26$). As such, these tendons and ligaments are considered to have adapted a pronounced stress withstanding behavior relative to their intrinsic elasticity when compared with the average tendon and ligament biomechanics. On the other hand, PT in young and elderly people has modulus-stress pairs above the correlation line. This suggests that PT tissue is adapted to a higher elasticity than average tendon tissue and also relative to its inherent ultimate stress property.

## 2.3 Intrinsic factors: Gender and age

Not only positional changes lead to differences in structural and material mechanical properties, gender also plays a pivotal role. For example, ACL injuries do occur much more often in females than in males. This is not only due to the smaller size of the female ACL compared to the male ACL, but also due to its lower biomechanical properties (Table 2.1). It was found that female ACL only reached 85% failure stress, 77% elastic modulus, 70% failure load, and 65% stiffness values of their male equivalents (Chandrashekar et al., 2006). Especially the significantly lower stiffness may be a core factor for predisposing females involved in sports activities to injured ACLs.

Other gender differences were found for the tendinous inscription (TI) in the human *semitendinosus* muscle (Kellis and Balidou, 2014). The *semitendinosus* muscle is a part of the hamstring muscle group. It is divided into two parts by a diagonal TI. In an ultrasound (US) study with 76 young participants (all aged around 20 years), the TI varied not only with different hip and knee positions which was assessed by three angles of 0, 45, and 90 degrees, but also significantly with gender; while female participants had a higher normalized TI long arm, males had higher surface, apex, and deep angles. As such, the authors speculated that the TI increases the resistance to stretch the muscle because the *semitendinosus* muscle has a maximum strain of 19%, which is higher than the rest of the hamstring muscles (Magnusson et al., 2000). While it is accepted that free tendons store and release energy, the aponeurosis ensures effective transmission of forces (Kjaer et al., 2008) (see Chapter 3). However, after Huijing (2009), a series of muscle-tendon-muscle more affects the force transmission to the tendon part. As such, females with higher TI long arms than males may exhibit a higher resistance to stretch the semitendinosus muscle than males.

Depending on the age of the donor and also on the test conditions such as the strain rates during tensile testing, load until failure values for human ATs are reported to range between 843 and 5579 N (Giza et al., 2011). In terms of age, the failure stresses of human AT decrease with increasing age—as found in a linear correlation with a correlation coefficient of 0.474 (negatively correlated) (Lewis and Shaw, 1997). Accordingly, ATs of donors aged 79–100 years have failure stresses of only 65% of those found for donors aged 36–50 years (Table 2.1). Moreover, also modulus of elasticity and linear stiffness decline with increasing age, although not as that expressed as the failure stresses (Lewis and Shaw, 1997). With respect to strain rates during tensile testing of human ATs, increasing strain rate from 1 to 10%/s increases the elastic modulus by 0.7%, while it increases the failure stress by 21% (Wren et al., 2001), explaining partly the wide range of values found in the literature for that kind of specimen (Table 2.1). Another reason can be attributed to the measurement technique which may have a high impact on the biomechanical data acquired (Butler et al., 1984).

Furthermore, the impact of age on tendon properties was assessed by RNA extraction of human ATs of young and old participants (young group: $19 \pm 5.8$ years, $n=4$, 4 males; old group: $69 \pm 7.3$ years, $n=5$, 3 females and 2 males) (Peffers et al., 2015). Interestingly, in their transcriptome analysis, the authors found 325 transcribed elements that changed between the young and old groups, of which 191 had increased in the old group and 134 had decreased in the old group. The changed transcribed elements included protein-coding as well as noncoding transcripts such as small noncoding RNAs, pseudo-genes, long noncoding RNAs, and one miRNA. The top-scoring networks derived from the 325 transcribed elements differentially expressed in aging included (i) cellular function and maintenance, (ii) cellular growth and proliferation, (iii) cell cycle and skeletal and muscular system development, and (iv) cellular development. Although tendon disease is clearly associated with aging, this study investigating participants with throughout healthy tendons, obviously showed, that aging and disease-associated processes are completely different in terms of gene expression. While various tendinopathy-related studies demonstrated for example an upregulation in collagen 1 alpha 1 (COL1A1) expression and also collagen 2 alpha 1 (COL2A1), aggrecan, SOX 9 (human transcription factor), and various MMPs (matrix metalloproteinases) (Corps et al., 2004, 2012), the study by Peffers et al. (2015) showed contrarily a very low expression of COL2A1, aggrecan, SOX 9, and MMPs—indicating a very important fact: tendon degeneration is not an inevitable consequence of aging!

ACL was another human tendon biomechanically investigated in terms of age effect (Noyes and Grood, 1976). The young group consisted of 6 participants aged 16–26 years, while the old group had 20 participants aged 48–86 years. Not only the biomechanics changed significantly with 1.4 times lower stiffness in the old group compared to the young one, 2.4 times lower maximum force, 2.6 times lower energy to failure, 1.7 times lower elastic modulus, and 2.8 times lower maximum stress, but also the failure mode was different in the old and young group. While the ACLs showed a clear ligament disruption during tensile testing in the young group, it was an avulsion of bone beneath the ligament insertion in the old group. Histological analyzes revealed that in the old group there was a significantly reduced bone mass which corroborated the failure mode at the insertion site in the old group rather than in the ligament tissue as found for the young participants.

In contrast to these findings with clear aging effects in terms of biomechanics, the central parts of the PT showed only minimal differences when stress-relaxation curves, cyclic stress-relaxation, and load to failure tests were performed in two groups (group I aged 29–50 years and group II aged 64–93 years; Johnson et al., 1994). Of all the acquired biomechanical properties, only the ultimate load was significantly lower in the old group compared to the young group, with an average value that was 17% lower. No other differences in tensile or viscoelastic behavior were found.

## 2.4 Extrinsic factors: Physical activity and exercise

How do tendons adapt to loading? When tendons are subjected to high stress, they show a viscoelastic behavior and they store and release elastic strain energy (Biewener and Roberts, 2000). In this regard, a long and thin tendon would serve best because strain energies would be accordingly high. However, a thicker tendon having less strain energy stored and released would have a lower stress in the tendon tissue and therefore provide a higher safety in terms of possible rupture caused by overload. Tendon tissue adapts to repetitive force through alterations of the tissue composition and the biomechanical properties as shown for tendons in runners compared to nonrunners after a regular exercise during five years; the cross-sectional area (CSA) in runners was associated with a significantly higher value (+22%) leading to a lower stress if forces obtained are assumed to be the same (Rosager et al., 2002).

In a study including young men and women aged 18–30 years, biomechanical data of ATs were assessed using an US-based method for three groups; at rest, after a 10-min. treadmill walk, and after a fatigue protocol of 100 toe jumps performed in a Smith machine (load was equal to 20% of the body weight) (Joseph et al., 2014). Force and stress values were lowered to approximately 85% after the "jumping" regime compared to the "rest" and "walk" groups, for both, men and women, respectively. This can be explained by the fatiguing nature of the loading protocol. When stiffness and elastic modulus were compared, only women showed a significant reduction with 70% stiffness and 68% modulus after "jumping" compared to the "rest" and "walk" situations, which had very similar values. As for men, no significant changes occurred to these biomechanical properties. Joseph et al. concluded that women respond to an acute loading protocol with a clear increase in tendon compliance, in other words: The significantly reduced stiffness and modulus determined in women's tendons after high loading may represent a protective effect—which can explain the large discrepancy in tendinopathy, tendon disorders, and ruptures found between the sexes.

In another study, the length and the strain of young men's and women's ATs ($28 \pm 4$ years) was significantly increased after an eccentric heal drop exercise compared to subjects without exercise (Obst et al., 2016). As such, a single bout of eccentric heal drop exercise in healthy young adults had an immediate effect on AT biomechanical properties, which can be explained by mechanical creep of tendon tissue under load.

The impact of exercise was also studied in elderly people (Grosset et al., 2014). Two groups of participants with group 1 aged $74 \pm 5$ years ($n = 9$) and group 2 aged $68 \pm 6$ years ($n = 8$). The mean anthropometric and habitual physical activity of the two groups was very similar. The two groups performed different training regimen during 12 weeks; while group 1 performed 40% of repetition maximum (low regimen), group 2 reached 80% of repetition maximum (high regimen). In other words, group 1 had one supervised gym-based class and two home-based sessions per week, while group 2 had two supervised gym-based classes and one home-based session per week. There was a significant increase from pretest situation to posttest situation only in group 2: the stiffness and the young modulus of the patella tendon increased by a factor of 1.6 and 1.5, as assessed by US. Moreover, when the two groups were compared, group 2 also had 1.6 and 1.5 times higher stiffness and young modulus compared to group 1 after 12 weeks, because in group 1 the biomechanical properties did not change. As a conclusion, Grosset et al. claim that low intensity exercise does not result in tendon adaption in elderly people. There is a clear need for high intensity resistance training in order to increase tendon stiffness as well as young modulus (Grosset et al., 2014).

## 2.5 Which tendon is the best (allo)graft in terms of material properties?

Besides the length of a graft as discussed above, the tendon material properties as such vary from tendon to tendon, whereby an adequate choice is only possible, if the characteristics of different alternatives are known. Therefore baseline biomechanical studies are of great importance. For example, anterior *tibialis* tendon (ATT) is often used for ACL reconstruction. As an alternative, Delcroix et al. (2013) suggest a fan-folded iliotibial band (FITB) as an off-the-shelf allograft because they were able to show that FITBs were similar with respect to tensile strength, creep, and stress-relaxation viscoelastic properties as ATTs. In addition, compared to ACL, both, FITB and ATT had around three times higher ultimate loads providing a strong basis for the inhibition of re-rupture after implantation (Table 2.1). A further option for ACL reconstruction is reported by Hunziker et al. where the *semitendinosus* ligaments are used for this purpose (Hunziker et al., 2006). In order to get higher loading resistance, *semitendinosus* ligaments may be doubled or even quadrupled. Similarly, the AT is also used for ACL reconstructions, however, not in a doubled mode. And donor site morbidity as well as weakness go along with this procedure, which has to be judged precisely before AT grafting.

For AT augmentation, *peritoneus longus* (PL) or *flexor hallucis longus* (FHL) were tested (Sebastian et al., 2007). Although native PL has a failure load of only 65% of that of FHL (Table 2.1), when used as an AT augmentation, PL turned out to perform better with a 144% higher failure load compared to an AT reconstruction with FHL.

According to the authors, the more important point, however, was not the choice of the type of tendon graft, but that the grafts are carefully attached to the stumps of the AT to prevent pullout (Sebastian et al., 2007).

## 2.6 Animal models

To study tendon reconstructions in a preclinical setting, animal models are often used. Rabbits are very commonly used as they have ATs with similar biomechanical properties and dimensions as found in human hands (Goodman and Choueka, 2005; Buschmann et al., 2013) and therefore the translation from the laboratory to the clinical setting is rendered easier compared to smaller animal models such as mice. Biomechanical properties of native tendons and ligaments in animals are given in Table 2.2.

### 2.6.1 Rabbit

In order to design new functional prostheses for regenerative purposes, Kahn et al. (2013) determined the biomechanical properties of the rabbit AT meticulously. They determined the young modulus of the tendon fibers, of the interfibrils and the fibrils, finding that the fibrils had by far the highest modulus (10 times higher than the fibers themselves; Table 2.2). Moreover, the effect of fibroblast infiltration into acellular rabbit PTs for 3 and 6 weeks was determined with respect to tangent moduli of fascicles and bundles with a micro-tensile testing machine; only the fascicles were significantly affected by this ex vivo cellular infiltration (Ikema et al., 2007), lowering the modulus to around half compared to unseeded tendons after 6 weeks in culture.

As a whole specimen, rabbit ATs usually do exhibit failure loads of around 300 N (Buschmann et al., 2011) and failure stresses of 116–228 MPa in Lewis and Shaw (1997). Depending on the training level, however, linear stiffness of rabbit ATs may vary; rabbits that were trained three times per day for 1 hour and for 40 weeks exhibited 13% higher linear stiffness compared to untrained rabbits (Viidik, 1969), although absolute values of stiffness of rabbit ATs may vary with study conditions by factors of up to 5 (Ouyang et al., 2003). As a matter of fact, such inter-literature comparisons should encourage researchers to closely check their reference values with respect to how they have been determined.

Similar to AT, failure loads of ACL in rabbits have been reported to be $291 \pm 73$ N (Tischer et al., 2010; Table 2.2). Although only having half the failure loads of ACLs, rabbit STT allografts have been reported to be useful as grafts after decellularization by sodium dodecyl sulfate (Tischer et al., 2007, 2010).

### 2.6.2 Sheep

ACL reconstruction in the hind limb of sheep using a STT autograft (without decellularization) has been reported to be successful with respect to biomechanics. At 52 weeks postoperation, they had 40% of the control ultimate strength and 82% of the primary stiffness, (Goradia et al., 2000), which is far beyond in vivo peak forces and stiffness. However, a similar study also using STT autografts for ACL reconstruction in sheep suggested that vigorous activity should not be permitted for patients in the early periods after ACL reconstruction using STT autografts, which are necrotized and weakened after surgery; indeed, the authors determined at 52 weeks postoperation that the maximum load of the femur–graft–tibia complex was significantly lower than that of the femur–ACL–tibia complex (Kondo et al., 2012; Table 2.2).

### 2.6.3 Mouse

The mouse model is not very frequently used for biomechanical tendon studies (Table 2.2). However, Palmes et al. (2002) reported an AT mouse model where long-term effects of postoperative immobilization versus the mobilization were explored. The ATs were fully transected and sutured with a Kirchmayr–Kessler technique. While the mobilization group reached failure loads of healthy tendons after 35 days, immobilized tendons had only 67% of ultimate loads. Interestingly, by day 112 postsurgery, the mobilized ATs had significantly higher ultimate loads compared to controls, with 121% of the control tendons. As for the immobilized tendons at 112 days postinjury, they had very similar ultimate loads as the control and were only then considered to be recovered. In terms of stiffness, the mobilization also had a clear positive effect, with significantly higher stiffness (+171% compared to the immobilized tendons). This effect even remained at 112 days postsurgery, where the mobilization group showed the same stiffness as the control, whereas the immobilization group was only reaching around 45% of the control values. The mouse model to study AT defects and different regeneration methods is a suitable option in terms of biomechanical assessment, however, absolute ultimate loads are quite low, not exceeding 10 N (Mikic et al., 2001; Palmes et al., 2002).

Nevertheless, also tendinopathy murine AT models were successfully established (Bell et al., 2013). A single injection of TGF-β1 translating mechanical force into a biological response (see Chapter 3) and stimulating chondrogenesis (accumulation of glycosamin-rich chondroid matrix) (Diederichs et al., 2012) was applied at the AT mid-substance of mice. Typical features of tendinopathy occurred thereafter at 2 weeks postsurgery: the maximum stress of the tendon was reduced by 66%, there were abundant chondrocytes in the tendon tissue and an increased expression of Acan, Col1A1, Col2A1 and Col3A1, Fn, and MMP3. However, a remarkable restoration of the tensile properties was achieved when the mice were subjected to 4 weeks of treadmill exercise. Maximum load, stiffness, stress relaxation, maximum stress, and elastic modulus were again very similar as the data assessed for the native ATs.

Furthermore, the mouse model was also used to study tendon tears with the highest incidence of re-tear, the *supraspinatus* tendon (Bell et al., 2015). For that purpose, the RC was dissected and then, a tunnel was created at the insertion site of the *supraspinatus*. The suture was ligated and the tendon was brought back to its original insertion site. Two weeks postsurgery, biomechanics were assessed and compared to the contralateral healthy side. The maximum load of the RCs was 2.7 times smaller than that of the control which was $1.2 \pm 0.5$ N, six times smaller stiffness than the control being $2.37 \pm 1.6$ N/mm, and the displacement at maximum load was increased in the surgically repaired RCs by a factor of 2.3 (Bell et al., 2015). As the authors used different mouse strains (among them wild types, arthritis mice, joint laxity strains, and a strain where healing is investigated) and all the strains revealed very similar results in terms of morphology and biomechanics 2 weeks postsurgery, they concluded to have developed a mouse model for *supraspinatus* detachment and repair aimed at elucidating biomechanical mechanisms that are responsible for the clinically observed poor *supraspinatus* insertion healing.

## 2.6.4 Rat

Compared to murine ATs, rat ATs have 4 to 8 times higher ultimate failure loads and a 5 to 14 times higher stiffness when native healthy tendon biomechanics are compared (Majewski et al., 2008, 2012; Table 2.2). Therefore, the rat AT model is used more often compared to the mouse AT model. For example, in order to study tissue engineered AT grafts based on poly glycolic acid seeded with tenocytes or stem cells, the rat AT model revealed tenocytes to be favorable over stem cells; not only were the failure loads higher compared to stem cell-seeded grafts, but also the side effect of ossification was only found in case stem cells were applied (Pietschmann et al., 2013).

Moreover, rats not only serve as AT models, but are also often used for RC studies; in contrast to murine RCs that are biomechanically quite weak with loads until failure around 1 N only (Bell et al., 2015), the rat RC has at least an ultimate load of $18 \pm 3$ N (Ide et al., 2009). For example, Ide et al. used adult male Sprague–Dawley rats in order to reconstruct RC defect with an acellular dermal matrix patch and compared the ultimate failure load with nontreated specimen. Twelve weeks postoperation, the patch-treated rats showed 73% of failure load compared to the control, while nontreated rats had only 52% (Ide et al., 2009). Finally, the supraspinatus was investigated in the rat model and adjacent to a supraspinatus defect of $2 \times 2$ mm$^2$ either an in situ freezing or not was performed, in order to mimic an intrinsically reduced healing capacity of the supraspinatus. Although the freezing had no effect on biomechanical properties 12 weeks postsurgery, the maximum stress and the elastic modulus were an order of magnitude lower than the control values (Carpenter et al., 1998; Fig. 2.2).

Fig. 2.2 gives an overview of the modulus-stress ranges of typical tendons of different animals and compares them to the human tendon and ligament biomechanical "window." While the *supraspinati* and *infraspinati* of rats and rabbits, respectively, cover a wide range of human tendons and ligaments biomechanics, the rabbit AT is usually used to mimic tendons of the hand (Buschmann et al., 2013), as for example flexor tendons, because rabbit ATs are very similar in terms of their biomechanical properties (Goodman and Choueka, 2005). The choice of an appropriate xenograft can also be made consulting Fig. 2.2—reinforcements and complete replacements with decellularized animal xenografts for human tendon and ligament repair are further discussed in Chapter 6.

## 2.6.5 Dog

Adult mongrel dogs were used to study the effectiveness of decellularized porcine small intestinal submucosa (SIS) for replacement of a completely resected infraspinatus tendon (Dejardin et al., 2001). Biomechanical analysis 6 months postoperatively showed an ultimate failure load of only 45% of native canine *infraspinatus* (187 ± 31 N, Table 2.2). However, compared to the sham-operated shoulders (reattachment to the greater tubercle with sutures), the SIS treated specimen were not performing any better—having no significantly different values. In another study, the canine infraspinatus was used as a reference material in order to compare the biomechanical behavior of GraftJacket and others in the corresponding stress–strain curves, allowing to deduce an elastic modulus of the dog infraspinatus of 405 ± 86 MPa (Derwin et al., 2006). Moreover, ultimate load and stiffness of poly-L-lactic acid (PLLA)-augmented canine RC tendons were reported to reach failure loads of 77%, while the nonaugmented reached only 62% at 12 weeks postsurgery (Derwin et al., 2009). The stiffness was 47% of the native RC tendon in the augmentation group and 39% in the nonaugmented group. As such, the canine RC model revealed the beneficial biomechanical effect of such a device. Interestingly, also at time point 0, biomechanics were assessed. While the stiffness was very similar in the nonaugmented and the augmented group, the primary repair strength was significantly higher in the augmented group (120%).

Compared to the rat RC model, the canine RC model offers considerably higher ultimate loads, being 88 times higher (Ide et al., 2009)—which is on the one hand easier to measure as stress–strain curves get more accurate the higher the ultimate loads are in absolute terms; however, on the other hand, it is much more expensive and not even allowed to perform experiments with dogs in any animal facilities, may it be hospitals or universities.

### 2.6.6 Pig

Porcine models are rarely used to study tendon and ligament reconstructions, probably caused by the high costs associated with those experiments. Moreover, as pigs do not inherently jump as do rabbits, tendons such as the AT are mimicking less the functionalities of human tendons and ligaments. Nevertheless, Liu et al. performed a *flexor digitorum superficialis* (FDS) replacement in the porcine model, using either a dermal fibroblast-seeded scaffold or a dermal tenocyte-seeded scaffold and compared the biomechanical outcome at 6, 14, and 26 weeks postoperatively to the cell-free dermal graft (Liu et al., 2006). While the two cell-seeded dermal grafts behaved very similar, the cell-free graft was significantly less strong at all three time points with ultimate stress of around 85% of the cell-seeded specimen. However, even 26 weeks postsurgery, the cell-seeded grafts reached only around 75% of the ultimate stress of the native pig FDS (Table 2.2).

In another study, ultimate load and stiffness of ACLs treated either with or without a collagen-platelet composite material in addition to a suture were measured in immature pigs after transection of the ACL at 4, 6, and 12 weeks (Joshi et al., 2009). Although the application of the collagen-platelet composite significantly improved failure load and stiffness compared to the suture-alone group, even at 12 weeks postsurgery, it did not reach values found for native ACLs (Table 2.2); only 18% of the ultimate load was reached and only 26% of the stiffness. Compared to sheep, native porcine ACLs have very similar ACLs in terms of ultimate loads with 780 ± 60 and

713 ± 118 N for sheep and pigs respectively, while the stiffness is higher in sheep (210 ± 60 and 154 ± 20 N/mm for sheep and pigs, respectively; Yoshikawa et al., 2006).

### 2.6.7 Monkey

Compared to sheep or pigs, rhesus monkeys have ACL biomechanical characteristics that are similar to humans with respect to ultimate loads and stiffness (Noyes and Grood, 1976). However, although monkeys would be an ideal model to mimic human beings at first sight, their ACLs have a twofold ultimate stress when compared with young humans and even a fivefold higher ultimate stress when compared to old people (Tables 2.1 and 2.2, Fig. 2.2). At the same time, rhesus monkeys have a twofold and threefold higher elastic modulus compared to young and old people, respectively. When the ultimate loads and the stiffness are compared, however, rhesus monkeys bear only half the ultimate load of young people and around the same as old people. As for the stiffness, it is very similar to that of young people and around 1.5 times higher than that for old people. As such, the structural properties (ultimate load and stiffness) of the monkeys ACL are similar to those of human beings, while the material properties (ultimate stress and modulus) exhibit clearly higher values.

### 2.6.8 Horse

Horses are rather seldomly used in tendon tissue engineering studies; rather they are subjected to ultrasonic diagnostic investigations delivering baseline values especially for the FDS (Denoix and Busoni, 1999) or for the establishment of methods to analyze collagen organization in equine tendons (Sivaguru et al., 2014; see also Chapter 5). Nevertheless, biomechanical assessment of FDS in foals and adult horses has been reported (Table 2.2 and Fig. 2.2). While ultimate stress for foals aged 5 and 11 months, respectively, is very similar (Dowling and Dart, 2005) and around double as much as found for pigs (Liu et al., 2006), ultimate stress values reported for adult horses differ significantly depending on the reference with 65 ± 4 MPa (Dowling et al., 2002) or 109 ± 8 MPa (Crevier et al., 1996), the latter being similar to the values found for foals (Dowling and Dart, 2005). Moreover, maximum loads were assessed with 7–9 kN in 5-month old foals and 11–12 kN in 11-months old ones (Dowling and Dart, 2005). Similar values as for the older foals were also found for adult horses (Crevier et al., 1996).

Also the effect of exercise has been investigated in foals (Cherdchutham et al., 2001). From 1 week until 5 months of age, foals in the nonexercised group were confined to a small box stall, foals in the exercised group were housed in stalls but were exercised (special training program), and foals in the pastured group were maintained at pasture (free movement). Eight foals in each group were euthanatized at 5 months. The remaining foals were housed together in a loose stall and paddock until euthanatized at 11 months. The right fore SDFT (superficial digital flexor tendon) was collected from each foal immediately after euthanasia and subjected to biomechanical testing. Interestingly, although the pastured group showed significantly higher ultimate force at 5 months, this effect faded at 11 months, when the three groups had obtained the same motion regimen for the past 6 months.

## 2.7 Summary

In this chapter, we have shown the wide range of human and animal tendons and ligaments in terms of biomechanical properties. Some tendons are extremely strong and their modulus/ultimate stress ratio lies below the positive correlation line found in a modulus versus ultimate stress plot. Others lie above this line, implying a higher stiffness compared to their inherent strength. Furthermore, the impact of gender, age, and exercise was discussed in terms of biomechanics. It was shown that females are more injury-prone than males because of their weaker structural and mechanical tendon properties. Doubtless, age-related degeneration of tendon and ligament tissue can be observed, however, exercise—also for the elderly population—is highly recommended because the tendons' biomechanical properties are significantly improved by exercise which has been found with many animal models as well as in human studies. Moreover, we also presented a large set of animal models that may be used in preclinical studies to elucidate new interventions after tendon rupture or laceration such as surgical techniques or novel tendon and ligament grafts and augmentations (for different grafts, see Chapters 6–8).

## References

Angelidis, I.K., Thorfinn, J., Connolly, I.D., Lindsey, D., Pham, H.M., Chang, J., 2010. Tissue engineering of flexor tendons: the effect of a tissue bioreactor on adipoderived stem cell-seeded and fibroblast-seeded tendon constructs. J. Hand Surg. 35, 1466–1472.

Badylak, S., Arnoczky, S., Plouhar, P., Haut, R., Mendenhall, V., Clarke, R., Horvath, C., 1999. Naturally occurring extracellular matrix as a scaffold for musculoskeletal repair. Clin. Orthop. Relat. Res., 367, S333–S343.

Bell, R., Li, J., Gorski, D.J., Bartels, A.K., Shewman, E.F., Wysocki, R.W., Cole, B.J., Bach, J.B.R., Mikecz, K., Sandy, J.D., Plaas, A.H., Wang, V.M., 2013. Controlled treadmill exercise eliminates chondroid deposits and restores tensile properties in a new murine tendinopathy model. J. Biomech. 46, 498–505.

Bell, R., Taub, P., Cagle, P., Flatow, E.L., Andarawis-Puri, N., 2015. Development of a mouse model of supraspinatus tendon insertion site healing. J. Orthop. Res. 33, 25–32.

Biewener, A.A., Roberts, T.J., 2000. Muscle and tendon contributions to force, work, and elastic energy savings: a comparative perspective. Exerc. Sport Sci. Rev. 28, 99–107.

Burks, R.T., Desio, S.M., Bachus, K.N., Tyson, L., Springer, K., 1998. Biomechanical evaluation of lateral patellar dislocations. Am. J. Knee Surg. 11, 24–31.

Buschmann, J., Müller, A., Feldman, K., Tervoort, T.A., Fessel, G., Snedeker, J.G., Giovanoli, P., Calcagni, M., 2011. Small hook thread (quill) and soft felt internal splint to increase the primary repair strength of lacerated rabbit Achilles tendons: biomechanical analysis and considerations for hand surgery. Clin. Biomech. 26, 626–631.

Buschmann, J., Meier Buergisser, G., Bonavoglia, E., Neuenschwander, P., Milleret, V., Giovanoli, P., Calcagni, M., 2013. Cellular response of healing tissue to DegraPol tube implantation in rabbit Achilles tendon rupture repair: an in vivo histomorphometric study. J. Tissue Eng. Regen. Med. 7, 413–420.

Butler, D.L., Grood, E.S., Noyes, F.R., Zernicke, R.F., Brackett, K., 1984. Effects of structure and strain-measurement technique on the material properties of young human tendons and fascia. J. Biomech. 17, 579–596.

Cao, Y., Liu, Y., Liu, W., Shan, Q., Buoncuore, S., Cui, L., 2002. Bridging tendon defects using autologous tenocyte engineered tendon in a Hen model. Plast. Reconstr. Surg. 110, 1280–1289.

Carpenter, J.E., Thomopoulos, S., Flanagan, C.L., Debano, C.M., Soslowsky, L.J., 1998. Rotator cuff defect healing: a biomechanical and histologic analysis in an animal model. J. Shoulder Elb. Surg. 7, 599–605.

Chamberlain, C.S., Duenwald-Kuehl, S.E., Okotie, G., Brounts, S.H., Baer, G.S., Vanderby, R., 2013. Temporal healing in rat Achilles tendon: ultrasound correlations. Ann. Biomed. Eng. 41, 477–487.

Chandrashekar, N., Mansouri, H., Slauterbeck, J., Hashemi, J., 2006. Sex-based differences in the tensile properties of the human anterior cruciate ligament. J. Biomech. 39, 2943–2950.

Cherdchutham, W., Meershoek, L.S., Van Weeren, P.R., Barneveld, A., 2001. Effects of exercise on biomechanical properties of the superficial digital flexor tendon in foals. Am. J. Vet. Res. 62, 1859–1864.

Chong, A.K.S., Riboh, J., Smith, R.L., Lindsey, D.P., Pham, H.M., Chang, J., 2009. Flexor tendon tissue engineering: acellularized and reseeded tendon constructs. Plast. Reconstr. Surg. 123, 1759–1766.

Corps, A.N., Robinson, A.H.N., Movin, T., Costa, M.L., Ireland, D.C., Hazleman, B.L., Riley, G.P., 2004. Versican splice variant messenger RNA expression in normal human Achilles tendon and tendinopathies. Rheumatology 43, 969–972.

Corps, A.N., Robinson, A.H.N., Harrall, R.L., Avery, N.C., Curry, V.A., Hazleman, B.L., Riley, G.P., 2012. Changes in matrix protein biochemistry and the expression of mRNA encoding matrix proteins and metalloproteinases in posterior tibialis tendinopathy. Ann. Rheum. Dis. 71, 746–752.

Crevier, N., Pourcelot, P., Denoix, J.M., Geiger, D., Bortolussi, C., Ribot, X., Sanaa, M., 1996. Segmental variations of in vitro mechanical properties in equine superficial digital flexor tendons. Am. J. Vet. Res. 57, 1111–1117.

Criscenti, G., De Maria, C., Sebastiani, E., Tei, M., Placella, G., Speziali, A., Vozzi, G., Cerulli, G., 2016. Material and structural tensile properties of the human medial patello-femoral ligament. J. Mech. Behav. Biomed. Mater. 54, 141–148.

Datta, B., Salleh, R., Neil, M., Bulter, A., Walsh, W.R., 2006. Mechanical properties of human flexor hallucis longus, peroneus brevis and tendo Achillies tendons. In: 52nd Annual Meeting of the Orthopaedic Research Society. Orthopaedic Research Society, Chicago, IL, p. 1889.

Defrate, L.E., Van Der Ven, A., Gill, T.J., Li, G., 2004. The effect of length on the structural properties of an Achilles tendon graft as used in posterior cruciate ligament reconstruction. Am. J. Sports Med. 32, 993–997.

Dejardin, L.M., Arnoczky, S.P., Ewers, B.J., Haut, R.C., Clarke, R.B., 2001. Tissue-engineered rotator cuff tendon using porcine small intestine submucosa. Histologic and mechanical evaluation in dogs. Am. J. Sports Med. 29, 175–184.

Delcroix, G.J.R., Kaimrajh, D.N., Baria, D., Cooper, S., Reiner, T., Latta, L., D'Ippolito, G., Schiller, P.C., Temple, H.T., 2013. Histologic, biomechanical, and biological evaluation of fan-folded iliotibial band allografts for anterior cruciate ligament reconstruction. Arthroscopy 29, 756–765.

Denoix, J.M., Busoni, V., 1999. Ultrasonographic anatomy of the accessory ligament of the superficial digital flexor tendon in horses. Equine Vet. J. 31, 186–191.

Derwin, K.A., Baker, A.R., Spragg, R.K., Leigh, D.R., Iannotti, J.P., 2006. Commercial extracellular matrix scaffolds for rotator cuff tendon repair. Biomechanical, biochemical, and cellular properties. J. Bone Joint Surg. Am. 88, 2665–2672.

Derwin, K.A., Baker, A.R., Spragg, R.K., Leigh, D.R., Farhat, W., Iannotti, J.P., 2008. Regional variability, processing methods, and biophysical properties of human fascia lata extracellular matrix. J. Biomed. Mater. Res. A 84, 500–507.

Derwin, K.A., Codsi, M.J., Milks, R.A., Baker, A.R., Mccarron, J.A., Iannotti, J.P., 2009. Rotator cuff repair augmentation in a canine model with use of a woven poly-L-lactide device. J. Bone Joint Surg. Am. 91, 1159–1171.

Diederichs, S., Baral, K., Tanner, M., Richter, W., 2012. Interplay between local versus soluble transforming growth factor-beta and fibrin scaffolds: role of cells and impact on human mesenchymal stem cell chondrogenesis. Tissue Eng. A 18, 1140–1150.

Dowling, B.A., Dart, A.J., 2005. Mechanical and functional properties of the equine superficial digital flexor tendon. Vet. J. 170, 184–192.

Dowling, B.A., Dart, A.J., Hodgson, D.R., Rose, R.J., Walsh, W.R., 2002. Recombinant equine growth hormone does not affect the in vitro biomechanical properties of equine superficial digital flexor tendon. Vet. Surg. 31, 325–330.

Funakoshi, T., Majima, T., Iwasaki, N., Suenaga, N., Sawaguchi, N., Shimode, K., Minami, A., Harada, K., Nishimura, S.-I., 2005. Application of tissue engineering techniques for rotator cuff regeneration using a chitosan-based hyaluronan hybrid fiber scaffold. Am. J. Sports Med. 33, 1193–1201.

Gilday, S.D., Chris Casstevens, E., Kenter, K., Shearn, J.T., Butler, D.L., 2014. Murine patellar tendon biomechanical properties and regional strain patterns during natural tendon-to-bone healing after acute injury. J. Biomech. 47, 2035–2042.

Giza, E., Frizzell, L., Farac, R., Williams, J., Kim, S., 2011. Augmented tendon Achilles repair using a tissue reinforcement scaffold: a biomechanical study. Foot Ankle Int. 32, S545–S549.

Goodman, H.J., Choueka, J., 2005. Biomechanics of the flexor tendons. Hand Clin. 21, 129–149.

Goradia, V.K., Rochat, M.C., Grana, W.A., Rohrer, M.D., Prasad, H.S., 2000. Tendon-to-bone healing of a semitendinosus tendon autograft used for ACL reconstruction in a sheep model. Am. J. Knee Surg. 13, 143–151.

Grosset, J.-F., Breen, L., Stewart, C.E., Burgess, K.E., Onambélé, G.L., 2014. Influence of exercise intensity on training-induced tendon mechanical properties changes in older individuals. Age (Dordr.) 36, 9657.

Halder, A., Zobitz, M.E., Schultz, F., An, K.N., 2000a. Mechanical properties of the posterior rotator cuff. Clin. Biomech. 15, 456–462.

Halder, A., Zobitz, M.E., Schultz, F., An, K.N., 2000b. Structural properties of the subscapularis tendon. J. Orthop. Res. 18, 829–834.

Henn, R.F., Kuo, C.E., Kessler, M.W., Razzano, P., Grande, D.P., Wolfe, S.W., 2010. Augmentation of zone II flexor tendon repair using growth differentiation factor 5 in a rabbit model. J. Hand. Surg. [Am.] 35A, 1825–1832.

Hosey, G., Kowalchick, E., Tesoro, D., Balazsy, J., Klocek, J., Pederson, B., Wertheimer, S.J., 1991. Comparison of the mechanical and histologic properties of Achilles tendons in New zealand white rabbits secondarily repaired with marlex mesh. J. Foot Surg. 30, 214.

Huijing, P.A., 2009. Epimuscular myofascial force transmission: a historical review and implications for new research. International society of biomechanics muybridge award lecture, Taipei, 2007. J. Biomech. 42, 9–21.

Hunziker, E., Spector, M., Libera, J., Gertzman, A., Woo, S.L.Y., Ratcliffe, A., Lysaght, M., Coury, A., Kaplan, D., Vunjak-Novakovic, G., 2006. Translation from research to applications. Tissue Eng. 12, 3341–3364.

Huri, G., Bicer, O.S., Ozgozen, L., Ucar, Y., Garbis, N.G., Hyun, Y.S., 2013. A novel repair method for the treatment of acute Achilles tendon rupture with minimally invasive approach using button implant: a biomechanical study. Foot Ankle Surg. 19, 261–266.

Ide, J., Kikukawa, K., Hirose, J., Iyama, K.-I., Sakamoto, H., Mizuta, H., 2009. Reconstruction of large rotator-cuff tears with acellular dermal matrix grafts in rats. J. Shoulder Elbow Surg. 18, 288–295.

Ikema, Y., Tohyama, H., Yamamoto, E., Kanaya, F., Yasuda, K., 2007. Ex vivo infiltration of fibroblasts into the tendon deteriorates the mechanical properties of tendon fascicles but not those of tendon bundles. Clin. Biomech. (Bristol, Avon) 22, 120–126.

Ilhami, K., Eray, B.M., Gokhan, M., Ulukan, I., Levent, A., 2004. Comparative effects of monopolar radiofrequency energy and conservative management of mechanical properties of elongated lateral collateral ligament in rabbits: an experimental study. Clin. Biomech. 19, 184–189.

Irie, T., Majima, T., Sawaguchi, N., Funakoshi, T., Nishimura, S., Minami, A., 2011. Biomechanical and histologic evaluation of tissue engineered ligaments using chitosan and hyaluronan hybrid polymer fibers: a rabbit medial collateral ligament reconstruction model. J. Biomed. Mater. Res. A 97A, 111–117.

Itoi, E., Berglund, L.J., Grabowski, J.J., Schultz, F.M., Growney, E.S., Morrey, B.F., An, K.-N., 1995. Tensile properties of the supraspinatus tendon. J. Orthop. Res. 13, 578–584.

Johnson, G.A., Tramaglini, D.M., Levine, R.E., Ohno, K., Choi, N.Y., Woo, S.L., 1994. Tensile and viscoelastic properties of human patellar tendon. J. Orthop. Res. 12, 796–803.

Joseph, M., Lillie, K.R., Bergeron, D.J., Cota, K.C., Yoon, J.S., Kraemer, W.J., Denegar, C.R., 2014. Achilles tendon biomechanics in response to acute intense exercise. J. Strength Cond. Res. 28, 1181–1186.

Joshi, S.M., Mastrangelo, A.N., Magarian, E.M., Fleming, B.C., Murray, M.M., 2009. Collagen-platelet composite enhances biomechanical and histologic healing of the porcine anterior cruciate ligament. Am. J. Sports Med. 37, 2401–2410.

Juncosa-Melvin, N., Boivin, G.P., Galloway, M.T., Gooch, C., West, J.R., Butler, D.L., 2006a. Effects of cell-to-collagen ratio in stem cell-seeded constructs for Achilles tendon repair. Tissue Eng. 12, 681–689.

Juncosa-Melvin, N., Boivin, G.P., Gooch, C., Galloway, M.T., West, J.R., Dunn, M.G., Butler, D.L., 2006b. The effect of autologous mesenchymal stem cells on the biomechanics and histology of gel-collagen sponge constructs used for rabbit patellar tendon repair. Tissue Eng. 12, 369–379.

Kahn, C.J.F., Dumas, D., Arab-Tehrany, E., Marie, V., Tran, N., Wang, X., Cleymand, F., 2013. Structural and mechanical multi-scale characterization of white New-zealand rabbit Achilles tendon. J. Mech. Behav. Biomed. Mater. 26, 81–89.

Kellis, E., Balidou, A., 2014. In vivo examination of the morphology of the tendinous inscription of the human semitendinosus muscle: gender and joint position effects. J. Morphol. 275, 57–64.

Kennedy, M.I., Claes, S., Fuso, F.A., Williams, B.T., Goldsmith, M.T., Turnbull, T.L., Wijdicks, C.A., Laprade, R.F., 2015. The anterolateral ligament: an anatomic, radiographic, and biomechanical analysis. Am. J. Sports Med. 43, 1606–1615.

Kjaer, M., Langberg, H., Bojsen-Møller, J., Koskinen, S.O., Mackey, A., Heinemeier, K., Holm, L., Skovgaard, D., Døssing, S., Hansen, M., Hansen, P., Haraldsson, B., Carøe, I., Magnusson, S.P., 2008. Novel methods for tendon investigations. Disabil. Rehabil. 30, 1514–1522.

Kondo, E., Yasuda, K., Katsura, T., Hayashi, R., Kotani, Y., Tohyama, H., 2012. Biomechanical and histological evaluations of the doubled semitendinosus tendon autograft after anterior cruciate ligament reconstruction in sheep. Am. J. Sports Med. 40, 315–324.

Lalley, A.L., Dyment, N.A., Kazemi, N., Kenter, K., Gooch, C., Rowe, D.W., Butler, D.L., Shearn, J.T., 2015. Improved biomechanical and biological outcomes in the MRL/MpJ murine strain following a full-length patellar tendon injury. J. Orthop. Res. 33, 1693–1703.

Leong, N.L., Kabir, N., Arshi, A., Nazemi, A., Wu, B., Petrigliano, F.A., McAllister, D.R., 2015. Evaluation of polycaprolactone scaffold with basic fibroblast growth factor and fibroblasts in an athymic rat model for anterior cruciate ligament reconstruction. Tissue Eng. Part A 21, 1859–1868.

Lewis, G., Shaw, K.M., 1997. Tensile properties of human tendo achillis: effect of donor age and strain rate. J. Foot Ankle Surg. 36, 435–445.

Li, X., Snedeker, J.G., 2013. Wired silk architectures provide a biomimetic ACL tissue engineering scaffold. J. Mech. Behav. Biomed. Mater. 22, 30–40.

Liang, R., Woo, S.L.Y., Takakura, Y., Moon, D.K., Jia, F., Abramowitch, S.D., 2006. Long-term effects of porcine small intestine submucosa on the healing of medial collateral ligament: a functional tissue engineering study. J. Orthop. Res. 24, 811–819.

Liu, W., Chen, B., Deng, D., Xu, F., Cui, L., Cao, Y., 2006. Repair of tendon defect with dermal fibroblast engineered tendon in a porcine model. Tissue Eng. 12, 775–788.

Maganaris, C.N., Paul, J.P., 1999. In vivo human tendon mechanical properties. J. Physiol. 521 (Pt 1), 307–313.

Magnusson, S.P., Aagaard, P., Simonsen, E.B., Bojsen-Møller, F., 2000. Passive tensile stress and energy of the human hamstring muscles in vivo. Scand. J. Med. Sci. Sports 10, 351–359.

Majewski, M., Betz, O., Ochsner, P.E., Liu, F., Porter, R.M., Evans, C.H., 2008. Ex vivo adenoviral transfer of bone morphogenetic protein 12 (BMP-12) cDNA improves Achilles tendon healing in a rat model. Gene Ther. 15, 1139–1146.

Majewski, M., Porter, R.M., Betz, O.B., Betz, V.M., Clahsen, H., Fluckiger, R., Evans, C.H., 2012. Improvement of tendon repair using muscle grafts transduced with TGF-beta1 cDNA. Eur. Cell. Mater. 23, 94–101. discussion 101–2.

Meier Buergisser, G., Calcagni, M., Muller, A., Bonavoglia, E., Fessel, G., Snedeker, J.G., Giovanoli, P., Buschmann, J., 2014. Prevention of peritendinous adhesions using an electrospun DegraPol polymer tube: a histological, ultrasonographic, and biomechanical study in rabbits. Biomed Res. Int. 2014, 656240.

Meimandi-Parizi, A., Oryan, A., Moshiri, A., Silver, I.A., 2013. Novel application of a tissue-engineered collagen-based three-dimensional bio-implant in a large tendon defect model: a broad-based study with high value in translational medicine. Tissue Cell 45, 282–294.

Mikic, B., Schalet, B.J., Clark, R.T., Gaschen, V., Hunziker, E.B., 2001. GDF-5 deficiency in mice alters the ultrastructure, mechanical properties and composition of the Achilles tendon. J. Orthop. Res. 19, 365–371.

Moshiri, A., Oryan, A., Meimandi-Parizi, A., 2013. Role of tissue-engineered artificial tendon in healing of a large Achilles tendon defect model in rabbits. J. Am. Coll. Surg. 217. 421–441.e8.

Mountney, J., Senavongse, W., Amis, A.A., Thomas, N.P., 2005. Tensile strength of the medial patellofemoral ligament before and after repair or reconstruction. J. Bone Joint Surg. (Br.) 87, 36–40.

Musahl, V., Abramowitch, S.D., Gilbert, T.W., Tsuda, E., Wang, J.H.C., Badylak, S.F., Woo, S.L.Y., 2004. The use of porcine small intestinal submucosa to enhance the healing of the medial collateral ligament—a functional tissue engineering study in rabbits. J. Orthop. Res. 22, 214–220.

Nagumo, A., Yasuda, K., Numazaki, H., Azuma, H., Tanabe, Y., Kikuchi, S.-I., Harata, S., Tohyama, H., 2005. Effects of separate application of three growth factors (TGF-beta1, EGF, and PDGF-BB) on mechanical properties of the in situ frozen-thawed anterior cruciate ligament. Clin. Biomech. (Bristol, Avon) 20, 283–290.

Ni, T., Senthil-Kumar, P., Dubbin, K., Aznar-Cervantes, S.D., Datta, N., Randolph, M.A., Cenis, J.L., Rutledge, G.C., Kochevar, I.E., Redmond, R.W., 2012. A photoactivated nanofiber graft material for augmented Achilles tendon repair. Lasers Surg. Med. 44, 645–652.

Nirmalanandhan, V.S., Rao, M., Sacks, M.S., Haridas, B., Butler, D.L., 2007. Effect of length of the engineered tendon construct on its structure-function relationships in culture. J. Biomech. 40, 2523–2529.

Noyes, F.R., Grood, E.S., 1976. The strength of the anterior cruciate ligament in humans and rhesus monkeys. J. Bone Joint Surg. Am. 58, 1074–1082.

Noyes, F.R., Butler, D.L., Grood, E.S., Zernicke, R.F., Hefzy, M.S., 1984. Biomechanical analysis of human ligament grafts used in knee-ligament repairs and reconstructions. J. Bone Joint Surg. Am. 66, 344–352.

Obst, S.J., Newsham-West, R., Barrett, R.S., 2016. Changes in Achilles tendon mechanical properties following eccentric heel drop exercise are specific to the free tendon. Scand. J. Med. Sci. Sports 26 (4), 421–431.

Oryan, A., Moshiri, A., Parizi Meimandi, A., Silver, I.A., 2013. A long-term in vivo investigation on the effects of xenogenous based, electrospun, collagen implants on the healing of experimentally-induced large tendon defects. J. Musculoskelet. Neuronal Interact. 13, 353–367.

Oryan, A., Moshiri, A., Meimandi-Parizi, A., Maffulli, N., 2015. Role of xenogenous bovine platelet gel embedded within collagen implant on tendon healing: an in vitro and in vivo study. Exp. Biol. Med. (Maywood) 240, 194–210.

Ouyang, H.W., Goh, J.C.H., Thambyah, A., Teoh, S.H., Lee, E.H., 2003. Knitted poly-lactide-co-glycolide scaffold loaded with bone marrow stromal cells in repair and regeneration of rabbit Achilles tendon. Tissue Eng. 9, 431–439.

Palmes, D., Spiegel, H.U., Schneider, T.O., Langer, M., Stratmann, U., Budny, T., Probst, A., 2002. Achilles tendon healing: long-term biomechanical effects of postoperative mobilization and immobilization in a new mouse model. J. Orthop. Res. 20, 939–946.

Peffers, M.J., Fang, Y., Cheung, K., Wei, T.K.J., Clegg, P.D., Birch, H.L., 2015. Transcriptome analysis of ageing in uninjured human Achilles tendon. Arthritis Res. Ther. 17, 544.

Pietschmann, M.F., Frankewycz, B., Schmitz, P., Docheva, D., Sievers, B., Jansson, V., Schieker, M., Mueller, P.E., 2013. Comparison of tenocytes and mesenchymal stem cells seeded on biodegradable scaffolds in a full-size tendon defect model. J. Mater. Sci. Mater. Med. 24, 211–220.

Pridgen, B.C., Woon, C.Y.L., Kim, M., Thorfinn, J., Lindsey, D., Pham, H., Chang, J., 2011. Flexor tendon tissue engineering: acellularization of human flexor tendons with preservation of biomechanical properties and biocompatibility. Tissue Eng. Part C Methods 17, 819–828.

Prud'homme, J., Budoff, J.E., Nguyen, L., Hipp, J.A., 2008. Biomechanical analysis of medial collateral ligament reconstruction grafts of the elbow. Am. J. Sports Med. 36, 728–732.

Quapp, K.M., Weiss, J.A., 1998. Material characterization of human medial collateral ligament. J. Biomech. Eng. 120, 757–763.

Rosager, S., Aagaard, P., Dyhre-Poulsen, P., Neergaard, K., Kjaer, M., Magnusson, S.P., 2002. Load-displacement properties of the human triceps surae aponeurosis and tendon in runners and non-runners. Scand. J. Med. Sci. Sports 12, 90–98.

Saber, S., Zhang, A.Y., Ki, S.H., Lindsey, D.P., Smith, R.L., Riboh, J., Pham, H., Chang, J., 2010. Flexor tendon tissue engineering: bioreactor cyclic strain increases construct strength. Tissue Eng. A 16, 2085–2090.

Sahin, H., Tholema, N., Petersen, W., Raschke, M.J., Stange, R., 2012. Impaired biomechanical properties correlate with neoangiogenesis as well as VEGF and MMP-3 expression during rat patellar tendon healing. J. Orthop. Res. 30, 1952–1957.

Sakai, T., Yasuda, K., Tohyama, H., Azuma, H., Nagumo, A., Majima, T., Frank, C.B., 2002. Effects of combined administration of transforming growth factor-beta 1 and epidermal growth factor on properties of the in situ frozen anterior cruciate ligament in rabbits. J. Orthop. Res. 20, 1345–1351.

Santoni, B.G., Mcgilvray, K.C., Lyons, A.S., Bansal, M., Turner, A.S., Macgillivray, J.D., Coleman, S.H., Puttlitz, C.M., 2010. Biomechanical analysis of an ovine rotator cuff repair via porous patch augmentation in a chronic rupture model. Am. J. Sports Med. 38, 679–686.

Schlegel, T.F., Hawkins, R., Lewis, C.W., Motta, T., Turner, A.S., 2006. The effects of augmentation with swine small intestine submucosa on tendon healing under tension: histologic and mechanical evaluations in sheep. Am. J. Sports Med. 34, 275–280.

Sebastian, H., Datta, B., Maffulli, N., Neil, M., Walsh, W.R., 2007. Mechanical properties of reconstructed Achilles tendon with transfer of peroneus brevis or flexor hallucis longus tendon. J. Foot Ankle Surg. 46, 424–428.

Seitz, H., Pichl, W., Matzi, V., Nau, T., 2013. Biomechanical evaluation of augmented and non-augmented primary repair of the anterior cruciate ligament: an in vivo animal study. Int. Orthop. 37, 2305–2311.

Shearn, J.T., Kinneberg, K.R.C., Dyment, N.A., Galloway, M.T., Kenter, K., Wylie, C., Butler, D.L., 2011. Tendon tissue engineering: progress, challenges, and translation to the clinic. J. Musculoskelet. Neuronal Interact. 11, 163–173.

Sivaguru, M., Eichorst, J.P., Durgam, S., Fried, G.A., Stewart, A.A., Stewart, M.C., 2014. Imaging horse tendons using multimodal 2-photon microscopy. Methods 66, 256–267.

Smith, M.J., Cook, J.L., Kuroki, K., Jayabalan, P.S., Cook, C.R., Pfeiffer, F.M., Waters, N.P., 2012. Comparison of a novel bone-tendon allograft with a human dermis–derived patch for repair of chronic large rotator cuff tears using a canine model. Arthroscopy 28, 169–177.

Tischer, T., Vogt, S., Aryee, S., Steinhauser, E., Adamczyk, C., Milz, S., Martinek, V., Imhoff, A.B., 2007. Tissue engineering of the anterior cruciate ligament: a new method using acellularized tendon allografts and autologous fibroblasts. Arch. Orthop. Trauma Surg. 127, 735–741.

Tischer, T., Aryee, S., Wexel, G., Steinhauser, E., Adamczyk, C., Eichhorn, S., Milz, S., Martinek, V., Gansbacher, B., Imhoff, A.B., Vogt, S., 2010. Tissue engineering of the anterior cruciate ligament-sodium dodecyl sulfate-acellularized and revitalized tendons are inferior to native tendons. Tissue Eng. A 16, 1031–1040.

Van Der Meijden, O.A., Wijdicks, C.A., Gaskill, T.R., Jansson, K.S., Millett, P.J., 2013. Biomechanical analysis of two-tendon posterosuperior rotator cuff tear repairs: extended linked repairs and augmented repairs. Arthroscopy 29, 37–45.

Viidik, A., 1969. Tensile strength properties of Achilles tendon system in trained and untrained rabbits. Acta Orthop. 40, 261–272.

Wilson, T.W., Zafuta, M.P., Zobitz, M., 1999. A biomechanical analysis of matched bone-patellar tendon-bone and double-looped semitendinosus and gracilis tendon grafts. Am. J. Sports Med. 27, 202–207.

Woo, S.L., Hollis, J.M., Adams, D.L., Lyon, R.M., Takai, S., 1991. Tensile properties of the human femur-anterior cruciate ligament-tibia complex. The effects of specimen age and orientation. Am. J. Sports Med. 19, 217–225.

Wren, T.A.L., Yerby, S.A., Beaupré, G.S., Carter, D.R., 2001. Mechanical properties of the human Achilles tendon. Clin. Biomech. 16, 245–251.

Yoshikawa, T., Tohyama, H., Katsura, T., Kondo, E., Kotani, Y., Matsumoto, H., Toyama, Y., Yasuda, K., 2006. Effects of local administration of vascular endothelial growth factor on mechanical characteristics of the semitendinosus tendon graft after anterior cruciate ligament reconstruction in sheep. Am. J. Sports Med. 34, 1918–1925.

Zalavras, C.G., Gardocki, R., Huang, E., Stevanovic, M., Hedman, T., Tibone, J., 2006. Reconstruction of large rotator cuff tendon defects with porcine small intestinal submucosa in an animal model. J. Shoulder Elb. Surg. 15, 224–231.

Zens, M., Feucht, M.J., Ruhhammer, J., Bernstein, A., Mayr, H.O., Sudkamp, N.P., Woias, P., Niemeyer, P., 2015. Mechanical tensile properties of the anterolateral ligament. J Exp Orthop. 2, 7.

# Mechanobiology of tendons and ligaments

# 3

## Abstract

Tendons and ligaments are load-bearing connective tissues. As such, they are dynamic structures that respond in various manners to mechanical loading having different magnitudes, directions and frequencies. They adapt to mechanical loading by altering their tissue composition resulting in changed biomechanical properties. On the cellular level, load-induced alterations include mechanotransduction pathways such as transforming growth factor beta signaling pathways. In terms of extracellular matrix, collagen synthesis may be increased by the application of load. The mechanism by which force is exactly transmitted in the tendon tissue is suggested to go via interfibrillar shear of the discontinuous fibrils in the tendon fascicles.

**Keywords:** Mechanotransduction, Force, Shear stress, Growth factors, Metalloproteinases.

## Abbreviations

| | |
|---|---|
| **AT** | Achilles tendon (tendons and ligaments) |
| **C1,2C** | collagen degradation marker (element of living cells) |
| **cDNA** | complementary deoxyribonucleic acid (element of living cells) |
| **CLEC3B** | gene that encodes tetranectin (element of living cells) |
| **COL11A1** | gene that encodes collagen alpha-1 (XI) chain (element of living cells) |
| **COL1A1** | collagen, type I, alpha 1, also known as alpha-1 type I collagen (element of soft tissues) |
| **COMP** | cartilage oligomeric matrix protein (protein in living cells) |
| **COX-2** | cyclooxygenase (enzyme in living cells) |
| **CTNNA1** | gene that encodes αE-catenin, also known as catenin alpha-1 (element of living cells) |
| **Cx** | connexin (protein in living cells) |
| **DMEM** | Dulbecco's minimal essential medium (medium used in cell culture technique) |
| **dyn** | dyne, $1\ \text{dyn} = 10^{-5}\ \text{N} \equiv 1\ \text{g cm/s}^2$ (unit of force) |
| **ECM** | extra cellular matrix (matrix in living cells) |
| **GAGs** | glycosaminoglycans (protein in living cells) |
| **IL-*n*** | interleukin-*n* ($n = 1$–17) (protein in living cells) |
| **ITGA4** | integrin, alpha 4 (protein in living cells) |
| **ITGA6** | integrin alpha-6 (protein in living cells) |
| **LAMA2** | laminin alpha 2 (protein in living cells) |
| **miRNA, miR (-*n*)** | microRNA (element of living cells) |
| **MMP (-*n*)** | matrix metalloproteinase-*n* (enzymes in living cells) |
| **mRNA** | messenger ribonucleic acid (element of living cells) |

**Biomechanics of Tendons and Ligaments. http://dx.doi.org/10.1016/B978-0-08-100489-0.00003-X**

| | |
|---|---|
| **Pa, MPa, GPa** | pascal, mega-pascal ($10^3$ Pa), giga-pascal ($10^6$ Pa) (unit of pressure, internal pressure, stress, young's modulus and ultimate tensile strength, defined as one $N/m^2$) |
| **PCR** | polymerase chain reaction (technical feature of gene technology lab) |
| **rpm** | revolutions per minute (frequency of rotation) (technical feature of laboratory automation) |
| **TGF-β*n* or R$n_1$** | transforming growth factor beta-*n* or -R ($n=1,2,3$, $n_1=$I–II) (protein in living cells) |
| **TIMPs** | tissue inhibitors of metalloproteinases (enzymes in living cells) |

## 3.1 Introduction

Tendons and ligaments change their composition and structure as a response to mechanical forces (Wang, 2006). Tenocytes and tendon fibroblasts (tenoblasts) are primarily involved in this mechanical adaptation. They react towards force application by biochemical signals ending up in physiological and also pathological changes, which include mechanisms on the tendon tissue level as well as on the cell and molecular level (Wang and Thampatty, 2006). Discussion of mechanobiology of tendons and ligaments therefore covers a diverse set of effects of mechanical forces during development, homeostasis as well as healing of the tendon and ligament tissue.

In the musculoskeletal system, tendons allow transmitting force from muscles to bones. They provide stability as well as efficient motion. This transmittance of force acts as a mechanical stimulus on tenocytes. Roughly, they "receive" this stimulus via the tendon tissue, in other words via various matrix components including collagen type I, elastin, glycoproteins, proteoglycans, glycolipids and water among others. As such, the tendon cell experiences force on its surface, and from the exterior the force is transmitted to the interior of the cell by a series of biochemical pathways including transmembrane structures.

Based on their structure and cellular organization, whole tendons behave in a nonlinear viscoelastic manner when load is applied (see Chapters 1 and 4). Mainly, the predominant component collagen I is responsible for the typical stress-strain curves found in tendon tensile stretching experiments (see Fig. 4.4), with a primary toe region where the crimps of the collagen fibers are stretched, followed by a linear region provided by the high strength of covalent bonds between collagen molecules (intermolecular collagen-crosslinking), being also responsible for the sliding between fibers (Fessel et al., 2014). Also water ads up to the visco-elasticity found for typical tendon tissue. Under physical load, water is transferred from the central to the periphery of the tendon and thereby changes its biomechanical behavior (Wellen et al., 2005).

Within the tendon tissue, the tenocytes are arranged in parallel linear arrays (see Chapter 1). When load is applied, the extracellular matrix of the cells is deformed and a combination of compressive and tensile forces as well as shear and strain stresses act on the cells. Thereby noteworthy to mention is the fact, that local tissue strains are only 25–30% of grip-to-grip strains (Butler et al., 1984) and that the applied

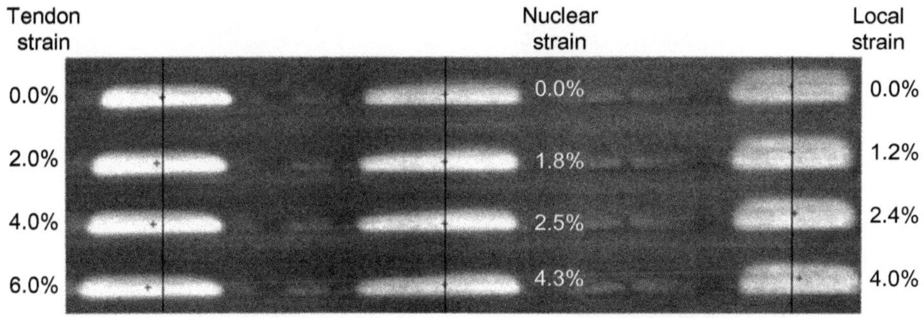


**Fig. 3.1** Composite photo illustrating the measurement technique for determining local strain relative to tendon (grip-to-grip) strain and cell nucleus strain. Images of three cells in line and in the same focal plane were evaluated at 0%, 2%, 4%, and 6% of tendon grip-to-grip strain. The distance between centroids (+) of nuclei on either side of the nuclei (center) used for deformation measurements are used to determine local tissue strain. In this series the local strain is less than the grip-to-grip strain and the cell nuclei strain are greater than the local tissue strain.
From Arnoczky, S.P., Lavagnino, M., Whallon, J.H., Hoonjan, A., 2002. In situ cell nucleus deformation in tendons under tensile load; a morphological analysis using confocal laser microscopy. J. Orthop. Res. 20, 29–35. © by Journal of Orthopaedic Research, with permission from Wiley.

load is always smaller than the received one (Screen et al., 2003), which can be made visible by the deformation of the localized cells in response to a given deformation of the whole tendon (Fig. 3.1; Arnoczky et al., 2002).

## 3.2 Impact of loading on tendon cells

### 3.2.1 Gene expression

Tendon cells respond to shear forces as well as mechanical load. In a simple in vitro experiment using tenocytes isolated from adult male rats in a culture plate equipped with a rotating cone, it has been shown by complementary deoxyribonucleic acid (cDNA) microarray and Northern blotting analysis led to an induced "antifibrotic" expression pattern of genes (Fong et al., 2005).

Several pro-fibrotic molecules were down regulated in different signaling pathways (including platelet-derived growth factor, insulin-like growth factor and fibroblast growth factor signaling pathway, respectively). Moreover, the shear forces induced down regulation of transforming growth factor beta 2 (TGF-β2), TGF-β3 and the receptors TGF-RI and TGF-RII, however, TGF-β1 was upregulated; this is noteworthy and interesting because TGF-β1 has been demonstrated to be mechanosensitive (Heinemeier et al., 2003). The TGF-β family is involved in the wound healing and in inducing collagen production via Smads and scleraxis (Fig. 3.2). While the mechanoresponsiveness of TGF-β2, TGF-β3 differed from the one observed for TGF-β1, it is interesting to mention that adult scar-mediated tendon

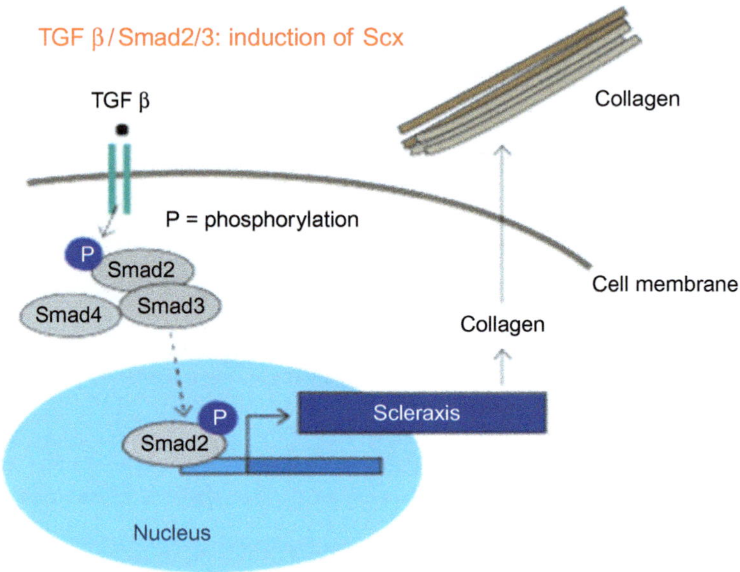


**Fig. 3.2** Upon binding of TGF-β to its receptor, Smads are phosphorylated and translocated to the nucleus, where they act as transcription factors to activate the expression of scleraxis. Scleraxis promotes the synthesis and secretion of collagen and other components of the tendon extra cellular matrix.
Modified from Sharir, A., Zelzer, E., 2011. Tendon homeostasis: the right pull. Curr. Biol. 21, R472–R474. Reprinted from Current Biology, Copyright 2011, with permission from Elsevier.

wound healing is also characterized by a different responsiveness of TGF-β2, TGF-β3 compared to TGF-β1; however, just in the opposite direction (Thomopoulos et al., 2015). In the case of adult wound healing, high levels of TGF-β2 and TGF-β3 are found and low levels of TGF-β1. Furthermore, adult tendon wound healing differs from fetal tendon wound healing, where the expression of the mentioned growth factors resembles the one induced by shear forces (low expression of TGF-β2 and TGF-β 3 and high expression of TGF-β1). Finally, it is noteworthy to mention that TGF-β2 plays a pivotal role in the tendon development which can be used for example in the tissue engineering of tendons following a paradigm called "developmental tissue engineering" (Glass et al., 2014).

With this overall down regulation of the TGF-β family induced by shear forces, the well-known improvement of flexor tendon healing and digital motion after early passive motion can be explained on the altered pathways, including signaling molecules in the cascade of fibrotic tissue formation (Gupta, 2005). Going along with the general down regulation of the TGF-β signaling molecules, also an increased expression of matrix metalloproteinases (MMPs) as well as a decrease in tissue inhibitors of metalloproteinases were measured (Fong et al., 2005), corroborating the overall antifibrotic effect of shear stress on flexor tenocytes in vitro.

An interesting study compared the effect of TGF-β applied on isolated tenocytes to the effect found by mechanical loading of a rat Achilles tendon (AT) analyzing of mRNA and microRNA (miRNA; Mendias et al., 2012). MiRNAs belong to the small noncoding RNAs and are involved in the posttranscriptional regulation of many mRNA molecules. They bind to the untranslated part of an mRNA and induce thereby its rapid degradation. As such, they may regulate the adaptation of tissues to mechanical loading as well as to diseases. Mendias et al. were able to show that not only the typical tendon genes including collagen type I, proliferating cell nuclear antigen, scleraxis and tenomodulin were very similar upregulated in both, the TGF-β treated tenocytes as well as the loaded rat ATs, but that in an analysis of over 50 miRNAs the ones involved in the binding to those genes were decreased. For example, miRNAs that were predicted to bind to the untranslated part of scleraxis like miR-338, miR-381 miR-743a and miR-743b were decreased—leading to a reduced degradation of that gene and hence to an upregulation.

From a gene regulation point of view, tendon fibroblasts extracted from fetal mice responded to linear shear stress (0.1 dyn/cm$^2$ for 14 hours) partly similar and partly different compared to fetal murine cornea and skin fibroblasts (Mackley et al., 2006). The authors analyzed 344 genes, of which only 15% were regulated by all three cell types, while one-third was at least regulated by two of the three—showing that fibroblasts originating from tendon tissue (tenocytes or tenoblasts) do react quite differently towards mechanical shear stress compared to skin and cornea fibroblasts. This implies that tenocytes and tenoblasts are highly specialized fibroblasts. Moreover, it was reported that the spectrum of genes affected by the mechanical stimulus was not primarily the one expected. For example, fluid shear stress evoked the down regulation of genes implicated for adipogenesis—as such underlining the fact that mechanical load stabilizes a more fibroblast-like phenotype compared to the adipocyte (which may be achieved by some fibroblast precursors differentiating towards adipocytes). Although it was expected that genes involved in the cytoskeleton-extra cellular matrix (ECM) linkage which is affected during mechanotransduction would be upregulated, the number of genes in this category was surprisingly low; in this regard, noteworthy to mention was the nonexpected down regulation of lumican (a small proteoglycan typically found in the ECM of tendons) in the tendon fibroblasts (−2.5*x*) opposed to the expected upregulation of the chondroitin sulfate proteoglycan 4 (+3.4*x*).

### 3.2.2 Gap junctions

Another effect of tensile loading on tenocytes is the regulation of the gap junction permeability (Maeda et al., 2012). Communication between tenocytes is an essential part of mechanotransduction during shear stress and load. Maeda et al. reported experiments with viable tenocytes in rat tail tendon fascicles that were labeled and subjected to a fluorescent loss induced by photobleaching. As such, they were able to record the fluorescent intensity in the neighboring tenocytes. After application of a 1 N static load for 10 minutes, no effect was observed on the gap junction communication, however, if the duration of load was extended to 1 hour, a significant reduction in gap junction permeability was found. These findings were further corroborated by the fact that connexin 43 (Cx 43) protein expression was reduced in the 1 hour-loaded samples and furthermore by a significant reduction in the permeability parameters. In contrast, on the mRNA-level, Cx 43 mRNA was upregulated. This concomitant upregulation of Cx 43 mRNA suggests that the tenocytes respond to the reduced permeability and the disruption of the gap junction communication by an increased connexin synthesis. Hence, tenocytes do react towards mechanical load by two mechanisms, involving both breakdown as well as remodeling of their gap junctions.

### 3.2.3 Calcium levels

How to study the effects of shear stress or strain on tendon cells? In an attempt to study the mechanotransductional effect of tensile strain, fluid shear stress or the combination of both, Maeda et al. developed a micro-grooved membrane and a flow unit (Fig. 3.3) where they analyzed the calcium levels of tenocytes originating from male bovine foot extensors (Maeda et al., 2013).

As a result, tenocytes showed no increase in calcium levels during the 5-minute fluorescence imaging period when they were not stimulated, while under fluid flow, tensile strain or the combination of both, elevations of calcium levels were found. Although calcium levels were significantly higher compared to nonstimulated tenocytes, the calcium levels under combined stimulation (flow and strain) were only tendentially higher than the calcium levels under only fluid flow or tensile strain.

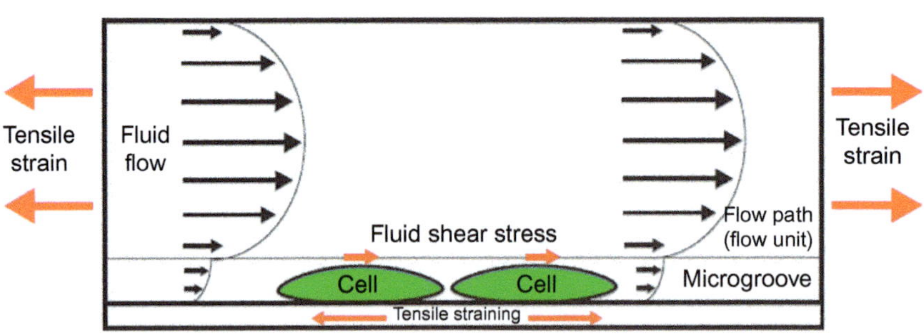


**Fig. 3.3** Fluid can be introduced into the device from the inlet to the outlet, which applies fluid shear stress to tenocytes seeded within the microgrooves. As the device is made from stretchable soft material (PDMS), both fluid shear stress and cyclic tensile strain can be applied to the cells simultaneously.
Based on Maeda, E., Hagiwara, Y., Wang, J.-C., Ohashi, T., 2013. A new experimental system for simultaneous application of cyclic tensile strain and fluid shear stress to tenocytes in vitro. Biomed. Microdevices 15, 1067–1075. © by Biomedical Microdevices, with permission from Springer.

### 3.2.4 Degenerative tendon tissue

So far, the effects of mechanical load or fluid shear stress on gene expression, gap junction, and calcium levels have been discussed for healthy tendons and ligaments or tenocytes and tendon fibroblasts isolated from healthy tissues. However, what about the mechanotransductional effects on cells isolated from degenerative tendon tissue? A study by Choi et al. (2014) compared the gene expression of normal, healthy human tendon cells with the one of degenerative tendon cells when cyclic strain was applied. For that purpose, tendon tissue of three donors having a tendinopathy for more than 6 months (two ATs and one *tibialis posterior* tendon) were collected. From the same three patients, also healthy tendon tissue biopsies were received (trimmed from the distal end of the residual tendon after debridement of degenerative tendon). Cyclic mechanical strain was applied during 15 and 60 minutes, respectively. While both, healthy and degenerative tendon cells, responded to strain by an increased proliferation, the cell viability of the degenerative cells was significantly lower compared the healthy cells. As for gene expression of typical mechanotransduction genes, there was an obvious nonresponsiveness of the degenerated tendon cells to mechanical stimulation (Fig. 3.4).

This pattern of gene expression indicates that the cytoskeletal tensional balance is impaired in the degenerative tendon because key cytoskeletal mediators are reduced or absent (time point 0, no mechanical stimulation). Also the mechanotransduction is different when cyclic strain is applied; while in healthy tendons, a 15-minute mechanical stimulation led to a significant initial increase in gene expression, there was no such effect in the degenerative cells. At 60 minutes, healthy tendon cells still had a significantly higher gene expression, while the degenerative cells did not show any response to the stimulation.

From these findings, the authors concluded that during tendinopathy, the most important genes of tensional balance and recovery are suppressed or remain inactive (Choi et al., 2014).

### 3.2.5 Finite element model

The influence of ECM strains and fluid-induced shear stress on tenocytes was modeled with a finite element model using a multiscale approach (Lavagnino et al., 2008). In this model, the geometry and composition was based on the rat tail tendon, with 70–80% of collagen I of the dry weight, an extrafibrillar matrix and water (60–80% of the wet weight). There was an axisymmetric global poroelastic model (Fig. 3.5A) and a nonaxisymmetric submodel (Fig. 3.5B) that was located in the mid-portion along the length of the global model.

In this model, collagen fibers were simulated using axially oriented nonlinear spring elements attached at the nodes of the matrix elements, with the largest strain needed to uncrimp the fibers in the center of the tendon and the smallest strain at the outer boundary. The collagen stiffness was set to 0 N/mm during straightening of the fiber and 20 N/mm after its straightening. As tendons have a Poisson's ratio (coefficient of expansion on the transverse axial) of >0.5 reflecting their orthotropy, the

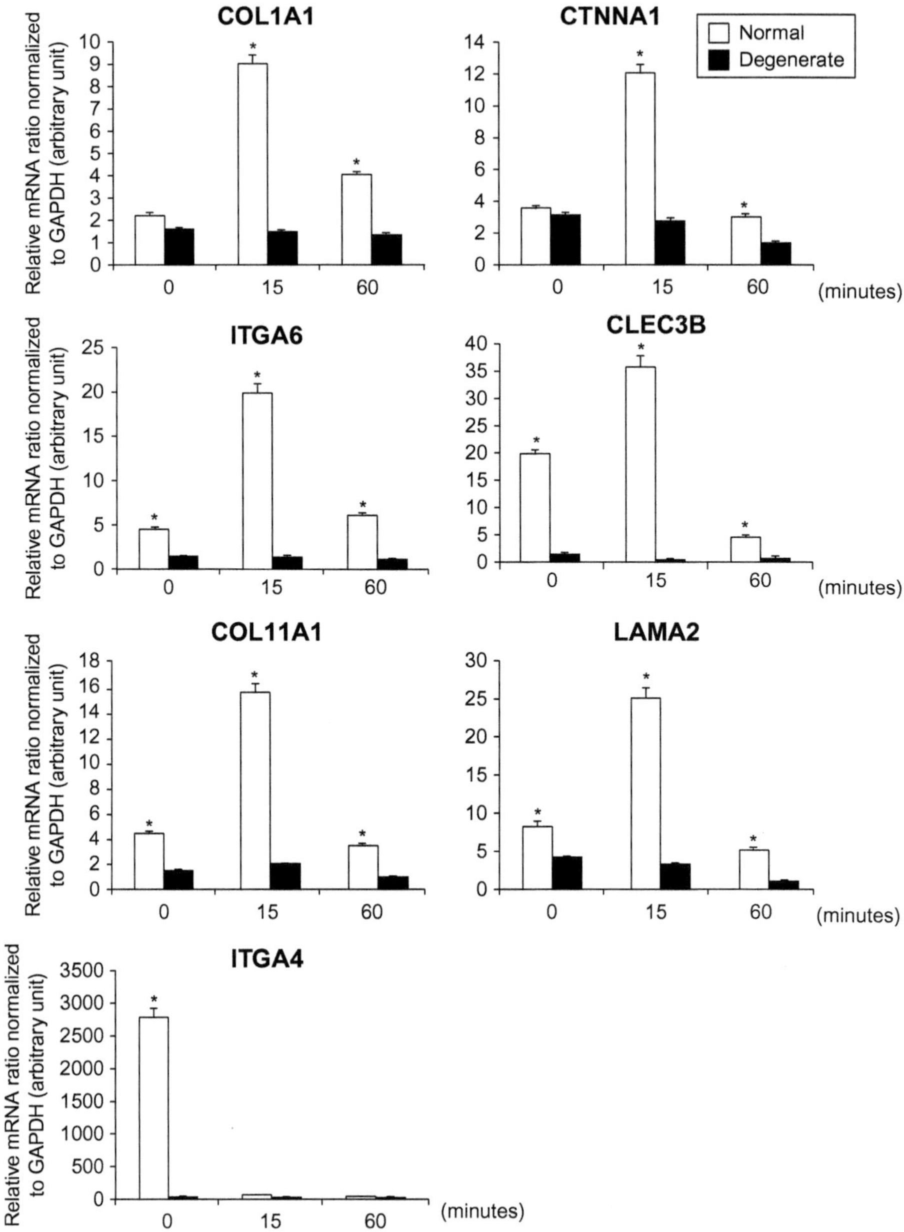


**Fig. 3.4** Normal (healthy) and degenerative tendon cells: real-time polymerase chain reaction analysis of 7 candidate gene expression. Gene expression was analyzed at different time points after cyclic strain (15 and 60 minutes, respectively). Values represent the mean data from three separate experiments with the same batch of cells. *COL1A1*, collagen, type I, alpha 1, also known as alpha-1 type I collagen; *CTNNA1*, gene that encodes αE-catenin, also known as catenin alpha-1; *ITGA6*, integrin alpha-6; *CLEC3B*, encodes tetranectin; *COL11A1*, encodes collagen alpha-1 (XI) chain; *LAMA2*, laminin alpha 2; *ITGA4*, integrin, alpha 4.
From Choi, W.J., Park, M.S., Park, K.H., Courneya, J.-P., Cho, J.S., Schon, L.C., Lee, J.W., 2014. Comparative analysis of gene expression in normal and degenerative human tendon cells: effects of cyclic strain. Foot Ankle Int. 35, 1045–1056. Copyright © 2014 by Foot & Ankle International, Reprinted by Permission of Sage Publications, Inc.

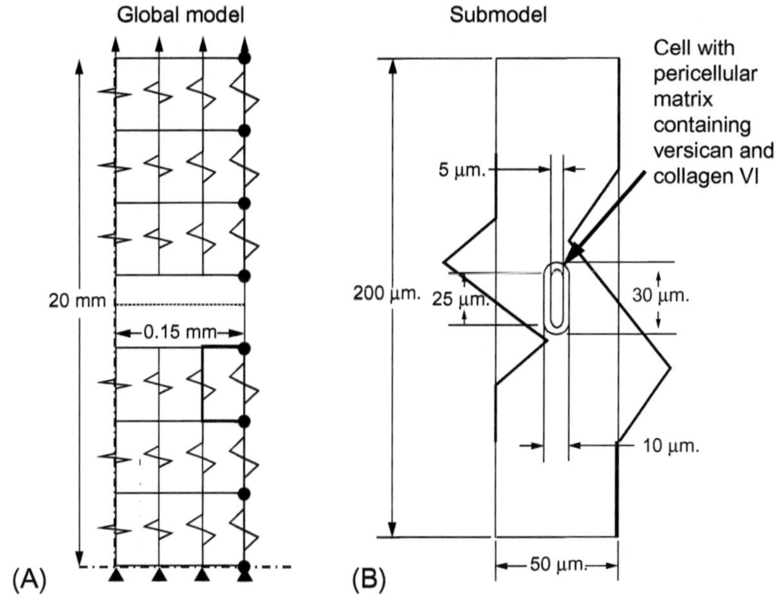


**Fig. 3.5** (A) Axisymmetric global poroelastic model of the rat tail tendon (20 × 0.15 mm), divided into 300 four-noded axisymmetric elements (0.2 × 0.05 mm) with radially variant nonlinear spring elements attached at the nodes of the matrix elements (*springs*), zero pore pressure on the outer boundary (*circles*), constrained at the tendon center (*triangles*), and loaded at the tendon end as per previous experimental conditions (*arrows*). The darkened element boundary indicates the location of the submodel. (B) Submodel of the rat tail tendon, the size of a global element, composed of an ovoid-shaped cell, cell membrane, peri-cellular matrix, extracellular matrix, and collagen fibers.
Based on Lavagnino, M., Arnoczky, S.P., Kepich, E., Caballero, O., Haut, R.C., 2008. A finite element model predicts the mechanotransduction response of tendon cells to cyclic tensile loading. Biomech. Model. Mechanobiol. 7, 405–416. © by Biomechanics and Modeling in Mechanobiology, with permission from Springer.

Poisson's ratio in the model was chosen to reflect the fluid exudation effect of the whole tendon. Although permeability of the tendon is not homogenous throughout the tendon tissue (the major determinants of permeability being strain, porosity, and/or void ratios), the model was based on a uniform permeability. Furthermore, cells were modeled to be situated at the boundary of the tendon because there they are subjected to the largest matrix deformations and the highest fluid flow. Finally, a peri-cellular matrix was included in the model (in addition to the extracellular matrix), with versican and collagen VI, which play a pivotal role in mechanotransduction and alter the mechanical environment of the tendon and cartilage cells (Guilak and Mow, 2000). For further detailed information on submodel components, see Lavagnino et al. (2008).

In order to compare the predicted stresses and strains with the interstitial collagenase (MMP-13) mRNA inhibition of tenocytes in situ, Lavagnino et al. performed experiments where rat tail tendons in Dulbecco's minimal essential medium (DMEM) underwent cyclic strain at different velocities (strains/minutes) and frequencies and quantitative polymerase chain reaction. As a result, they found a direct negative correlation between collagenase mRNA and the levels of membrane strain and shear stress that were predicted by the model. In other words, tenocytes exposed to higher strain rates and strain amplitudes expressed significantly less collagenase mRNA. Furthermore, it was shown that the applied strain rate and the strain amplitude were able to independently alter catabolic gene expression through enhanced fluid-flow-induced shear stress and matrix-induced tenocyte deformation (Lavagnino et al., 2008).

## 3.3 Effects of mechanical stimulation on ECM

### *3.3.1 Collagen*

As collagen I is the main component of tendon extracellular matrix (ECM), this suggests starting with collagen when effects of mechanical stimulation on ECM are discussed. Generally speaking, the formation as well as the degradation of collagen in tendons increases with loading. There are several experimental approaches to study collagen synthesis in tendon ECM—either indirectly by the determination of enzymes that are involved in collagen synthesis such as prolyl-4-hydroxylase (Kovanen, 1989), or directly by microdialysis assessing the interstitial concentrations of the procollagen propeptides. These propeptides are cleaved off during the transformation of procollagen to collagen (Langberg et al., 1999, 2001). Because the catheters are placed around the tendon tissue or directly in the tendon tissue, microdialysis allows assessing the local concentrations of such propeptides (compared to the rather small increase in the circulating blood for example). Furthermore, collagen synthesis can be determined also with stable isotopes that are not radioactive; such tracers are used to label amino acids that are incorporated in the living tissue. From biopsies, the amount of incorporated stable isotopes can be measured (Miller et al., 2006). It has been shown that mechanical loading increases the AT collagen synthesis after a 36 km run by a factor of 2 (human) (Langberg et al., 1999). Further results concerning collagen synthesis in tendons after mechanical stimulation (acute and chronic exercise included) are summarized in Kjaer et al. (2009).

One central question in mechanobiology of tendons is how the load applied is actually transmitted through the tendon tissue, or more specifically, whether the collagen fibrils bear the load independently or whether the load is transferred between the collagen fibrils through interfibrillar shear forces. The exact mechanism of load transmission has been discussed and investigated including different aspects. In 2006, load transmission in tendons has been attributed to the collagen fibril morphology and also to its organization (Provenzano and Vanderby, 2006); in contrast to early fetal tendons, mature tendon tissue was reported to have continuous collagen fibrils with only few fibril ends as revealed by the comparison of medial collateral ligament of fetal and mature rats under scanning electron microscopy (Fig. 3.6). Therefore, it was concluded that the majority of collagen fibrils are long (in the range of millimeters) and were considered as functionally continuous (Provenzano and Vanderby, 2006).

**Fig. 3.6** Collagen fibril continuity in mature rat ligament: no collagen fibril ends discovered by scanning electron microscopy.
Based on Provenzano, P.P., Vanderby, R., 2006. Collagen fibril morphology and organization: implications for force transmission in ligament and tendon. Matrix Biol. 25, 71–84. Reprinted from Matrix Biology, Copyright 2006, with permission from Elsevier.

In 2009, the concept of force transmission through continuously functional collagen fibrils along their "length" was re-discussed because of obvious collagen fibril branching. Starborg et al. showed that collagen fibrils fused in Y-shaped branches with an angle of 4–10 degrees using an embryonic mouse tail tendon model and transmission electron microscopy (Starborg et al., 2009). With such findings, the concept of force transmission along the "length" could be the sole underlying mechanism of mechanotransduction. Moreover, local tissue strains have been shown to be significantly lower than whole tendon tissue strains (Screen et al., 2003) and to be inhomogeneous (Snedeker et al., 2006). For example, the collagen fibril strains under tension were shown to be only half of the applied tissue strains. Therefore, there must be another underlying mechanism for force transfer than solely along a continuous collagen fibril. Relative sliding between collagen fibrils can account for different strains of tendon tissue and fibrils under tension (Rigozzi et al., 2011). In a multiscale experimental study using fascicles of mature rat tail tendons, Szczesny and Elliott (2014) were able to assess simultaneously macroscale mechanical behavior as well as microscale deformations in uniaxial tension. They reported decreasing fibril:tissue strain ratios with increasing tissue strain (no strain had a ratio of 1, while the ratio was

0.5 for around 9% of strain) and increasing degrees of interfibrillar sliding with increasing tissue strain. The fibril:tissue strain ratios were also plotted against the degree of sliding, as well as the macroscale biomechanical properties of elastic modulus and stress relaxation (Fig. 3.7).

In addition, these data were modeled with a shear lag model to test whether indeed the interfibrillar shear load transfer is the loading mechanism in tendons. The model was first used to fit the macroscale experimental mechanical data and then to predict the microscale deformations. As a successful result, the model fitted the macroscale mechanics well, and also the prediction of the fibril:tissue strain ratios was sound. This ability to approximate the mechanics at both length scales suggested that interfibrillar shear was the loading mechanism of the discontinuous fibrils in tendon fascicles (Szczesny and Elliott, 2014).

Other studies including partially lacerated *flexor* tendons (energy storing) showed that the mechanical compromise of these tendons was not proportional to the area of laceration, corroborating the findings of Szczesny and Elliott (2014) and indicating that longitudinal loading of fibers and fascicles was not the only load-bearing mechanism (Pensalfini et al., 2014; Ahmadzadeh et al., 2013). Focusing on the shear transfer between the fibers and the fascicles, the shear transfer was claimed to be the mechanism of action (Kondratko-Mittnacht et al., 2015). In this in vitro study, porcine flexor tendons (energy storing) and rat tail tendons (positional) were used and lacerated with a ratio of laceration depth/tendon width ranging from 0.2 to 0.6. Both tendon types exhibited a linear decrease in post/prelaceration ratios of mechanical parameters (elastic and viscoelastic stress and elastic modulus) with increasing laceration depth/tendon width. However, the rat tail tendons had a more negative slope, indicating that positional tendons have a lower ability for shear load transfer than energy storing tendons, which is also suggested by the more linearly arranged and more densely packed fascicles in the positional tendons.

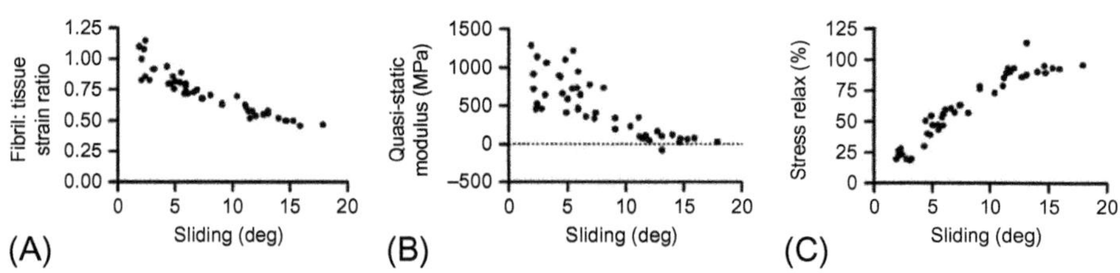


**Fig. 3.7** Relationship between fascicle macroscale mechanics and interfibrillar sliding. (A) The decrease in the fibril:tissue strain ratio with greater interfibrillar sliding suggests that relative sliding serves to unload the fibrils, which produces a (B) drop in the macroscale tissue modulus. These data suggest that interfibrillar sliding is responsible for the strain softening observed in the postyield behavior of tendon fascicles. (C) A strong correlation between the macroscale stress relaxation and interfibrillar sliding suggests that sliding is associated with a viscous interfibrillar shear stress that transmits load between fibrils.
From Szczesny, S.E., Elliott, D.M., 2014. Interfibrillar shear stress is the loading mechanism of collagen fibrils in tendon. Acta Biomater. 10, 2582–2590. Reprinted from Acta Biomaterialia, Copyright 2014, with permission from Elsevier.

### 3.3.2 The noncollagenous part in the ECM

Tendons can be roughly divided into positional tendons and energy storing tendons. Positional tendons are loaded along their long axis and enable muscles to move bones, while energy storing tendons (in addition to their positional role) store energy when they are stretched. The energy storing tendons can experience strains of up to 10% as opposed to the positional tendons where maximum strains are between 2% and 3% in vivo (Birch, 2007; Thorpe et al., 2012). Such different mechanical behaviors are reflected in the matrix composition of the tendon tissue. Energy storing tendons have higher levels of glycosaminoglycans (GAGs) than positional tendons. GAGs are part of proteoglycans, they are poly-carbohydrates, negatively charged, and coupled to a specific core protein. As for GAG's function in terms of mechanotransduction, it has been shown by Legerlotz et al. that tendon fascicles with GAG content reduced to 23% of normal content by chondoitinase digestion, behaved differently in mechanical loading experiments (Legerlotz et al., 2013). Either the fascicles were subjected to cyclic stretching (14%) at 1 Hz or they were statically strained for 30 minutes. Compared to natural tendons with full GAG content, the removal of GAGs led to more stress relaxation and to higher reductions in failure stress after loading, for both cyclic and static conditions, respectively. From these findings, the authors concluded that GAGs could have a protective effect by limiting the viscoelastic behavior of the tendon tissue and also by preventing tissue fatigue. In addition, they discussed these results in terms of tissue dehydration; both GAG depletion as well as cyclic loading have been shown to go along with a reduced tissue hydration (Sanchez-Adams et al., 2011). Therefore, the significantly lower failure stress found in the GAG-depleted fascicles might actually be due to tissue dehydration. However, as the stress-strain curves looked different for GAG-depleted and natural fascicles, another explanation would be that GAGs act as intermolecular bridges and after their removal lead to lower biomechanical failure properties.

Moreover, energy storing tendons have more cartilage oligomeric matrix protein (COMP) compared to positional tendons (Smith et al., 2002). COMP is a non-collagenous ECM protein primarily found in cartilage and stands for the cartilage turnover, however, it also plays a pivotal role in tendon development taking over an organizational role in collagenous matrix formation. It acts as a mediator in the growth of tendons. What does the higher amount of COMP in energy storing tendons compared to positional tendons imply? It has been shown that higher COMP concentrations in tendons of young horses have higher ultimate stress ($r^2=0.52$) and higher stiffness ($r^2=0.37$) (Smith et al., 2002). However, in adult horses, there was no correlation of biomechanical properties with COMP concentrations (ultimate stress: $r^2=0.01$ and stiffness: $r^2=0.02$). As such, the higher COMP concentration in energy storing tendons compared to positional ones remains to be elucidated because the mechanical properties such as ultimate stress or stiffness are usually higher in positional tendon fascicles than in energy storing tendon fascicles, as has been shown for the bovine extensor (positional) and flexor (energy storing) tendon fascicles (Shepherd et al., 2014b).

Furthermore, collagen crosslinking is different with the mature crosslink found only in positional tendons (Screen et al., 2015). This stands in accordance with their usually higher strength and stiffness (Shepherd et al., 2014b).

Proteoglycans located between the fibrils enable fibril sliding, which is found in a higher amount in positional tendons compared to energy storing tendons. In the latter, other molecules such as lubricin and elastin between the fascicles enable a more elastic, more recoverable sliding of the fascicles (explaining the high strechability of the energy storing tendons; Screen et al., 2015).

### 3.3.3 Inflammatory response

The impact of mechanical loading ending up in ECM damage such as collagen fiber kinks and interfascicular matrix disruption was also studied in terms of inflammation (Spiesz et al., 2015). For this purpose, bovine digital flexor tendons were used. Matrix response to cyclic loading for 300 and 1800 cycles, respectively, at 1 Hz from 1% to 10% of uniaxial strain was studied by immunostaining for inflammatory markers cyclooxygenase (COX-2) and interleukin (IL-6). Moreover, also matrix degradation markers such as MMP-1, MMP-3, and MMP-13 were assessed as well as collagen degradation marker C1,2C. Two phases in the tendon tissue were focused, the rather cell-poor fascicular matrix and the endotenon that has a higher cell density compared to the fascicular part (Buschmann et al., 2014). The intensity of immunostained inflammatory markers as well as matrix remodeling markers increased with increasing number of loading cycles (Spiesz et al., 2015). In terms of localization, both inflammatory and ECM-damage induced remodeling were in or near the cell-rich endotenon, indicating the predominant cell-responsiveness to overload in this part of the tendon tissue compared to the fascicular part. In some cases, tenocytes in endotenons are more sensitive towards mechanical loading compared to tenocytes in the fascicular tissue. As endotenons are between fascicles and allow sliding of fascicles with respect to each other, their role in mechanotransduction is to allow a higher tendon deformation compared to the fascicles as such (Thorpe et al., 2012). However, as this soft connective tissue can undergo higher shear stress than the fascicular part, it is also more prone to damage and therefore the cells in the endotenon react by a significantly higher expression of inflammatory markers compared to the cells in the fascicular matrix. To insure proper tendon function, the endotenon provides a higher matrix turnover rate (as implied by higher MMPs and C1,2C).

### 3.3.4 Changes of fiber strain and sliding under load

Although it has been clearly shown that cyclic creep loading experiments on fascicles from digital *extensor* tendon (a positional tendon), to one fourth of failure load (for 300 and 900 cycles at 1 Hz of two functionally different bovine tendon fascicles), led to a partial structural breakdown of the ECM and also to a widening of the interfiber space, there was no change in biomechanical properties (Shepherd et al., 2014b). The biomechanics of these early stage fatigue experiments were assessed on the macroscopic level, with the fascicles undergoing typical stress-strain tests. There, a significant difference was found between the two tendon types, with the *extensor* having higher ultimate stress and higher modulus than the *flexor* tendon; however, these properties did not change for each tendon type after different numbers of loading cycles under the mentioned conditions, respectively. Interestingly, also on the micro level, the authors did not find any mechanical alterations when micro-strain analysis was performed either by nuclei cell tracking or by the deformation of a laser-induced photo-bleached grid on the tendon fascicle to assess fiber strain and fiber sliding.

It has to be noted, however, that error bars in fiber strain and sliding data were quite large; indicating a high variability between the samples and relating to the localized nature of the micro-strain analysis. Moreover, at the completely damaged sites, micro-strain analysis was problematic or not possible at all because cell nuclei tracking was not reliable. Hence, at the sites where the highest degrees of deformation might be expected to be caused by ECM breakdown, measurements were impaired. Finally, the micro-strain analysis was displacement controlled, and corresponding loads associated with these displacements were not assessed. Load controlled micro-strain analysis might have given other results in terms of deformation after cyclic creep loading than displacement control. Nevertheless, these interesting results implicate that fascicles of positional as well as energy storing tendons withstand a visible degree of ECM disruption in early fatigue experiments without obvious changes in macro and micro mechanical behavior.

In a subsequent study, Shepherd et al. found a different response of positional and energy storing tendons when loads and strains were high in cyclic loading experiments before the analysis of mechanical properties. This study showed that energy storing tendons were better able to withstand ECM damage than positional tendons (Shepherd et al., 2014a). Here, not only bovine tendons (again *flexor* and *extensor* tendons as mentioned above), but also human *semitendinous* hamstring (energy storing) tendon, were used for comparison. Stress or strain controlled cyclic fatigue loading was performed (loading frequency 1 Hz, with constant load of 60% of the ultimate stress or 60% of the strain to failure, resulting in 14% of tendon strain = 1.4 mm displacement in 10 mm specimen). The experiments were carried out for various cycling regimes (300–1800 cycles), respectively. Like mentioned in the previous paragraph, static stress-strain and micro-strain analyzes were performed. While the more positional bovine extensor tendon demonstrated a poor fatigue response with a significantly lower number of cycles to failure compared to the human hamstring, the two energy storing tendons (bovine flexor and human hamstring) showed a similar mechanical behavior after the cyclic stress and strain controlled fatigue experiments. The ECM damage under the destructive loading conditions applied were clearly visible with confocal microscopy depicting impaired collagen organization and misalignment of tenocyte nuclei, with increasing damage caused by increasing number of cyclic loading. Although all three tendon types exhibited ECM damage, the energy storing tendons were better able to keep their biomechanical properties and function than the positional tendon. Again, the reason for such findings may be found in the energy storing tendons having more lubricin and elastin between their fascicles, which enables the fascicles to recover better under fatigue conditions (Screen et al., 2015). It might furthermore be speculated that tenocytes in the endotenon of positional tendons and of energy storing tendons are different in terms of their mechnosensitity; as has been shown for tenocytes in the endotenon versus tenocytes in the fascicle, tenocytes in endotenons of different tendons may be differently responsive towards tendon loading in terms of their remodeling capacity (Spiesz et al., 2015). However, the mechanobiology of tenocytes isolated from endotenon of different tendon types still has to be elucidated.

## 3.4 Summary

Tendons and ligaments are able to respond to mechanical stimuli by altering their structure, composition, and mechanical properties as well as gene expression in the tenocytes. The tenocytes sense mechanical forces in the form of matrix deformations caused by load and they convert the forces into biochemical signals. In this chapter, we have focused on altered gene expression with a focus on the TGF-β family, and also the impact of loading on the gap junction permeability and calcium levels in tenocytes. Interestingly, tenocytes of healthy tendon tissue do react differently in terms of gene expression compared to tenocytes of degenerative tendon tissue, showing a clear non-responsiveness to mechanotransduction genes in cells of damaged tissue. Furthermore, we presented a finite model that accurately predicted the tenocyte response towards varying strain rates and strain amplitudes. Finally, the impact of loading on extracellular matrix composition, including collagen, proteoglycans, and GAGs was discussed as well as the impact on inflammatory markers.

## References

Ahmadzadeh, H., Connizzo, B.K., Freedman, B.R., Soslowsky, L.J., Shenoy, V.B., 2013. Determining the contribution of glycosaminoglycans to tendon mechanical properties with a modified shear-lag model. J. Biomech. 46, 2497–2503.

Arnoczky, S.P., Lavagnino, M., Whallon, J.H., Hoonjan, A., 2002. In situ cell nucleus deformation in tendons under tensile load; a morphological analysis using confocal laser microscopy. J. Orthop. Res. 20, 29–35.

Birch, H.L., 2007. Tendon matrix composition and turnover in relation to functional requirements. Int. J. Exp. Pathol. 88, 241–248.

Buschmann, J., Puippe, G., Meier Buergisser, G., Bonavoglia, E., Giovanoli, P., Calcagni, M., 2014. Correspondence of high-frequency ultrasound and histomorphometry of healing rabbit Achilles tendon tissue. Connect. Tissue Res. 55, 123–131.

Butler, D.L., Grood, E.S., Noyes, F.R., Zernicke, R.F., Brackett, K., 1984. Effects of structure and strain-measurement technique on the material properties of young human tendons and fascia. J. Biomech. 17, 579–596.

Choi, W.J., Park, M.S., Park, K.H., Courneya, J.-P., Cho, J.S., Schon, L.C., Lee, J.W., 2014. Comparative analysis of gene expression in normal and degenerative human tendon cells: effects of cyclic strain. Foot Ankle Int. 35, 1045–1056.

Fessel, G., Cadby, J., Wunderli, S., Weeren, R.V., Snedeker, J.G., 2014. Dose and time dependent effects of genipin cross-linking on cell viability and tissue mechanics—toward clinical application for tendon repair. Acta Biomater. 10 (5), 1897–1906.

Fong, K.D., Trindade, M.C., Wang, Z., Nacamuli, R.P., Pham, H., Fang, T.D., Song, H.J.M., Smith, L., Longaker, M.T., Chang, J., 2005. Microarray analysis of mechanical shear effects on flexor tendon cells. Plast. Reconstr. Surg. 116, 1393–1404.

Glass, Z.A., Schiele, N.R., Kuo, C.K., 2014. Informing tendon tissue engineering with embryonic development. J. Biomech. 47, 1964–1968.

Guilak, F., Mow, V.C., 2000. The mechanical environment of the chondrocyte: a biphasic finite element model of cell-matrix interactions in articular cartilage. J. Biomech. 33, 1663–1673.

Gupta, R., 2005. Microarray analysis of mechanical shear effects on flexor tendon cells. Plast. Reconstr. Surg. 116, 1405–1406.

Heinemeier, K., Langberg, H., Olesen, J.L., Kjaer, M., 2003. Role of TGF-beta(1) in relation to exercise-induced type I collagen synthesis in human tendinous tissue. J. Appl. Physiol. 95, 2390–2397.

Kjaer, M., Langberg, H., Heinemeier, K., Bayer, M.L., Hansen, M., Holm, L., Doessing, S., Kongsgaard, M., Krogsgaard, M.R., Magnusson, S.P., 2009. From mechanical loading to collagen synthesis, structural changes and function in human tendon. Scand. J. Med. Sci. Sports 19, 500–510.

Kondratko-Mittnacht, J., Duenwald-Kuehl, S., Lakes, R., Vanderby, J.R., 2015. Shear load transfer in high and low stress tendons. J. Mech. Behav. Biomed. Mater. 45, 109–120.

Kovanen, V., 1989. Effects of aging and physical-training on rat skeletal-muscle—an experimental-study on the properties of collagen, laminin, and fiber types in muscles serving different functions. Acta Physiol. Scand. 135, 9–56.

Langberg, H., Skovgaard, D., Petersen, L.J., Bulow, J., Kjaer, M., 1999. Type I collagen synthesis and degradation in peritendinous tissue after exercise determined by microdialysis in humans. J. Physiol. Lond. 521, 299–306.

Langberg, H., Rosendal, L., Kjaer, M., 2001. Training-induced changes in peritendinous type I collagen turnover determined by microdialysis in humans. J. Physiol. Lond. 534, 297–302.

Lavagnino, M., Arnoczky, S.P., Kepich, E., Caballero, O., Haut, R.C., 2008. A finite element model predicts the mechanotransduction response of tendon cells to cyclic tensile loading. Biomech. Model. Mechanobiol. 7, 405–416.

Legerlotz, K., Riley, G.P., Screen, H.R.C., 2013. GAG depletion increases the stress-relaxation response of tendon fascicles, but does not influence recovery. Acta Biomater. 9, 6860–6866.

Mackley, J.R., Ando, J., Herzyk, P., Winder, S.J., 2006. Phenotypic responses to mechanical stress in fibroblasts from tendon, cornea and skin. Biochem. J. 396, 307–316.

Maeda, E., Ye, S., Wang, W., Bader, D.L., Knight, M.M., Lee, D.A., 2012. Gap junction permeability between tenocytes within tendon fascicles is suppressed by tensile loading. Biomech. Model. Mechanobiol. 11, 439–447.

Maeda, E., Hagiwara, Y., Wang, J.-C., Ohashi, T., 2013. A new experimental system for simultaneous application of cyclic tensile strain and fluid shear stress to tenocytes in vitro. Biomed. Microdevices 15, 1067–1075.

Mendias, C.L., Gumucio, J.P., Lynch, E.B., 2012. Mechanical loading and TGF-β change the expression of multiple miRNAs in tendon fibroblasts. .

Miller, B.F., Hansen, M., Olesen, J.L., Flyvbjerg, A., Schwarz, P., Babraj, J.A., Smith, K., Rennie, M.J., Kjaer, M., 2006. No effect of menstrual cycle on myofibrillar and connective tissue protein synthesis in contracting skeletal muscle. Am. J. Physiol. Endocrinol. Metab. 290, E163–E168.

Pensalfini, M., Duenwald-Kuehl, S., Kondratko-Mittnacht, J., Lakes, R., Vanderby, R., 2014. Evaluation of global load sharing and shear-lag models to describe mechanical behavior in partially lacerated tendons. J. Biomech. Eng.-Trans. ASME. 136. http://dx.doi.org/10.1115/1.4027714.

Provenzano, P.P., Vanderby, R., 2006. Collagen fibril morphology and organization: implications for force transmission in ligament and tendon. Matrix Biol. 25, 71–84.

Rigozzi, S., Stemmer, A., Mueller, R., Snedeker, J.G., 2011. Mechanical response of individual collagen fibrils in loaded tendon as measured by atomic force microscopy. J. Struct. Biol. 176, 9–15.

Sanchez-Adams, J., Willard, V.P., Athanasiou, K.A., 2011. Regional variation in the mechanical role of knee meniscus glycosaminoglycans. J. Appl. Physiol. 111, 1590–1596.

Screen, H.R., Lee, D.A., Bader, D.L., Shelton, J.C., 2003. Development of a technique to determine strains in tendons using the cell nuclei. Biorheology 40, 361–368.

Screen, H.R.C., Berk, D.E., Kadler, K.E., Ramirez, F., Young, M.F., 2015. Tendon functional extracellular matrix. J. Orthop. Res. 33, 793–799.

Shepherd, J.H., Legerlotz, K., Demirci, T., Klemt, C., Riley, G.P., Screen, H.R., 2014a. Functionally distinct tendon fascicles exhibit different creep and stress relaxation behaviour. Proc. Inst. Mech. Eng. H J. Eng. Med. 228, 49–59.

Shepherd, J.H., Riley, G.P., Screen, H.R.C., 2014b. Early stage fatigue damage occurs in bovine tendon fascicles in the absence of changes in mechanics at either the gross or micro-structural level. J. Mech. Behav. Biomed. Mater. 38, 163–172.

Smith, R.K.W., Gerard, M., Dowling, B., Dart, A.J., Birch, H.L., Goodship, A.E., 2002. Correlation of cartilage oligomeric matrix protein (COMP) levels in equine tendon with mechanical properties: a proposed role for COMP in determining function-specific mechanical characteristics of locomotor tendons. Equine Vet. J. 34, 241–244.

Snedeker, J.G., Pelled, G., Zilberman, Y., Gerhard, F., Muller, R., Gazit, D., 2006. Endoscopic cellular microscopy for in vivo biomechanical assessment of tendon function. J. Biomed. Opt. 11, 8.

Spiesz, E.M., Thorpe, C.T., Chaudhry, S., Riley, G.P., Birch, H.L., Clegg, P.D., Screen, H.R.C., 2015. Tendon extracellular matrix damage, degradation and inflammation in response to in vitro overload exercise. J. Orthop. Res. 33, 889–897.

Starborg, T., Lu, Y., Huffman, A., Holmes, D.F., Kadler, K.E., 2009. Electron microscope 3D reconstruction of branched collagen fibrils in vivo. Scand. J. Med. Sci. Sports 19, 547–552.

Szczesny, S.E., Elliott, D.M., 2014. Interfibrillar shear stress is the loading mechanism of collagen fibrils in tendon. Acta Biomater. 10, 2582–2590.

Thomopoulos, S., Parks, W.C., Rifkin, D.B., Derwin, K.A., 2015. Mechanisms of tendon injury and repair. J. Orthop. Res. 33, 832–839.

Thorpe, C.T., Udeze, C.P., Birch, H.L., Clegg, P.D., Screen, H.R.C., 2012. Specialization of tendon mechanical properties results from interfascicular differences. J. R. Soc. Interface 9, 3108–3117.

Wang, J.H.C., 2006. Mechanobiology of tendon. J. Biomech. 39, 1563–1582.

Wang, J.H.C., Thampatty, B.P., 2006. An introductory review of cell mechanobiology. Biomech. Model. Mechanobiol. 5, 1–16.

Wellen, J., Helmer, K.G., Grigg, P., Sotak, C.H., 2005. Spatial characterization of T1 and T2 relaxation times and the water apparent diffusion coefficient in rabbit Achilles tendon subjected to tensile loading. Magn. Reson. Med. 53, 535–544.

# Experimental methods for measuring tendon and ligament biomechanics

4

## Abstract

Biomechanical characterization of tendons and ligaments on all levels of structural hierarchy demands different experimental methods. Whole ex vivo specimen may be biomechanically characterized in classic static tensile testing machines, giving load–displacement or stress–strain curves where ultimate load, stiffness, maximum stress, and elastic modulus can be assessed. Collagen fiber and fascicle characterization may need additional techniques such as atomic force microscopy or transmission electron microscopy. Besides static tensile (and also compressive) testing, dynamic tests are in use in order to determine the fatigue behavior of the tissue under view. Finally in vivo biomechanical characterization is presented with ultrasound being one of the most often used noninvasive imaging methods.

**Keywords:** Tensile testing machine, Static and dynamic tests, Preconditioning, Loading, Biomechanical parameters.

## Abbreviations

| | |
|---|---|
| **ACL** | anterior cruciate ligament (tendons and ligaments) |
| **AFM** | atomic force microscopy (imaging technique) |
| **CCD** | charge-coupled device camera (device in digital imaging) |
| **CLM, CLsM** | confocal laser microscopy (technical feature of LM) |
| **CSA** | cross-sectional area (feature of tendons and ligaments) |
| **DDFT** | deep digital flexor tendon (tendons and ligaments) |
| **GAGs** | glycosaminoglycans (also called mucopolysaccharide) (polysaccharides in living cells) |
| **HPT** | human patellar tendon (tendons and ligaments) |
| **Hz** | hertz, 1/s (unit of frequency) |
| $L_0$ | length at the time point zero (unit of measurement) |
| **N** | Newton (unit of force) |
| **Pa, MPa, GPa** | pascal, mega-pascal ($10^3$ Pa), giga-pascal ($10^6$ Pa) (unit of pressure, internal pressure, stress, young's modulus, and ultimate tensile strength, 1 Pa defined as 1 N/m$^2$ = 1 kg/m s$^2$) |
| **PBS** | phosphate-buffered saline (physiological conditioned solution) |
| **ROIs** | regions of interest (unit of measurement) |
| **RTT** | rat tail tendon (tendons and ligaments) |

Biomechanics of Tendons and Ligaments. http://dx.doi.org/10.1016/B978-0-08-100489-0.00004-1

| | |
|---|---|
| **SD** | standard deviation (measure in statistics) |
| **SHG** | second harmonic generation (imaging technique) |
| **SWE** | shear wave elastography (imaging technique) |
| **TEM** | transmission electron microscopy (imaging technique) |
| **US** | ultrasound (imaging technique) |

## 4.1 Introduction

The choice of a specific biomechanical testing system or method depends on what the examiner wants to know. Therefore, whole specimen of tendons or ligaments are investigated differently than only single fibers (Rigozzi et al., 2010), fascicles (Maeda et al., 2012), or even tenocytes (Lavagnino et al., 2011, 2013). First, classic static tensile testing systems and their typical read-outs are presented in this chapter, with tensile testing systems allowing the determination of load until failure, stiffness, maximum stress, and elastic modulus. Testing conditions such as the clamps needed to fix the samples, the loading rate, and the preconditioning with static and cyclic preloading are discussed in terms of their effect on the biomechanical outcome. Moreover, compressive tests are also elucidated (Fang et al., 2014; Bah et al., 2016), which are less common in tendon and ligament characterization than the classic tensile tests. Second, dynamic testing like fatigue tests are presented, where the cycle number influences the peak strain, the stiffness, the laxity as well as the crimp pattern of the tendon. Third, besides the ex vivo biomechanical tests, in vivo biomechanical assessment is presented, with ultrasound (US) as one of the most often used imaging techniques. Finally, the methods to investigate smaller entities such as fibers, fascicles, or single cells are shown, among them for example atomic force microscopy (AFM), second harmonic generation (SHG), or transmission electron microscopy (TEM).

## 4.2 Classic tensile testing

### 4.2.1 Load–displacement and stress–strain

A common approach that is widely used to assess biomechanical properties is based on ex vivo measurements. After extraction of the tendon or ligament of interest, it is placed in a tensile testing machine. The two ends of the tendon or ligament are fixed with clamps which may be based on physical force through screws (Wiig et al., 2011) (Fig. 4.1, *left*), by adding glue, wrapping the ends in cloth (Rigozzi et al., 2009) and then also using screws (Fig. 4.1, *right*) or by cryogenic fixation assemblies where liquid nitrogen as well as ice containers provide secure fixation (Trudel et al., 2007). The principle of freezing clamps was reported to be efficient in other studies, too (Herbort et al., 2008, 2014; Zantop et al., 2006).

If tendons are measured biomechanically, the end including the bone is often fixed in a different way compared to the end facing the muscle; special devices such as that shown in Fig. 4.2 on the right allow the bone to be placed at a right angle to the tendon

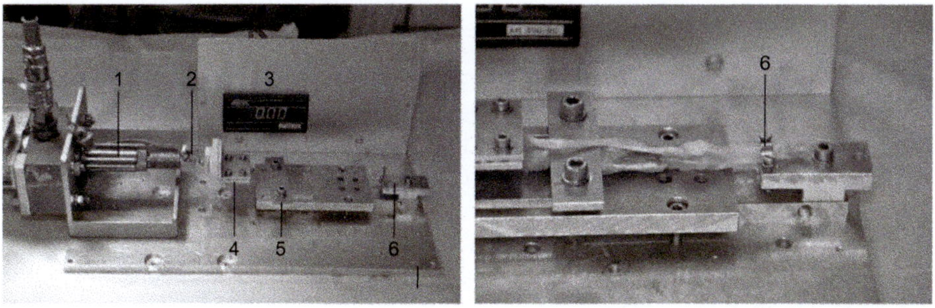


**Fig. 4.1** The testing equipment to assess biomechanical properties of rabbit flexor tendons. A servo-hydraulic actuator (1), designed for applying controlled force or displacement is pulling the specimen. Simultaneous force readings are recorded by a load aluminum frame (2) and are displayed on a digital monitor (3). The tendon is fixed in a clamp (4) and the hind paw is secured to the testing set-up by a similar construction (5). For "load-to-failure" studies, another clamp (6) is used to prevent digit flexion.
From Wiig, M., Olmarker, K., Håkansson, J., Ekström, L., Nilsson, E., Mahlapuu, M., 2011. A lactoferrin-derived peptide (PXL01) for the reduction of adhesion formation in flexor tendon surgery: an experimental study in rabbits. J. Hand Surg. (Eur. Vol.) 36E, 656–662. Copyright © 2011 by Journal of Hand Surgery (European Volume), Reprinted by Permission of SAGE Publications, Inc.

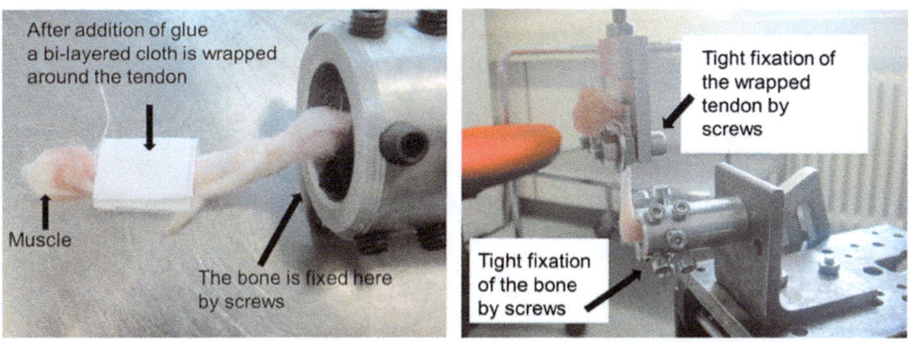


**Fig. 4.2** Details of testing equipment to assess biomechanical properties: a cloth is wrapped around the tendon after the addition of glue (e.g., *locktight* glue, cyanoacrylate) before it is tightly fixed with screws on the muscle side *(left)*. Special devices might be used for the fixation of the bony part; here, a rabbit Achilles tendon is shown where the calcaneus is tightly fixed with 12 screws in metal cylinder *(right)*. To prevent from dehydration, tendons are sprayed by phosphate-buffered solution.

and are themselves connected to the base of the tensile testing machine. Basically, the tendon or ligament is then distracted with a certain distraction rate (see below for the influence of this rate on the outcome measure) and force–displacement curves are recorded [note that load–displacement, force–elongation, and force–deformation are other terms for the same graph with force in newtons (N) on the *y*-axis and displacement in millimeter (mm) on the $x$-axis]. Depending on how the tendon or ligament tissue breaks under tension, the force–displacement curves may look different. Very often, after a toe region and a linear region, there is an abrupt decrease in force while displacement is increasing—many researchers define this force as the "load until gap formation" because sutured tendons show this behavior under tensile force when the threads used for suturing eat themselves through the tendon tissue, which leads to a gap between the tendon stumps. Other breaking patterns include mid-substance failure or interface-failures that may occur at the interface to the bone as well as at the interface to the muscle (Fig. 4.3) and that often depend on the age of the donor (Woo et al., 1991).

#### 4.2.1.1 Load until failure

The *load until failure* or *peak load* is defined as the maximum force a tendon or ligament bears until it breaks. As discussed above, some micro-damage may occur in the tendon tissue during loading, accompanied by a *load until gap formation*; nevertheless, *peak load* may be higher than *load until gap formation*. Therefore, researchers often define *peak load* by referring to the force drop after reaching a certain peak force

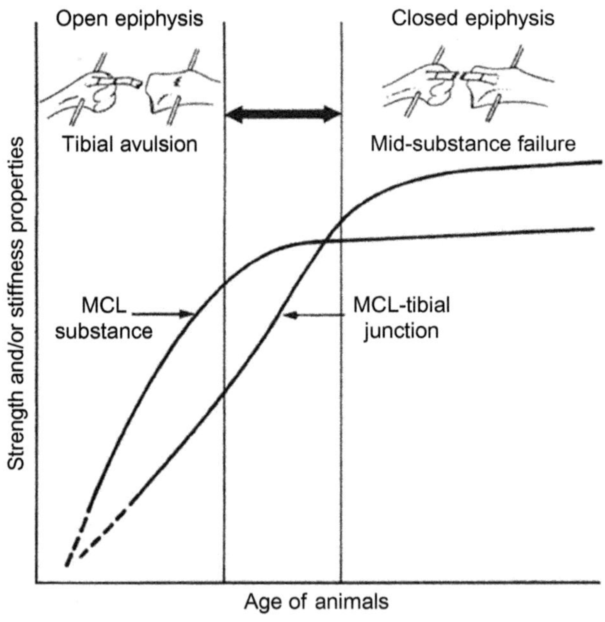


**Fig. 4.3** A schematic diagram hypothesizing the asynchronous rates of maturation from both the structural characteristics of the F-M-T (femur-MCL-tibia) complex and the mechanical properties of medial collateral ligament (MCL) substance.
From Woo, S.L.Y., Orlando, C.A., Gomez, M.A., Frank, C.B., Akeson, W H., 1986. Tensile properties of the medial collateral ligament as a function of age. J. Orthop. Res. 4, 133–141. © by Journal of Orthopaedic Research, with permission from Wiley.

value. For example, Trudel et al. defined the *peak load* of extracted rabbit Achilles tendons (AT) when a 50% drop in peak load is measured. Thus, decreases in load in the force–elongation curves that are less than 50% are not associated with loads until failure values (Trudel et al., 2007).

#### 4.2.1.2 Stiffness

The *stiffness* in N/mm is another structural property of the tendon and ligament besides the *load until failure*. It is the slope in the load–elongation curve. If not otherwise mentioned, the slope in the linear region of the load–elongation curve is used, however, some researchers define exactly which part of this curve they use for stiffness assessment. For example, linear regression to the load–deformation data between 30% and 90% of the peak load was used for the calculation of the stiffness (Trudel et al., 2007).

*Material properties* such as *maximum stress* as well as the *elastic modulus* (both in Pa) can also be assessed via force–elongation data; however, it has to be taken into account that the cross-sectional area (CSA) of the specimen has to be measured, too, because the definition of maximum stress and elastic modulus is load until failure divided by CSA and the slope in the stress–strain curve, respectively (Fig. 4.4). As the CSA varies along the corresponding specimen, the maximum stress as well as the elastic modulus changes along the segments of the specimen. For example, the CSA of human AT differs by more than 50% along their lengths, with the most proximal segment being significantly smaller compared to the most distal segment—leading to the highest stress at the interface to the *calcaneus* (Kongsgaard et al., 2005; Magnusson and Kjaer, 2003). Therefore, an exact definition of WHERE the corresponding material properties are referred to is absolutely needed. As for the determination of the CSA, different methods are being used, for example the laser-assisted measurement of the two halves of the circumference by a linear laser scanner adapted by Vergari which is being calculated to give the area inside (=CSA) (Vergari et al., 2010).

### 4.2.2 Loading rate

The rate of distraction is given either in mm/s, N/s, or in %/s, the latter being the percentage of length at the time point zero ($L_0$) divided by time. Screening different protocols in the biomechanical literature reveals a wide range of distraction rates. For example, rabbit digital flexor tendons' biomechanics were assessed with 1 mm/s (Wiig et al., 2011), rabbit ATs with 18 N/s (Trudel et al., 2007), rat patellar tendons (PT) with 0.1 mm/s (Stange et al., 2015), human, pig, and sheep ATs with 5 mm/min (Gatt et al., 2015), bovine deep digital flexor tendons (DDFT) with 20 mm/min (Fang et al., 2014), and human flexor *digitorum profundus* with 120 mm/min (Fox et al., 2013).

What influence do different distraction rates have on the biomechanical read-out? In an ex vivo study by Wren et al., human ATs were subjected to tensile testing with strain rates ranging from 1%/s to 10%/s. This increase in strain rate led to an increase

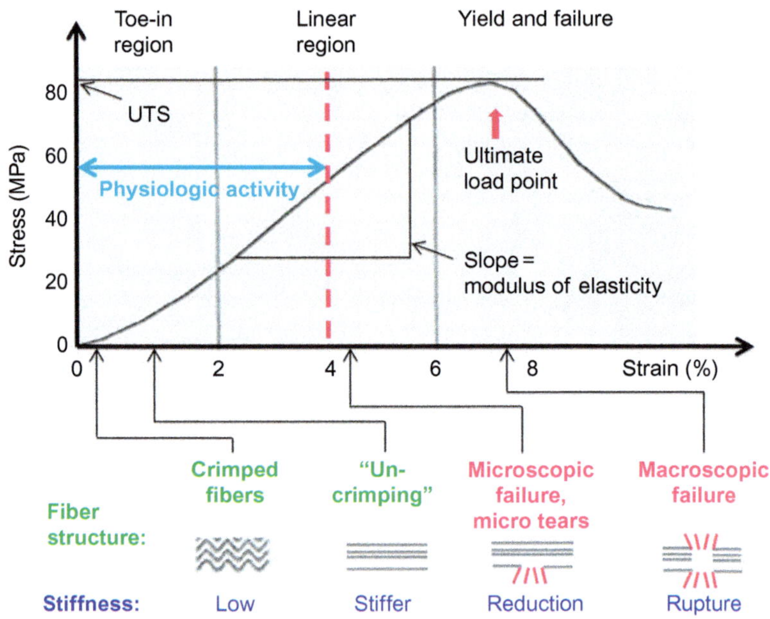


**Fig. 4.4** Typical stress–strain curve of a tendon or ligament showing three regions—the toe-in region where the fibers get uncrimped and strain is increased to around 2%, the linear region where some micro-damages to the fibers may occur and where usually the slope is measured to assess the elastic modulus of the specimen (between around 2–6% of strain), and the failure region where the maximum stress is measured and where the stress is decreased abruptly upon further straining of the specimen. This macroscopic failure is accompanied by the tendon being completely ruptured (~>6%). *UTS*, ultimate tensile strength.
Based on Lewis, G., Shaw, K.M., 1997. Tensile properties of human tendo achillis: effect of donor age and strain rate. J. Foot Ankle Surg. 36, 435–445; Sharma, P., Maffulli, N., 2006. Biology of tendon injury: healing, modeling and remodeling. J. Musculoskelet. Neuronal Interact. 6, 181–190.

in the elastic modulus by 0.7%, while it increased the failure stress by 21% (Wren et al., 2001). Loading rates were also varied in an in vivo study by Gerus et al., where healthy young males ($n=8$) were seated on the bench of a custom ergometer with the knee fully extended and the sole of the foot perpendicular to the shank. Then, the subjects had to perform isometric contraction for 3 s, followed by ramp up contractions either in 1.5 s or in minimal time (as fast as possible), having on-screen as feedback. These motions were US-imaged with a 50 mm US probe at 10 Hz (Gerus et al., 2011). The resulting force–strain curves were significantly different for force–strain values above 20% of maximum forces; the contractions performed at the highest rate possible resulted in steeper force–strain curves, implying higher values for stiffness at higher rates. As also reported by Pearson et al., highly dynamic motion patterns as found for running or hopping led to (apparently) stiffer tendons (Pearson et al., 2007). Therefore, the distraction rates, for in vivo as well as ex vivo measurements, is of high importance, especially in terms of data comparability.

### 4.2.3 Preconditioning

Preconditioning of the tendon or ligament specimen may include a preloading to a certain force (N) for a defined time or a prestretching to a certain strain (% of full length), it may also be performed by cyclic loading to specified forces or strains at a certain frequency (Hz) for a defined time. No doubt, preconditioning has an impact on the biomechanical measurements that follow. For example, preconditioning of rat ATs to a strain of 2% for different times was investigated and had an impact on the load until failure measured just right after the preconditioning (Fig. 4.5). Obviously, the load until failure was significantly higher than the control (no preconditioning) for 30, 100, 300, and 600 s (Teramoto and Luo, 2008). According to the authors, this preconditioning regimen induced progressive collagen fiber recruitment and subsequent fatigues at the micro-level; the latter being found in a 1000 s-preconditioning, where the failure load was smaller than at 600 s. Moreover, as in the same study stiffness was not affected by the 2%-preconditioning, it was concluded that failure load was more sensitive towards preconditioning than stiffness (Teramoto and Luo, 2008). Other static preconditioning examples are to stretch sheep rotator cuffs to 40 N for 2 min (Santoni et al., 2010) or human rotator cuffs to 25 N for only 10 s to remove possible creep (van der Meijden et al., 2013).

Preconditioning can include a cyclic loading regimen between a lower and an upper border (force) value, such as shown in a human cadaver study where the isolated medial patellofemoral ligament was preconditioned in an uniaxial tensile testing machine with 10 cycles between 5 and 20 N at a rate of 200 mm/min. Afterwards, load until failure and stiffness were assessed with a rate of 200 mm/min, too (Herbort et al., 2014). Also Baker et al. used a cyclic preconditioning when testing rotator cuffs augmented and reinforced by fascia patches, however, not just 10 cycles, but 100 cycles.

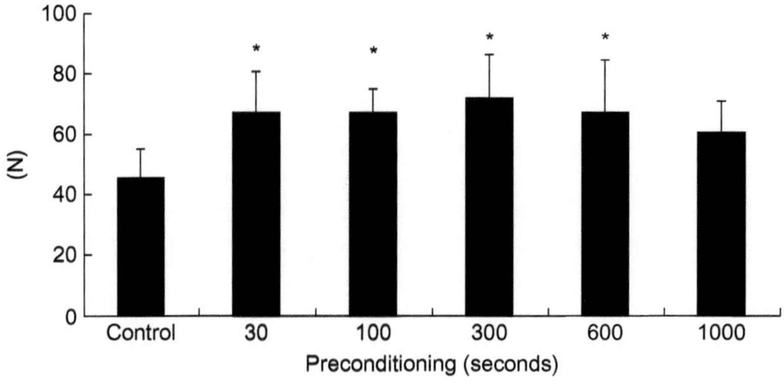


**Fig. 4.5** Investigation of the preconditioning parameters in rat Achilles tendon: ultimate failure load in control and preconditioning groups. Values are means ± SD. *Significant difference was found comparable to the control group.
From Teramoto, A., Luo, Z.-P., 2008. Temporary tendon strengthening by preconditioning. Clin. Biomech. 23, 619–622. Reprinted from Clinical Biomechanics, Copyright 2008, with permission from Elsevier.

The range of those "prefailure" loading cycles was 5 and 100 N with 30 mm/min (Baker et al., 2012). Others also include cyclic preconditioning, however, without mentioning the "border" values; may they be expressed by force (most often), elongation, or percentage of length (Perry et al., 2007).

### 4.2.4 Other testing conditions

Not always is the mechanical behavior of tendon tissue being stretched is of interest, however, the compressive deformation under force can be important; for example, Bah et al. examined the mechanical changes of human ATs caused by insertional AT tendinopathy (Bah et al., 2016). Insertional AT tendinopathy is located at the insertion site to the *calcaneus* and is hypothesized to be caused by compression of the AT tendon-bone-insertion against the *calcaneus* (Cook and Purdam, 2012). In their study, small biopsies of the anterior deep AT were taken and subjected to compressive force applied perpendicular to the collagen fiber alignment—to mimic in vivo conditions where compression of the tissue is due to calcaneal impingement. After a preloading of −0.0050 N (the negative value standing for compression, while positive values usually stand for stretching), the specimen underwent 20 cycles of triangle wave compression to −0.10 N at a rate of −5% strain/s which was defined as a fast loading and followed by two cycles to −0.10 N at a rate of 0.50% strain/s (slow loading). Healthy specimens were compared to insertional AT tendinopathy patients' specimen. It was found that the pathological AT tissue stiffened earlier, in other words at lower compressive strains than healthy tissues, which is associated with functional loss.

Compressive loading was also used to assess tendon biomechanics in a study by Fang et al. (2014), because tendons are not only subjected to tensile loading (=stretching), but also to compression, shear, and/or torsion. Therefore, they studied compressive and shear loaded bovine DDFT in terms of biomechanics. For the compression protocol, they used −0.01 N as preload or contact force, followed by three incremental steps of −0.08 strain ($\varepsilon = d/t$, where $\varepsilon$ = compressive strain, $d$ = displacement, and $t$ = thickness of the specimen) at a rate of 20 mm/s, with 12 min of relaxation time between each step. This incremental stress-relaxation protocol was used to evaluate the viscoelastic response as a function of strain. Furthermore, the shear protocol included also a preload of −0.01 N, however, in horizontal direction, and also followed by three incremental steps of −0.08 strain at a rate of 20 mm/s, with 12 min of relaxation time between each step. The shear strain $\gamma$ was defined as $\gamma = \arctan(d/t)$. It was reported that distal and proximal parts of the DDFT showed different biomechanical response, in compression as well as shear loading experiments; distal samples experienced 2–3 times higher forces than proximal samples under the same strains (Fang et al., 2014).

Tendon tissue is widely considered to be a transversely isotropic material because the fibers and the fascicles are unidirectionally arranged along the force transmitting axis. However, direction-dependent experiments are needed to reveal the anisotropy of the tendon tissue. Therefore, Boel et al. not only did their compression tests in mode I along the uniaxially parallel aligned fibers and fascicles of cubic specimen punched from porcine *flexor digitorum longus* (Figs. 4.6 and 4.7), but also perpendicular to the

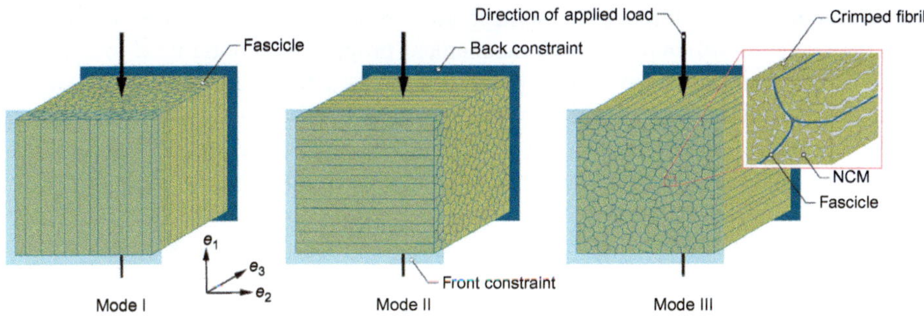


**Fig. 4.6** Idealized illustration of tissue testing with constraints in the ε3-direction: in mode I the fascicles are oriented in loading direction, in modes II and III fascicles are perpendicular to the loading direction, while in mode II fascicle extension is inhibited by the front and back plates, in mode III, the fascicles are kept at constant length by the two confining plates (back and front constraints).
Based on Boel, M., Ehret, A.E., Leichsenring, K., Ernst, M., 2015. Tissue-scale anisotropy and compressibility of tendon in semi-confined compression tests. J. Biomech. 48, 1092–1098. Reprinted from Journal of Biomechanics, Copyright 2015, with permission from Elsevier.

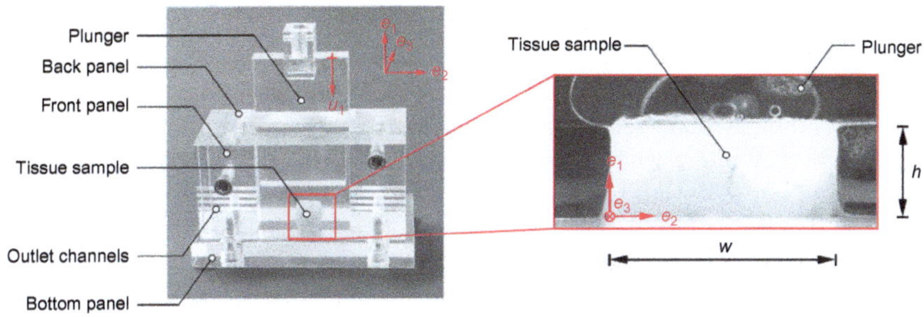


**Fig. 4.7** Custom-built pure shear compression device including an undeformed sample. Close-up view shows the deformed specimen recorded with a CCD-camera. In this example, the fibrils were oriented in the ε1-direction. For direction, see Fig. 4.6.
From Boel, M., Ehret, A.E., Leichsenring, K., Ernst, M., 2015. Tissue-scale anisotropy and compressibility of tendon in semi-confined compression tests. J. Biomech. 48, 1092–1098. Reprinted from Journal of Biomechanics, Copyright 2015, with permission from Elsevier.

fibers either with a semiconfining plate perpendicular to the fibers or parallel to the fibers (Boel et al., 2015). As such, they were able to show a completely different biomechanical behavior, when they followed a constant strain rate of −0.5% strain/s until a maximum compression of 30% of the cube. Load–displacement curves were assessed. Moreover, also stress–strain curves were obtained after dividing loads by the average CSAs of the cubes. While in mode III compression fascicles kept at constant length by two confining plates showed the softest response, mode II compression of the tendon tissue evoked the stiffest response—and mode I (along the force transmitting axis where usually the tensile stretching tests are performed) was in between the two modes. The differences in biomechanics between the three modes of compression tests were significant. Such biomechanical properties may be used to elucidate the load transfer mechanism that is based not only on fibers and fascicles, but also on proteoglycans where the core protein binds to fibrils, and glycosaminoglycans to other GAGs or to fibrils—ending up in a three dimensional network. From their results, the authors concluded that in mode I the thick "columns" composed of the tube-like sheaths (made of fascicles, fibrils, and subfibrils covered by the epitenon) are less prone for buckling than just the thin fibrils exposed to compressive force in mode III. Furthermore, they concluded that GAGs contribute to compressive resistance because the lateral stiffness (mode II) is significantly higher than the extensional stiffness (as assessed in mode III), where only in mode III (and not in mode II) the interconnecting GAG side chains are stretched, which is caused by the compressive force that induces a change from the primarily circular cylinder sample into a deformed elliptic one. Interestingly, large proteoglycans (such as aggrecan) are usually found in higher concentrations in tendon regions experiencing compression than in regions subjected to tensile stretching (regions with more small proteoglycans with less side chains as decorin, only 1 GAG) (Smith et al., 2002).

### 4.2.5 Fatigue tests

Besides static tensile loading tests, dynamic tests are also in use; however, there is less literature about structure–function relationships of tendon and ligament tissue under dynamic conditions than under static conditions. Fatigue loading tests under high loads are not only interesting in terms of elucidating fatigue-induced tendon or ligament injuries, but also with regard to the usually monotonic increase in peak strains found with increasing cycles and the structural changes causing this observed behavior. Combinations of fatigue tests with imaging methods such as polarized light imaging may help to reveal the structural changes that lead to increasing peak strains during fatigue loading, for example as discussed for the changes in the crimp pattern (Freedman et al., 2015). Murine PT were used and the changes in crimp frequency and amplitude were assessed as a function of time (cycle number) and of locality (mid-substance versus insertion site; center versus lateral) during fatigue loading and polarized light imaging. The experimental conditions and the loading protocol are summarized in Fig. 4.8. As a result, cycle number was a significant factor for peak strain, tangent stiffness, hysteresis, and laxity at all different localities tested. While fatigue loading, peak strain, tangent stiffness, and laxity increased, the hysteresis decreased. As for the crimp pattern, crimp frequencies decreased and crimp amplitudes increased with increasing cycles at 0.1 N (representing the toe region in a typical load–displacement curve). As such, nondestructive real-time monitoring during fatigue loading at low cost elucidated biomechanical changes in correlation to structural changes—which may be used as a tool in diagnostics (Freedman et al., 2015).

It is not just tendon and ligament tissues that may be subjected to fatigue tests. In order to test tendon replacement (bio)-materials for their suitability to act as substitutes at corresponding defect sites, dynamic biomechanical testing may be of interest.

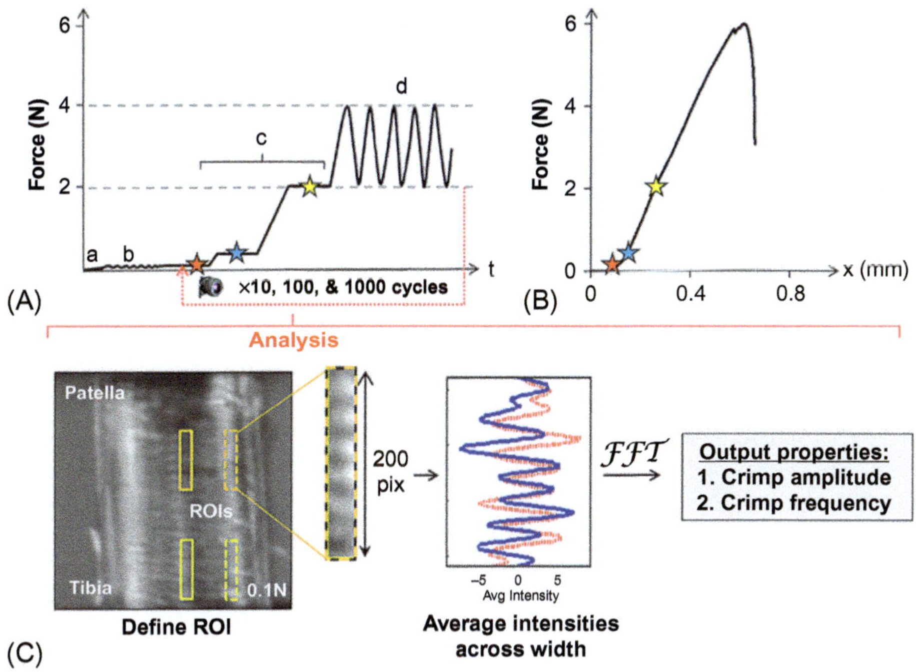


**Fig. 4.8** Mechanical testing and image capture protocol of mouse patellar tendon (A). Tendons were preloaded (a), preconditioned (b), imaged at three loads (0.1, 0.5, and 2.0 N) (c), and fatigue loaded (d). After 10, 100, and 1000 cycle intervals of fatigue loading, images were captured at these three loads to quantify tendon crimp properties in the toe, transition, and linear regions of a representative load–displacement curve (B). This process was repeated until tendons reached 1000 fatigue loading cycles (C). Four regions of interest (ROIs) were selected representing the mid-substance *(orange)*, insertion *(yellow)*, center *(solid)*, and lateral *(dashed)* regions of the tendon. ROIs were low pass filtered to enhance the visibility of light and dark bands, and intensities were averaged across the ROI width *(red dashed line)* before being high-pass filtered *(blue line)*. From these spectra, the crimp amplitude and frequency were computed. From Freedman, B.R., Zuskov, A., Sarver, J.J., Buckley, M.R., Soslowsky, L.J., 2015. Evaluating changes in tendon crimp with fatigue loading as an ex vivo structural assessment of tendon damage. J. Orthop. Res. 33, 904–910. © by Journal of Orthopaedic Research, with permission from Wiley.

For example, Altman et al. examined silk as an optional material for tissue engineering of anterior cruciate ligament (ACL). They assessed the fatigue performance of single silk fibers at different temperatures (37°C and 90°C-treatment in order to remove the glue-like sericin) as well as hierarchically fabricated 6-corded ACL matrix (Altman et al., 2002). Single pull-to-failure testing was performed at a strain rate of 100%/s. Cycles to failure at ultimate tensile strength, 1680 and 1200 N, were determined using a H-sine wave function at 1 Hz.

## 4.3 Other biomechanical tests

Besides the classical tensile testing, there are other biomechanical tests, especially when the tendon or ligament tissue is not evaluated macroscopically, meaning as a full organ, but on smaller entities of the hierarchically structured tissue—such as collagen molecules within tendon tissue, fibrils, fibers, fascicles, or even single tendon cells. Many imaging methods are applied for that purpose (see Chapter 5). Here, we present the rationales behind such tests and elucidate some exemplary read-outs.

Not only the variations in the elastic modulus of fibrils extracted from rat tail tendon (RTT) (Baldwin et al., 2014), but also stress–strain curves of RTT fibrils as well as human PT (HPT) fibrils (Svensson et al., 2013) have been assessed by AFM. In the latter study, tapping mode AFM imaging was performed in air to determine the dry CSA before mechanical testing. Afterwards, solitary fibrils that were rehydrated with PBS (phosphate-buffered saline) were tested mechanically. Only short fibril pieces (<40 μm) were used because the system was limited to 20 μm of deformation. Furthermore, a stiff AFM cantilever (measure needle with a very small spring having a size of 100–300 μm and a tip of a few nanometers) was taken. The corresponding stress–strain curve is shown in Fig. 4.9. In the first region of the curve, the collagen coils resist force to a certain extent, where unfolding starts (beginning of the plateau in region II). Stretching the fibrils even more leads finally to backbone stretching ending up in a second increase in force and in modulus in region III.

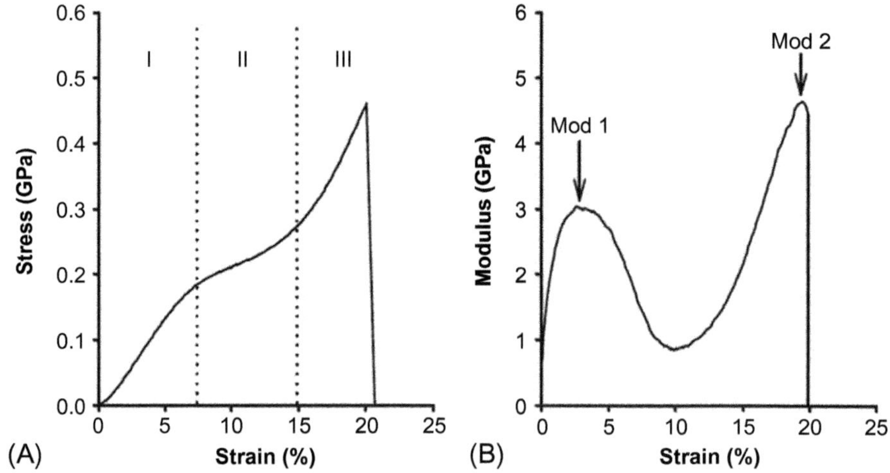


**Fig. 4.9** (A) Stress–strain curve of a representative human patellar tendon collagen fibril. Three distinct regions were observed (I, II, III). (B) Local modulus as a function of strain of the same curve. The three regions rise to two distinct peak modulus values (Mod 1 and Mod 2).
From Svensson, R.B., Mulder, H., Kovanen, V., Magnusson, S.P., 2013. Fracture mechanics of collagen fibrils: influence of natural cross-links. Biophys. J. 104, 2476–2484. Reprinted from Biophysical Journal, Copyright 2013, with permission from Elsevier.

Other AFM-fibril studies focused on adhesion forces in injured compared to healthy fibrils in rats (Lee et al., 2011); the changes in the ultrastructure and adhesion force of the collagen fibrils on the rat ATs having collagenase-induced tendinitis were observed, based on the diameter and D-banding of the collagen fibrils.

The viscoelastic properties of tendon fibers and their crimp pattern were assessed by a combination of AFM and SHG (Kahn et al., 2013). There, rabbit AT fibers, fibrils, and fascicles were investigated in a multiscale study, and it was found that the elastic modulus of the fibrils were 3.0 GPa. Furthermore the crimping pattern was identified by SHG revealing as a plane sine wave with $28.4 \pm 75.8$ μm of amplitude and $141 \pm 41$ μm of wavelength. Other methods to assess the mechanics of crimped collagen fibers as well as the angle of loaded fibers include confocal laser microscopy. RTT fibers were put in a tensile test rig (Fig. 4.10) that is usable for loading and unloading regimen of viable tendon explants (Screen et al., 2004): during stretching of the samples, the fluorescently stained cell nuclei were tracked with a confocal laser microscope, whereby the cells acted as discrete markers (Fig. 4.11). Principally, during stretching the cells move apart from each other and strain can therefore be determined by their relative position. More specifically, within-group strains were taken as the strain between the two extreme nuclei positioned along the same fiber at each strain increment (Screen et al., 2004).

There are also biomechanics-related assessments of tendon and ligament tissue by TEM. For example, strength gaining of rat AT fibers (Ehrlich et al., 2005), collagen fibril strength (Sluss et al., 2001; Lo et al., 2004), and collagen morphology in murine AT fibers (Rigozzi et al., 2010) were investigated by using TEM. Other imaging methods are thoroughly discussed in Chapter 5.

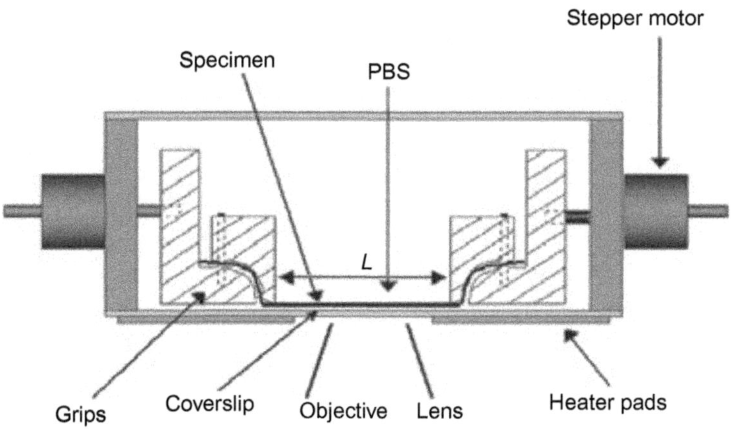


**Fig. 4.10** Schematic of tensile testing rig used on the stage of the inverted confocal laser microscope.
From Screen, H.R.C., Bader, D.L., Lee, D.A., Shelton, J.C., 2004. Local strain measurement within tendon. Strain, 40, 157–163. © by Strain, with permission from Wiley.

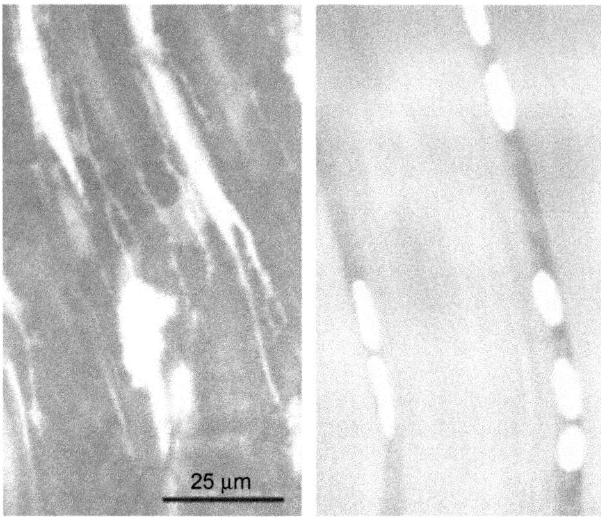


**Fig. 4.11** Confocal microscopy images of tenocytes within a tendon fascicle explant depicting *(left)* calcein AM staining of the whole cell, in which the complex cell morphology and processes are visible and *(right)* acridine orange staining of the cell nucleus, demonstrating the uniform morphology suitable for use as a strain marker.
From Screen, H.R.C., Bader, D.L., Lee, D.A., Shelton, J.C., 2004. Local strain measurement within tendon. Strain, 40, 157–163. © by Strain, with permission from Wiley.

## 4.4 In vivo biomechanical tests

The most often used method to assess in vivo biomechanics is ultrasound (US, see also Chapter 5 for further information about the imaging methods for tendons and ligaments). For the HPT, not only studies about in vivo elasticity (Hsiao et al., 2015), but also for in vivo creep behavior (Pearson et al., 2007) as well as for the influence of the loading rate on biomechanical outcome (Kösters et al., 2014) are reported. The in vivo elasticity of the HPT was studied as a function of age and with shear wave elastography (SWE) (Hsiao et al., 2015). Shear wave velocity and elastic modulus were measured at the proximal, middle, and distal areas of the PT of three groups of healthy participants aged 20–30, 40–50, and 60–70 years, respectively. As a result, the elastic modulus as well as the shear wave velocity were significantly lower in the group aged 60–70 years than the younger ones (Fig. 4.12).

Furthermore, the in vivo creep behavior of the PT in humans was assessed (Pearson et al., 2007). For this purpose, healthy men performed three short (3–4 s) and three long (10–12 s) maximal ramped isometric contractions on an isokinetic dynamometer, with real-time recordings of patellar excursions using B-mode ultrasonography (Fig. 4.13).

Deformation, strain, and stiffness were dependent on the contraction rate performed by the test person; independent of the force applied, the slow ramped contractions always led to smaller deformations, to higher strains, and to lower stiffness

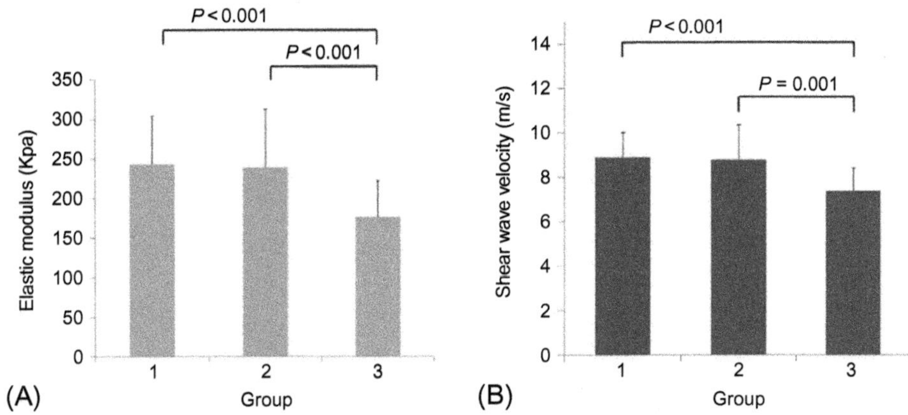


**Fig. 4.12** In vivo investigation of human patellar tendon with shear wave elastography: elastic modulus (A) and shear wave velocity (B) of the middle patellar tendons in three different age groups: group 1: 20–30 years old, group 2: 40–50 years old, and group 3: 60–70 years old. Error bars represent standard deviations.
From Hsiao, M.-Y., Chen, Y.-C., Lin, C.-Y., Chen, W.-S., Wang, T.-G., 2015. Reduced patellar tendon elasticity with aging: in vivo assessment by shear wave elastography. Ultrasound Med. Biol. 41, 2899–2905. Reprinted from Ultrasound in Medicine & Biology, Copyright 2015, with permission from Elsevier.

values when compared to the same force applied with a fast ramped contraction. As such, this study clearly showed that in vivo assessment of biomechanical properties is creep-dependent; in other words, the viscoelastic behavior of the tendon tissue leads to different read-outs depending on the loading rate under which they are assessed. Therefore, the loading rate is tremendously important in inter-literature comparisons, not only for ex vivo assessments (Section 2.2), but also for in vivo measurements. In another study also dealing with the influence of the loading rate for in vivo biomechanics, the HPT was analyzed in terms of deformation, strain, stiffness, stress, and the elastic modulus. The results clearly indicated substantial changes in stiffness and elastic modulus across various loading rates (50, 80, and 110 N m/s). Increasing loading velocities resulted in increasing stiffness and elastic modulus, respectively (Kösters et al., 2014). As for strain and stress, the loading rate did not appreciably affect them under these conditions; however, the authors noted that although there was on-line visual feedback for the participants engaged in the study, nevertheless interindividual differences in performing the protocol may have led to different forces reached within 2 s (the time of the ramp) such as 300 or 360 N, respectively; and such differences would lead to a difference of around 20% in the loading rate (=30 N m/s), which was the difference between the three groups of loading velocities.

Another clinically very relevant tendon is the AT which was also investigated in vivo with regard to its biomechanical properties (ultimate force, stiffness, elastic modulus, and maximum stress, Kongsgaard et al., 2011), in terms of echogenicity, elongation, and stiffness (Suydam and Buchanan, 2014) and also for resting length

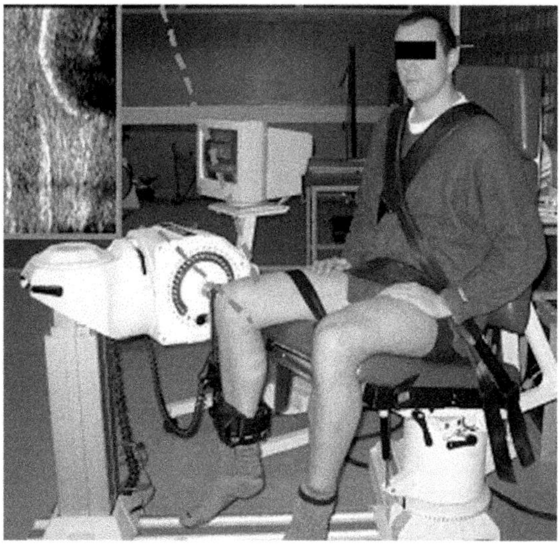

**Fig. 4.13** In vivo investigation of human patellar tendon with real-time B-mode ultrasound: isometric knee extension torque was determined by fixing the knee at 90 degree flexion (full extension = 0 degree) and hip at 85 degrees (supine = 0 degree). The knee joint center was aligned with the center of rotation of the dynamometer lever arm and straps were fixed across the chest, hip, and thigh of the test limb to prevent any extraneous movement. A padded cuff was strapped on the lower leg at ~3 cm above the medial malleolus. Inset—showing sagittal plane ultrasound image of patella and tendon.
From Pearson, S.J., Burgess, K., Onambele, G.N.L., 2007. Creep and the in vivo assessment of human patellar tendon mechanical properties. Clin. Biomech. 22, 712–717. Reprinted from Clinical Biomechanics, Copyright 2007, with permission from Elsevier.

defined as the length when the force–length curve is extrapolated to zero force (Morrison et al., 2015). In the latter study, gender effects on the resting length of the AT were elucidated by ultrasound; it was found that males had significantly longer resting length than females. Regarding stiffness, it was found that muscle strength rather than gender was the important parameter to induce differences.

## 4.5 Summary

In the first part of this chapter, some basic information about classic tensile testing machines and typical read-out parameters such as load until failure, stiffness, maximum stress, and elastic modulus and their corresponding stress–strain curves is given. Then, conditions during tensile testing are discussed and their effects on the biomechanical parameters are elucidated; for example, the loading rate does have an impact on the failure stress—an increase of a factor of 10 in strain rate may result in a 20% higher failure stress. Other conditions such as the preconditioning of the samples which may be cyclic or static preloading also do influence the final biomechanical characterization parameters. Also less typical biomechanical tendon and ligament tests are presented, as for example the compression tests performed at the AT insertion site to the bone, where usually compressive forces are dominant and not tensile stretching.

Besides static tests, dynamic biomechanical tests are also in use, which are presented in the second part of this chapter. Fatigue tests and high as well as low loading at different cycle numbers were reported to elucidate the typical viscoelastic behavior of tendons and ligaments. Moreover, some techniques to discover the biomechanics of smaller entities like the collagen fibril are presented. For example, the biomechanics of single fibrils might be determined by AFM, leading to a different stress–strain curve as that found for whole tendon or ligament specimen.

Finally, some clinically relevant noninvasive imaging techniques to determine the biomechanics in vivo are presented. In this last part of the chapter, especially ultrasound and its application are elucidated and the advantages as well as the limitations are discussed.

## References

Altman, G.H., Horan, R.L., Lu, H.H., Moreau, J., Martin, I., Richmond, J.C., Kaplan, D.L., 2002. Silk matrix for tissue engineered anterior cruciate ligaments. Biomaterials 23, 4131–4141.

Bah, I., Kwak, S.T., Chimenti, R.L., Richards, M.S., Ketz, P., Samuel Flemister, J., Buckley, M.R., Ketz, P., 2016. Mechanical changes in the Achilles tendon due to insertional Achilles tendinopathy. J. Mech. Behav. Biomed. Mater. 53, 320–328.

Baker, A.R., Mccarron, J.A., Tan, C.D., Iannotti, J.P., Derwin, K.A., 2012. Does augmentation with a reinforced fascia patch improve rotator cuff repair outcomes? Clin. Orthop. Relat. Res. 470, 2513–2521.

Baldwin, S.J., Quigley, A.S., Clegg, C., Kreplak, L., 2014. Nanomechanical mapping of hydrated rat tail tendon collagen I fibrils. Biophys. J. 107, 1794–1801.

Boel, M., Ehret, A.E., Leichsenring, K., Ernst, M., 2015. Tissue-scale anisotropy and compressibility of tendon in semi-confined compression tests. J. Biomech. 48, 1092–1098.

Cook, J.L., Purdam, C., 2012. Is compressive load a factor in the development of tendinopathy? Br. J. Sports Med. 46, 163–168.

Ehrlich, H.P., Lambert, P.A., Saggers, G.C., Myers, R.L., Hauck, R.M., 2005. Dynamic changes appearing in collagen fibers during intrinsic tendon repair. Ann. Plast. Surg. 54, 201–206.

Fang, F., Sawhney, A.S., Lake, S.P., 2014. Different regions of bovine deep digital flexor tendon exhibit distinct elastic, but not viscous, mechanical properties under both compression and shear loading. J. Biomech. 47, 2869–2877.

Fox, P.M., Farnebo, S., Lindsey, D., Chang, J., Schmitt, T., Chang, J., 2013. Decellularized human tendon-bone grafts for composite flexor tendon reconstruction: a cadaveric model of initial mechanical properties. J. Hand Surg. 38, 2323–2328.

Freedman, B.R., Zuskov, A., Sarver, J.J., Buckley, M.R., Soslowsky, L.J., 2015. Evaluating changes in tendon crimp with fatigue loading as an ex vivo structural assessment of tendon damage. J. Orthop. Res. 33, 904–910.

Gatt, R., Wood, M.V., Gatt, A., Zarb, F., Formosa, F., Azzopardi, K.M., Casha, K.M., Agius, A., Schembri-Wismayer, P., Attard, L., Chockalingam, N., Grima, J.N., 2015. Negative

Poisson's ratios in tendons: an unexpected mechanical response. Acta Biomater. 24, 201–208.

Gerus, P., Rao, G., Berton, E., 2011. A method to characterize in vivo tendon force–strain relationship by combining ultrasonography, motion capture and loading rates. J. Biomech. 44, 2333–2336.

Herbort, M., Haber, A., Zantop, T., Gosheger, G., Rosslenbroich, S., Raschke, M.J., Petersen, W., 2008. Biomechanical comparison of the primary stability of suturing Achilles tendon rupture: a cadaver study of Bunnell and Kessler techniques under cyclic loading conditions. Arch. Orthop. Trauma Surg. 128, 1273–1277.

Herbort, M., Hoser, C., Domnick, C., Raschke, M.J., Lenschow, S., Weimann, A., Kösters, C., Fink, C., 2014. MPFL reconstruction using a quadriceps tendon graft: Part 1: Biomechanical properties of quadriceps tendon MPFL reconstruction in comparison to the Intact MPFL. A human cadaveric study. Knee 21, 1169–1174.

Hsiao, M.-Y., Chen, Y.-C., Lin, C.-Y., Chen, W.-S., Wang, T.-G., 2015. Reduced patellar tendon elasticity with aging: in vivo assessment by shear wave elastography. Ultrasound Med Biol. 41, 2899–2905.

Kahn, C.J.F., Dumas, D., Arab-Tehrany, E., Marie, V., Tran, N., Wang, X., Cleymand, F., 2013. Structural and mechanical multi-scale characterization of white New-Zealand rabbit Achilles tendon. J. Mech. Behav. Biomed. Mater. 26, 81–89.

Kongsgaard, M., Aagaard, P., Kjaer, M., Magnusson, S.P., 2005. Structural Achilles tendon properties in athletes subjected to different exercise modes and in Achilles tendon rupture patients. J. Appl. Physiol. 99, 1965–1971.

Kongsgaard, M., Nielsen, C.H., Hegnsvad, S., Aagaard, P., Magnusson, S.P., 2011. Mechanical properties of the human Achilles tendon, in vivo. Clin. Biomech. 26, 772–777.

Kösters, A., Wiesinger, H.P., Bojsen-Møller, J., Müller, E., Seynnes, O.R., 2014. Influence of loading rate on patellar tendon mechanical properties in vivo. Clin. Biomech. 29, 323–329.

Lavagnino, M., Arnoczky, S.P., Gardner, K., 2011. In situ deflection of tendon cell-cilia in response to tensile loading: an in vitro study. J. Orthop. Res. 29, 925–930.

Lavagnino, M., Gardner, K., Sedlak, A.M., Arnoczky, S.P., 2013. Tendon cell ciliary length as a biomarker of in situ cytoskeletal tensional homeostasis. Muscles Ligaments Tendons J. 3, 118–121.

Lee, G.-J., Choi, S., Chon, J., Yoo, S., Cho, I., Park, H.-K., 2011. Changes in collagen fibril pattern and adhesion force with collagenase-induced injury in rat achilles tendon observed via AFM. J. Nanosci. Nanotechnol. 11, 773–777.

Lo, I.K., Burkhart, S.S., Chan, K.C., Athanasiou, K., 2004. Arthroscopic knots: determining the optimal balance of loop security and knot security. J. Arthrosc. Relat. Surg. 20, 489–502.

Maeda, E., Ye, S., Wang, W., Bader, D.L., Knight, M.M., Lee, D.A., 2012. Gap junction permeability between tenocytes within tendon fascicles is suppressed by tensile loading. Biomech. Model. Mechanobiol. 11, 439–447.

Magnusson, S.P., Kjaer, M., 2003. Region-specific differences in Achilles tendon cross-sectional area in runners and non-runners. Eur. J. Appl. Physiol. 90, 549–553.

Morrison, S.M., Dick, T.J., Wakeling, J.M., 2015. Structural and mechanical properties of the human Achilles tendon: sex and strength effects. J Biomech.

Pearson, S.J., Burgess, K., Onambele, G.N.L., 2007. Creep and the in vivo assessment of human patellar tendon mechanical properties. Clin. Biomech. 22, 712–717.

Perry, S.M., Gupta, R.R., Van Kleunen, J., Ramsey, M.L., Soslowsky, L.J., Glaser, D.L., 2007. Use of small intestine submucosa in a rat model of acute and chronic rotator cuff tear. J. Shoulder Elb. Surg. 16 (5 Suppl), S179–S183.

Rigozzi, S., Muller, R., Snedeker, J.G., 2009. Local strain measurement reveals a varied regional dependence of tensile tendon mechanics on glycosaminoglycan content. J. Biomech. 42, 1547–1552.

Rigozzi, S., Muller, R., Snedeker, J.G., 2010. Collagen fibril morphology and mechanical properties of the Achilles tendon in two inbred mouse strains. J. Anat. 216, 724–731.

Santoni, B.G., Mcgilvray, K.C., Lyons, A.S., Bansal, M., Turner, A.S., Macgillivray, J.D., Coleman, S.H., Puttlitz, C.M., 2010. Biomechanical analysis of an ovine rotator cuff repair via porous patch augmentation in a chronic rupture model. Am. J. Sports Med. 38, 679–686.

Screen, H.R.C., Bader, D.L., Lee, D.A., Shelton, J.C., 2004. Local strain measurement within tendon. Strain 40, 157–163.

Sluss, J.R., Liberti, J.P., Jiranek, W.A., Wayne, J.S., Zuelzer, W.A., 2001. pN collagen type III within tendon grafts used for anterior cruciate ligament reconstruction. J. Orthop. Res. 19, 852–857.

Smith, R.K.W., Birch, H.L., Goodman, S., Heinegård, D., Goodship, A.E., 2002. The influence of ageing and exercise on tendon growth and degeneration—hypotheses for the initiation and prevention of strain-induced tendinopathies. Comp. Biochem. Physiol. A Mol. Integr. Physiol. 133, 1039–1050.

Stange, R., Sahin, H., Wieskotter, B., Persigehl, T., Ring, J., Bremer, C., Raschke, M.J., Vieth, V., 2015. In vivo monitoring of angiogenesis during tendon repair: a novel MRI-based technique in a rat patellar tendon model. Knee Surg. Sports Traumatol. Arthrosc 23, 2433–2439.

Suydam, S.M., Buchanan, T.S., 2014. Is echogenicity a viable metric for evaluating tendon properties in vivo? J. Biomech. 47, 1806–1809.

Svensson, R.B., Mulder, H., Kovanen, V., Magnusson, S.P., 2013. Fracture mechanics of collagen fibrils: influence of natural cross-links. Biophys. J. 104, 2476–2484.

Teramoto, A., Luo, Z.-P., 2008. Temporary tendon strengthening by preconditioning. Clin. Biomech. 23, 619–622.

Trudel, G., Koike, Y., Ramachandran, N., Doherty, G., Dinh, L., Lecompte, M., Uhthoff, H.K., 2007. Mechanical alterations of rabbit Achilles' tendon after immobilization correlate with bone mineral density but not with magnetic resonance or ultrasound imaging. Arch. Phys. Med. Rehabil. 88, 1720–1726.

Van Der Meijden, O.A., Wijdicks, C.A., Gaskill, T.R., Jansson, K.S., Millett, P.J., 2013. Biomechanical analysis of two-tendon posterosuperior rotator cuff tear repairs: extended linked repairs and augmented repairs. Arthroscopy 29, 37–45.

Vergari, C., Pourcelot, P., Holden, L., Plumioan, R., Laugier, P., Mitton, D., Crevier Denoix, N., 2010. A linear laser scanner to measure cross-sectional shape and area of biological specimens during mechanical testing. J. Biomech. Eng. 132, 105001.

Wiig, M., Olmarker, K., Håkansson, J., Ekström, L., Nilsson, E., Mahlapuu, M., 2011. A lactoferrin-derived peptide (PXL01) for the reduction of adhesion formation in flexor tendon surgery: an experimental study in rabbits. J. Hand Surg. Eur. Vol. 36E, 656–662.

Woo, S.L., Hollis, J.M., Adams, D.L., Lyon, R.M., Takai, S., 1991. Tensile properties of the human femur-anterior cruciate ligament-tibia complex. The effects of specimen age and orientation. Am. J. Sports Med. 19, 217–225.

Wren, T.A.L., Yerby, S.A., Beaupré, G.S., Carter, D.R., 2001. Mechanical properties of the human achilles tendon. Clin. Biomech. 16, 245–251.

Zantop, T., Weimann, A., Schmidtko, R., Herbort, M., Raschke, M.J., Petersen, W., 2006. Graft laceration and pullout strength of soft-tissue anterior cruciate ligament reconstruction: in vitro study comparing titanium, poly-D,L-lactide, and poly-D,L-lactide-tricalcium phosphate screws. J. Arthrosc. Relat. Surg. 22, 1204–1210.

# Imaging of tendons and ligaments in animal models

**5**

## Abstract

There are many viable options to image tendons and ligaments. Depending on the structure under view, ultrasound, magnetic resonance imaging (MRI), light microscopy, fluorescence microscopy (FM), scanning electron microscopy (SEM), transmission electron microscopy (TEM), X-ray diffraction (X-ray-D), atomic force microscopy (AFM), or second-harmonic generation (SHG) is used. For macroscopic investigations including anatomical questions, ultrasound and MRI are preferred imaging methods, while for smaller entities such as fiber bundles, fascicles, or tenocytes, light microscopy, FM, and SEM are in use. Finally, on the collagen molecule level, X-ray-D, TEM, AFM, or SHG are the methods of choice. In this chapter, each methodology is introduced with respect to the physical background, and its application demonstrated in a large variety of reports dealing with tendon and ligaments. A special focus is set on biomechanical measurements based on imaging methods.

**Keywords:** Ultrasound, Magnetic resonance, Light microscopy, Electron microscopy, X-ray diffraction.

## Abbreviations

| | |
|---|---|
| **~Coll** | acellular collagen zone (tendon and ligament structure during healing) |
| **2-D** | two dimensions (technical feature of imaging techniques) |
| **2-D-PLM** | 2-D polarized light microscopy (technical feature of LM) |
| **3-D** | three dimensions (technical feature of imaging techniques) |
| **ACL** | anterior cruciate ligament (tendon and ligaments) |
| **ADC** | apparent diffusion coefficient (technical feature of MRI) |
| **AF** | anisotropy factor (factor in optical technologies) |
| **AFM** | atomic force microscopy (imaging technique) |
| **AGEs** | advanced glycation end (products) (substances in degenerative diseases) |
| **ALD** | *anterior latissimus dorsi* (muscle) |
| **ASC** | adipose-derived stem cell (cells for tissue engineering and cell therapy) |
| **AT** | Achilles tendon (tendon and ligaments) |
| **BF** | bright field (technical feature of LM) |
| **CCD** | charge-coupled device camera (device in digital imaging) |
| **CFM** | confocal fluorescence microscope (technical feature of LM) |
| **CLM** | confocal light microscopy (technical feature of LM) |
| **CLsM** | confocal laser microscopy (technical feature of LM) |
| **CM** | confocal microscopy (technical feature of LM) |
| **CR** | cell-rich tissue (tendon and ligament structure during healing) |
| **CSA** | cross-sectional area (feature of tendon and ligaments) |

**Biomechanics of Tendons and Ligaments. http://dx.doi.org/10.1016/B978-0-08-100489-0.00005-3**

| | |
|---|---|
| **CT** | computed tomography (imaging technique) |
| **DDFT** | *deep digital flexor* tendon (tendon and ligaments) |
| **DP** | DegraPol® polymer tube (synthetic polymer for tissue engineering) |
| **ECM** | extracellular matrix (matrix in living cells) |
| **EP** | epitenon (tendon and ligament intact structure) |
| **FDL** | *flexor digitorum longus* (tendon and ligaments) |
| **FDP** | *flexor digitorum profundus* (tendon and ligaments) |
| **FDS** | *flexor digitorum superficialis* (tendon and ligaments) |
| **FITC** | fluorescein isothiocyanate (agent in light microscopy) |
| **FM** | fluorescence microscopy (technical feature of LM) |
| **GAG** | glycosaminoglycan (polysaccharide and element of living cells) |
| **Gd-DTPA/Dimeg** | gadopentate dimeglumine (contrast agent in MRI technique) |
| **Gr** | granulation tissue (tendon and ligament structure during healing) |
| **H&E** | hematoxylin and eosin (histological staining technique) |
| **HF** | high frequency (technical feature of US) |
| **HIF (−*n*)** | hypoxia-inducible factor (transcription factor in living cells) |
| **Hum** | human (species classification) |
| **ICLsM** | inverted confocal laser microscopy (technical feature of LM) |
| **ILM** | interference light microscopy (technical feature of LM) |
| **LF** | low frequency (technical feature of US) |
| **LGP (−*n*)** | lubricating glycoprotein (protein of living cells) |
| **LGT** | lateral gastrocnemius muscle-tendon (tendon and ligaments) |
| **LM** | light microscope, light microscopy (imaging technique) |
| **MCL** | medial collateral ligament (tendon and ligaments) |
| **MGO** | methylglyoxal (organic compound in degenerative diseases) |
| **MMP (−*n*)** | matrix metalloproteinase-3/stromelysin-1 (enzyme in living cells) |
| **MPh-CM** | multiphoton confocal microscope (technical feature of LM) |
| **MR** | magnetic resonance (imaging technique) |
| **MRE** | phase-sensitive magnetic resonance elastography (technical feature of MRI) |
| **MRI** | magnetic resonance imaging (imaging technique) |
| **mRNA** | messenger ribonucleic acid (element of living cells) |
| **MRT** | magnetic resonance tomography (imaging technique) |
| **NMRI** | nuclear magnetic resonance imaging (imaging technique) |
| **NT** | no treatment (no treated tendon or tissue and mainly contralateral part of the body) |
| **OCT** | optical coherence tomography (imaging technique) |
| **P(LLA/CL)** | poly(L-lactic acid-*co*-caprolactone) synthetic polymer (element in tissue engineering) |
| **PD** | proton density (technical feature of MRI) |
| **PDWI** | proton density weighted (technical feature of MRI) |
| **PLD** | *posterior latissimus dorsi* (muscle) |
| **PLM** | polarized light microscopy (technical feature of LM) |
| **PS** | picrosirius red (histological staining technique) |
| **PS-OCT** | polarization-sensitive optical coherence tomography (imaging technique) |
| **PT** | patellar tendon (tendon and ligaments) |
| $R^2$ | correlation coefficient (element of statistics) |
| **RC** | rotator cuff (tendon and ligaments) |
| **RS** | Raman spectroscopy (technical feature of LM) |

| | |
|---|---|
| **rt-CM** | real-time confocal microscopy (technical feature of LM) |
| **SAXS** | small-angle X-ray scattering (technical feature of X-ray technology) |
| **SDFT** | superficial digital flexor tendons (tendon and ligaments) |
| **SE/T2WI** | spin-echo $T_2$-weighted sequence (technical feature of MRI) |
| **SEM** | scanning electron microscopy (imaging technique) |
| **SHG** | second-harmonic generation (imaging technique) |
| **SHG-cc** | SHG-collar correlation (element of second-harmonic generation) |
| **SLC** | synovial lining cells (cell type) |
| **SSCT** | subsynovial connective tissue (tissue) |
| **SSP** | *supraspinatus* tendon (tendon and ligaments) |
| **STT** | semitendinosus tendon (tendon and ligaments) |
| **T** | tesla (unit for magnetic flux density) |
| **T** | temperature (unit for hot and cold) |
| **T&L** | tendon and ligaments (tendon and ligaments) |
| $T_1$ | spin-lattice relaxation process (technical feature of MRI) |
| $T_2$ | spin-spin relaxation process (technical feature of MRI) |
| **TE** | echo time (technical feature of MRI) |
| **TEM** | transmission electron microscopy (imaging technique) |
| **TR** | repetition time (technical feature of MRI) |
| **US** | ultrasound, ultrasonography (imaging technique) |
| **USPIO** | ultrasmall superparamagnetic iron oxide (technical feature of MRI) |
| **UTC** | ultrasound tissue characterization (technical feature of US) |
| **UTE** | ultrashort echo imaging (technical feature of MRI) |
| **UV** | ultraviolet (technical feature of LM) |
| **VEGF** | vascular endothelial growth factor (protein in living cells) |
| **XANES-PEEM** | X-ray absorption near-edge structure and photoemission electron microscopy (imaging technique) |
| **X-ray-D** | X-ray diffraction (imaging technique) |

## 5.1 Introduction

Imaging of tendons and ligaments (Fig. 5.1) has been and is still being performed for different purposes; not only anatomical and diagnostic aims are followed but also studies to get insight into the structures and correlations to their mechanical behavior are based on imaging techniques (Table 5.1)—having under view of experimental animals or cadaver tissue and ranging from macroscopic to microscopic levels.

## 5.2 Ultrasonography

### *5.2.1 Overview*

Currently, a largely used imaging technique is ultrasonography (US) (Table 5.2). US offers the possibility to noninvasively study tendons and ligaments in vivo, which is an attractive feature. It is a portable device and provides good images of tendons and ligaments where the size and the structure can be assessed qualitatively (Avella et al., 2009)

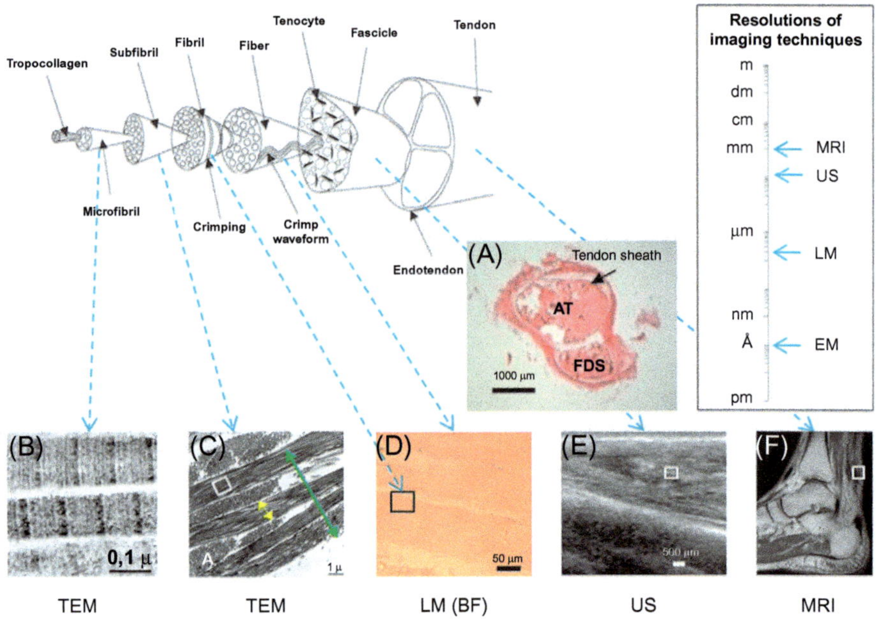


**Fig. 5.1** Tendon structure and morphology, drawing and correspondent microscopic photographs (with resolutions of microscopic techniques). Drawing from "anatomy of a normal tendon." (A) Intact rabbit Achilles tendon (AT) and flexor digitorum superficialis (FDS), image performed by light microscopy (LM; *BF*; bright field; histological staining, picrosirius red (PS); and 8 × magnification). (B) Collagen fibrils, AFM image (atomic force microscopy), and 67 nm periodicity. (C) Collagen fiber (*yellow arrow*), AFM image, and collagen subfascicle. (D) Intact rabbit Achilles tendon fascicle and subfascicle (frame), LM (Leica DM 6000 B; histological staining, *H&E*; hematoxylin and eosin; 100 × magnification; and image source in-house). (E) Rabbit Achilles tendon rupture (frame) and US image (ultrasonography). (F) Human calcaneus; Achilles tendon rupture (frame); and *MRI*, magnetic resonance imaging (figure source in-house). Resolutions of the imaging techniques, MRI, 1 mm; US, 150 µm; LM, 200 nm; *EM*, electron microscopy, 0.05 nm = 0.5 Å; and *X-ray-D*, X-ray diffraction, 0.01 nm.
Figure taken from Screen, H.R.C., Bader, D.L., Lee, D.A., Shelton, J.C., 2004. Local strain measurement within tendon. Strain 40, 157–163, © by Strain, with permission from Wiley. Part figure (A) taken from Meier Buergisser, G., Calcagni, M., Müller, A., Bonavoglia, E., Fessel, G., Snedeker, J.G., Giovanoli, P., Buschmann, J., 2014. Prevention of peritendinous adhesions using an electrospun DegraPol polymer tube: a histological, ultrasonographic, and biomechanical study in rabbits. Biomed. Res. Int. 2014, 656240, open access article. Part figure (B) taken from Venturoni, M., Gutsmann, T., Fantner, G.E., Kindt, J.H., Hansma, P.K., 2003. Investigations into the polymorphism of rat tail tendon fibrils using atomic force microscopy. Biochem. Biophys. Res. Commun. 303, 508–513, reprinted from Biochemical and Biophysical Research Communications, Copyright 2003, with permission from Elsevier. Part figure (C) taken from Venturoni, M., Gutsmann, T., Fantner, G.E., Kindt, J.H., Hansma, P.K., 2003. Investigations into the polymorphism of rat tail tendon fibrils using atomic force microscopy. Biochem. Biophys. Res. Commun. 303, 508–513, reprinted from Biochemical and Biophysical Research Communications, Copyright 2003, with permission from Elsevier. Part figure (E) taken from Buschmann, J., Puippe, G., Meier Buergisser, G., Bonavoglia, E., Giovanoli, P., Calcagni, M., 2014. Correspondence of high-frequency ultrasound and histomorphometry of healing rabbit Achilles tendon tissue. Connect. Tissue Res., 55, 123–131, reprinted by permission of the publisher Taylor & Francis Ltd., http://www.informaworld.com.

**Table 5.1 Overview on the imaging techniques in medical diagnostics und morphological investigations on tendons and ligaments**

| Tech | Resolution/ magnification, aimed structure in tendons | Technical advantages | Technical limitations |
|---|---|---|---|
| US | 150 μm tendon bundle | – In vivo application<br>– Noninvasive<br>– Portable advice | – 2-D imaging only<br>– Comparability<br>– Repeatability<br>– Low resolution<br>– Limited value comp. to MRI +X-ray |
| MRI/MRT/ NMR | 1 $mm^3$ tendon tissue | – In vivo application<br>– Noninvasive<br>– Accurate 3-D imaging<br>– More sensitive than X-ray<br>– Nonionizing radiations | – Manly nonportable advices<br>– Water sensitivity<br>– Low resolution |
| LM | 200 nm, 400 × collagen bundles (fibers) | – Portable advices<br>– Very sensitive | – Ex vivo application<br>– Invasive |
| TEM | 0.05 nm, 50,000,000 × | – Even more sensitive | – Ex vivo application<br>– Invasive |
| SEM | 0.1 nm, 500,000 × | | – Nonportable advice |
| X-ray-D, AFM, SHG OCT | 0.01 nm molecular level | – In vivo application<br>– Noninvasive | – Nonportable advice<br>– Ionizing radiations |

*US*, ultrasonography; *MRI*, magnetic resonance imaging; *MRT*, magnetic resonance tomography; *NMR*, nuclear magnetic resonance; *LM*, light microscopy; *TEM*, transmission electron microscopy; *SEM*, scanning electron microscopy; *X-ray-D*, X-ray diffraction; *AFM*, atomic force microscopy; *SHG*, second-harmonic generation; *OCT*, optical coherence tomography.

and quantitatively. The echogenicity can be quantified with first-order gray-level statistics (Schie et al., 2000). However, compared with other imaging techniques such as magnetic resonance imaging (MRI) or computed tomography (CT), it is limited to only two dimensions (2-D); comparisons of the echogenicity between a series of US images are often difficult because of minor changes in the amplifier gain, transducer tilt, and displacement, which all affect the repeatability (Schie et al., 1999). A further limitation of US is the resolution because US images are based on structural reflections and interfering echoes. Therefore, only relatively large tendon structures (bundles) generate informative reflections, whereas small structures (single fibrils or tenocytes) end up in interferences.

**Table 5.2 Overview on ultrasonographic studies presented in this chapter, grouped by thematic subchapters**

| Aimed T&L | Technical features | Medical findings | Lit. |
|---|---|---|---|
| Human AT | LF | US as diagnosis tool in rupture AT | Maner and Marsh (1981) |
| Horse limb | HF (5–7.5 MHz) | Focusing on managing the technique | Rantanen (1982) and Rantanen et al. (1983) |
| Rabbit/Hum AT | HF (8 MHz) | Development of higher frequencies | Dillehay et al. (1984) |
| Calve/ sheep/Hum AT | HF (7.5–15 MHz) | Examination of echotexture | Martinoli et al. (1993) |
| Horse SDFT | HF, 3-D/UTC | Evaluation of ultrastructural changes | Docking et al. (2012) and Schie et al. (2003) |
| Rat LGT | HF (40 MHz) | Evaluation of structural changes | Peixinho et al. (2014) |
| Rabbit AT | Power Doppler | Comparison with histomorphology | Buschmann et al. (2014) |
| Chicken FDP | HF, 3-D/UTC | Evaluation of adhesion extent | Hung et al. (2013) |
| Rat AT | HF (12 MHz) | Comparison with biomechanics | Chamberlain et al. (2013) |
| Horse T&L | Color Doppler | Investigation of (neo) vasculariz. | Kristoffersen et al. (2005) |
| Mouse AT | Col + pow Doppler | Invest. changes in microcirculation | Yeh et al. (2009) |
| Human AT | Color Doppler | Large variability in diagnostics | Snellenberg et al. (2007) and Yang et al. (2011) |
| Mouse AT | HF (25 MHz) | Approaches for accurate information | Wang et al. (2011) |
| Rabbit AT | Power Doppler | Comp. with histomorphology/healing | Meier Buergisser et al. (2014) |
| Rabbit + human AT | HF/MRI | Anatomical studies | Doherty et al. (2006) |
| Horse SDFT | HF (7.5–10 MHz) | Baseline, compare normal/injured T | Denoix and Busoni (1999) |
| Horse flexors | HF (7.5 MHz) | Focus on cross-sectional area | Smith et al. (1994b) |
| IV, Rabbit AT | HF (5–10 MHz) | Tendon elasticity | Kuo et al. (1999) |
| Rabbit AT | HF (6–15 MHz) | Sonoelastography/ASC cell therapy | Gehmert et al. (2012) |
| Horse DDFT | HF (10 MHz) | Appl. US contrast agent | Bertuglia et al. (2014) |

*T&L*, tendon and ligaments; *Hum*, human; *AT*, Achilles tendon; *SDFT*, superficial digital flexor tendons; *LGT*, lateral gastrocnemius muscle-tendon; *FDP*, flexor digitorum profundus; *DDFT*, deep digital flexor tendon. Technical features, *LF*, low frequency; *HF*, high frequency; *UTC*, ultrasound tissue characterization.

## 5.2.2 Development of US as a diagnostic tool

Until 1981, only one case report had dealt with US as a diagnostic tool to examine a ruptured Achilles tendon (AT) (Maner and Marsh, 1981). Then, some basic reports about tendon injuries using US diagnostics had been published, based on the five criteria: alterations in echogenicity, size, margination, shape, and position (Rantanen, 1982; Rantanen et al., 1983). In 1984, Dillehay et al. suggested US technique should be considered as a useful method for the evaluation of tendons and ligaments because of the development of higher megahertz transducers at that time, which enabled the visualization of these superficial structures (Dillehay et al., 1984). The authors examined a rabbit AT using a high-resolution, short-focus, 8 MHz real-time small-parts ultrasound transducer. They examined a hematoma within the tendon that was artificially created by injection of 5 cc of rabbit's blood into the tendon. Moreover, they also successfully examined a partial laceration induced by a scalpel. In addition, their US images of human ATs of a 24-year-old basketball player who suddenly felt a "pop" and very sharp pain during training clearly showed an interruption of the border of the AT (Dillehay et al., 1984). As such, diagnostic usefulness was shown not only in the animal model of the rabbit AT but also in human diagnostics.

A comprehensive study by Martinoli et al. (1993) examined the echotexture of tendons by US. A series of different frequencies were used including 7.5, 10, 13, and 15 MHz for in vitro tendon imaging in calves and sheep and in vivo tendon imaging in human patients with a clinical diagnosis of calcaneal tendon disease and patients undergoing surgery after rupture of the AT. For normal healthy tendons, fine parallel and linear fibrillar echoes were reported that became more numerous and thinner with increasing US frequency applied. The echoes were explained by specular reflectance at the interface between collagen bundles and endotenons. In contrast to healthy tendons, patients with tendon disease or rupture showed an increase in fibrillar thickness, interruptions, and fragmentations or complete disappearance of the echotexture in the US image. The comparison of different frequencies led the authors to conclude that high-resolution and high-frequency (HF) probes are hampered by poor tissue penetration of the US beam—because evaluation of deeply lying tissues such as *bursae* or fat pads became difficult or even impossible at high frequencies. These structures and tissues were considered to be very important in diagnosis as they are often affected by inflammation or hematoma. Thus, high-resolution HF US should not be considered as a general alternative to conventional 7.5 MHz (Martinoli et al., 1993). Nevertheless, the authors admitted that HF may be used as a prediction of subtle tendon degenerations caused by tiny changes in the fibrillar structure—which can be made visible only at high frequencies.

Twenty years later, studies dealing with the ultrastructure in the gray-scale B-mode US images of tendons and their histological equivalents as references were reported. Such information is very important with respect to diagnostic purposes as it allows noninvasive fast and easy diagnosis of diseases or abnormalities and insight into special conditions such as excessive training.

For example, an US study on superficial digital flexor tendons (SDFT) of race horses has been performed in order to evaluate the ultrastructural changes in the tendon tissue caused by mechanical stimuli (Docking et al., 2012). For this purpose, race horses were examined before and after maximum exercise (competitive race). Transverse images at even distances of 0.2 mm over a 12 cm scanning window were gathered to give a three dimensional (3-D) US data block after compounding these images. Calculation of respective pixels in the region of interest as a percentage of total pixels was performed and classified according to four echo patterns. This approach is called ultrasound tissue characterization (UTC) and was developed to overcome the limitation of 2-D US imaging; in the 3-D US data block UTC-specific algorithms quantify the dynamics of echo patterns over a contiguous series of images by the intensity and the relative gray level of the pixels. As a reference, histological sections of the tendon specimens are used (Schie et al., 2003).

In the racehorse study, UTC elucidated changes in the ultrastructural composition of the flexor tendons after the competitive race (Docking et al., 2012). A clear reduction in pixels of the aligned and integer collagen bundles was found one and two days post race. On the third day, however, the echo pattern turned back to baseline values. As such, short-term changes in the flexor tendons of horses were accurately visualized. Such information is valuable for the design of optimal exercise and training schedules for race horses—and even more importantly for an early detection of impending tendon injuries.

Another exercise-related study using US has been reported by Peixinho et al. in a rat lateral gastrocnemius muscle-tendon (LGT) complex study. They examined the effect of a stretching program on the muscle thickness and the tendon length. Although the training had been performed three times per week for 8 weeks, no changes in muscle thickness nor tendon length were found. Only the muscle fiber length increased significantly in the stretching group by 8.5% compared with the nontraining group (Peixinho et al., 2014).

### 5.2.3 Comparison of US with histology and investigation of adhesion

In an attempt to correlate exquisite zones in high-frequency US images (17.5 MHz) in terms of echogenicity and homogeneity to histomorphometric parameters such as collagen structure, cell density, and cell nuclei width, a study in our research group was performed (Buschmann et al., 2014). We were able to show in a rabbit AT model of completely dissected tendons, which were sutured by a four-strand Becker suture distinct zones having typical features of echogenicity and homogeneity at three and 6 weeks postsurgery, respectively. The exact one-to-one vis-à-vis posting of US gray-scale image and longitudinal section of paraffin-embedded tendon tissue (picrosirius red staining) revealed distinct tissue zones (Fig. 5.2).

First of all, intact tendon tissue was characterized by a hyperechogenic signal with tightly packed parallel echoes evoked by the fibrillar structure of the collagen bundles. This zone was histomorphometrically characterized as a zone with approximately 500 tenoblasts and tenocytes per mm$^2$, with low collagen fiber crimps and slender cell nuclei having widths of around 1 μm. At the repair site, a zone-denoted cell-rich tissue

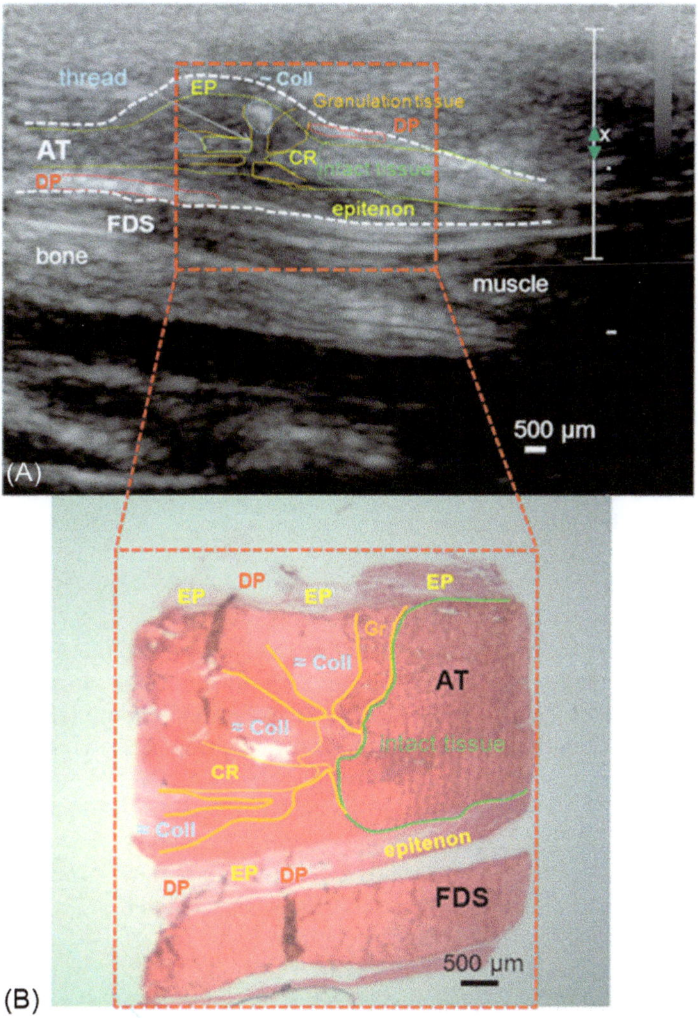


**Fig. 5.2** Ultrasound image (A) and corresponding picrosirius red-stained histological section at 8 × magnification using a light microscope (B) of rabbit Achilles tendon rupture repaired with a four-strand Becker suture and a polymer tube evaluated 3 weeks (A, B) post operation. *AT*, Achilles tendon (white dotted lines); *FDS*, flexor digitorum superficialis; *DP*, DegraPol polymer tube (*red dotted lines*); *EP*, epitenon (*yellow dotted lines*); *CR*, cell-rich tissue (*yellow solid lines*); *Gr*, granulation tissue (*orange dotted lines*); and ~*Coll*, acellular collagen zone (*light blue dotted lines*).

From Buschmann, J., Puippe, G., Meier Buergisser, G., Bonavoglia, E., Giovanoli, P., Calcagni, M., 2014. Correspondence of high-frequency ultrasound and histomorphometry of healing rabbit Achilles tendon tissue. Connect. Tissue Res. 55, 123–131, reprinted by permission of the publisher Taylor & Francis Ltd., http://www.informaworld.com.

(CR) was found, which was hypoechogenic and had tightly packed parallel echoes. The corresponding histological data consisted of a tendon cell density of approximately 1000 cell per $mm^2$, similar collagen crimp pattern as intact tissue, but morphologically less slender and more roundish cell nuclei with a significantly higher width of around 1.6 μm compared with the intact tissue. An even more hypoechogenic zone close to the repair site was found: granulation tissue (Gr) depicting unordered echoes, explained by the fact found in histology of highly unordered and crimped collagen structure. In another study, cell density was similar to the one found in the CR tissue; however, the widths of the cells were even wider, implying more roundish cells being immature tenoblasts (Moshiri and Oryan, 2011). An interesting aspect of this Gr was found as it looked like emerging from the tendon-surrounding epitenon (EP); the EP, 10 times swollen compared with intact tendons at 3 weeks postsurgery, showed very heterogeneous granulation zone "fingers" pointing toward the repair site suggesting that extrinsic healing took place by providing those cells migrating toward the repair site. Even an acellular zone could be recognized in the US image—as a hyperechogenic zone more than intact tissue and with tightly packed parallel echoes like in the intact tissue. As such, zones of ischemia, transection stitching (Wong et al., 2010), or stretch overload injuries leading to necrosis or apoptosis (Scott et al., 2005) can be fairly well visualized by US by our group (Buschmann et al., 2014).

The adhesion formation of the regenerating tendon tissue to the surroundings can also be accurately determined by US. Using a semiquantitative scoring system with 0 = no adhesion and good gliding, 1 = some adhesion and middle gliding, and 2 = maximum adhesion and no gliding, our research group performed dynamic ultrasound measurement before AT tectonomy and three and 6 weeks postsurgery in the rabbit model (Meier Buergisser et al., 2014). Short video films were thereby analyzed with respect to the gliding of the AT, while the *flexor digitorum superficialis* (FDS) of the rabbit hindpaw was actively moved. A good linear regression correlation to the findings determined in histological cross sections was obtained ($R^2 = 0.79$), although mere macroscopic evaluation by the surgeon during the tendon extraction correlated better to the histological findings ($R^2 = 0.86$).

Another study using US imaging to determine the extent of adhesion formation has been performed by Hung et al. The authors checked whether locally injected vitamin C (50 μL of either 5 or 50 mg/mL) reduced the adhesion formation of the *flexor digitorum profundus* (FDP) in chicken. At 2 and 6 weeks postsurgery, the fibrotic tissue between the FDS and the FDP was analyzed by UTC for a total length of 12 mm with a step size of 32 μm (Hung et al., 2013). The such produced 3-D data blocks of US images were analyzed with respect to fibrotic response, tendon adhesion, and peritendinous reactions at the injury site. In addition, gliding tests were performed using a testing jig with a mechanical testing machine. Maximum flexion angle and gliding resistance were determined and set in correlation to the findings of US imaging. The 5 mg/mL vitamin C group performed best; it showed a significant reduction of the fibrotic size at 6 weeks postsurgery, leading to the conclusion that an intraoperative application of a vitamin C solution during tendon repair reduces tendon adhesions.

### 5.2.4 Correspondence of US with biomechanics

Correlations not only between gray-scale images in US and histomorphometric information have been attempted but also between echo intensities in US images and biomechanical properties of healing tendons (Chamberlain et al., 2013). Such an approach is valuable because mechanical data are typically obtained by extracting tendons and subsequent in vitro biomechanical tensile testing. In contrast, a noninvasive method as US allows a direct in vivo longitudinal evaluation of the same specimen during the regeneration period. The theory behind the correspondence of US imaging and mechanical characteristics is found in acoustoelasticity, where the acoustic properties of a material are dependent of its stiffness and are changed when the material is deformed under load. Such changes in acoustic properties of a material result from the elastic deformation, which by itself can be measured as a change in the wave propagation velocity and amplitude of the reflected wave (Kobayashi and Vanderby, 2005, 2007).

Chamberlain et al. reported on a rat AT model, where the tendon was completely transected (no suture), the healing and gap filling were monitored as a function of time (7, 9, 14, and 29 days), and US, mechanical testing, and immunohistochemistry were performed at each time point (Chamberlain et al., 2013). As biomechanical readouts, ultimate stress, ultimate load, and normalized stiffness were assessed, echo intensities were correlated to ultimate stress and normalized stiffness and had strong positive linear correlations with $R^2$ values of 0.84 and 0.79, respectively. Echo intensities increased with increasing stress and stiffness, and the two mechanical properties also linearly correlated ($R^2=0.984$). These correlations allow a successful estimation of stress and stiffness in a tissue *via* US. As such, US may be able to track pathologies and healing over time.

In addition, Chamberlain et al. correlated biomechanics not only with US echo intensities but also with immunohistochemical data such as procollagen I and III in order to track healing markers. Interestingly, procollagen I correlated linearly with an $R^2$ of 0.99 to the normalized stiffness of the tendon. Intact tendons primarily consist of collagen type I, while after injury, collagen type III arises (Sharma and Maffulli, 2006). The authors showed that the decrease in procollagen type I after injury went along with the reduced normalized stiffness and strength of the tendon. As such, the biomechanical properties of the tendons correlated well with the biological aspects determined in the healing tendons. In sum, echo intensities, biomechanics, and histology had interrelated outcomes—and allowed for conclusions based on noninvasive US measurements alone.

### 5.2.5 Investigation of neovascularization after injury

Vascularization in tendon injuries has also been studied with US, the relevance lying in the fact that tendon pain is often coupled to (neo)vascularization, especially in chronic tendon injuries (Kristoffersen et al., 2005). In order to assess vascularity in flexor tendons and suspensory ligaments of horses with impaired function and chronic pain symptoms, Kristoffersen et al. used low-flow color Doppler without standoff and gray-scale US with a linear transducer of 10–13 MHz frequencies. Very clearly, abnormalities in the US images of injured tendons were observed, including local enlargement, irregular fiber structure, and hypoechogenic areas in contrast to the healthy tendons (Fig. 5.3). In the same injured tendons, color Doppler measurements revealed blood vessels coinciding with the regions of abnormalities found in the gray-scale US (Fig. 5.4A). As a conclusion, the authors mentioned the motivation to investigate the treatment of local sclerosing injections to target the zone of the vascularization.

Another research group also dealt with the imaging of blood flow using high-frequency US (Yeh et al., 2009). A tendinitis model in the mouse AT was based on local collagenase injection leading to an injury. The aim was to observe the morphology of the injury as well as to assess changes in the microcirculation of the mouse ATs before and post treatment with 5 min treatment of continuous mode low frequency (45 kHz) under the use of a linear transducer of 25 MHz (resolution 150 μm). Both power Doppler and color Doppler adjustments were performed and compared. As a main outcome, high-frequency Doppler imaging was suggested to

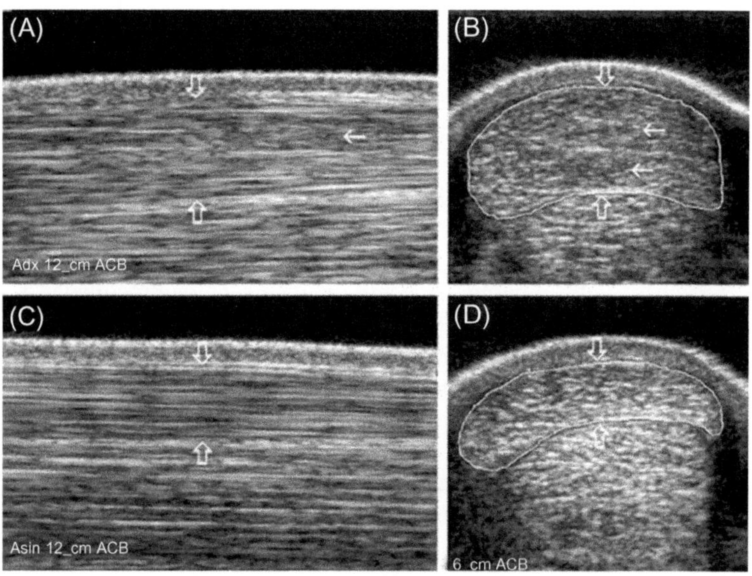


**Fig. 5.3** Gray-scale ultrasound images of a superficial digital flexor tendon (SDFT) in a horse. (A) Longitudinal plane of an injured tendon. The SDFT is between the *open arrows*; the *small arrow* shows an area with poor fiber quality. (B) Transversal plane of an injured tendon, the superficial digital flexor tendon is between the *open arrows*; the *small arrows* show an area with poor echogenicity. (C, D) Images from the normal tendon at the same level on the contralateral front limb.
From Kristoffersen, M., Ohberg, L., Johnston, C., Alfredson, H., 2005. Neovascularisation in chronic tendon injuries detected with colour Doppler ultrasound in horse and man: implications for research and treatment. Knee Surg. Sports Traumatol. Arthrosc. 13, 505–508, © by Knee Surgery, Sports Traumatology, Arthroscopy, with permission from Springer.

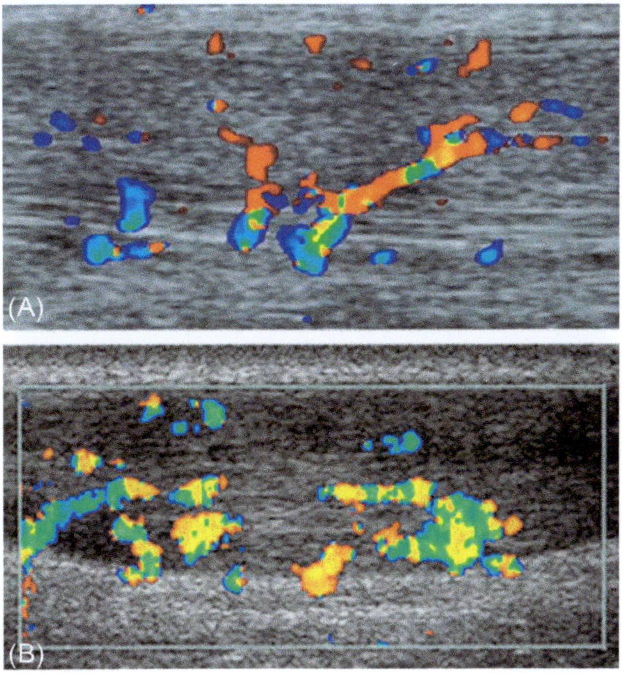


**Fig. 5.4** Longitudinal color Doppler ultrasound images demonstrating neovascularization in chronically injured tendons in the horse and man. (A) Image of the injured SDFT in the same horse and at the same level as demonstrated in Fig. 5.3A. (B) Human Achilles tendon suffering of tendinosis.
From Kristoffersen, M., Ohberg, L., Johnston, C., Alfredson, H., 2005. Neovascularisation in chronic tendon injuries detected with colour Doppler ultrasound in horse and man: implications for research and treatment. Knee Surg. Sports Traumatol. Arthrosc. 13, 505–508, © by Knee Surgery, Sports Traumatology, Arthroscopy, with permission from Springer.

be useful to monitor microcirculation; therapeutic 45 kHz US improved blood flow, even 30 min after its application, and this low-frequency therapeutic US should be favored over the conventional 1 MHz therapeutic US according to the authors, because of more nonthermal effects compared with the latter.

Other US studies dealing with the quantification of vascularization in tendons, however, deal with the obvious limits in accurately detecting the tiny blood flow in tendons, resulting in large variability in diagnostics (Snellenberg et al., 2007; Yang et al., 2011) or in the fact that only certain experimental approaches are leading to accurate information (Wang et al., 2011).

### 5.2.6 Extrinsic and intrinsic healing of tendons and ligaments

In our research group, we analyzed rabbit ATs after dissection and suturing with a four-strand Becker suture by power Doppler US in order to assess vascularization and relating the extent of vascularization to histologically determined adhesion

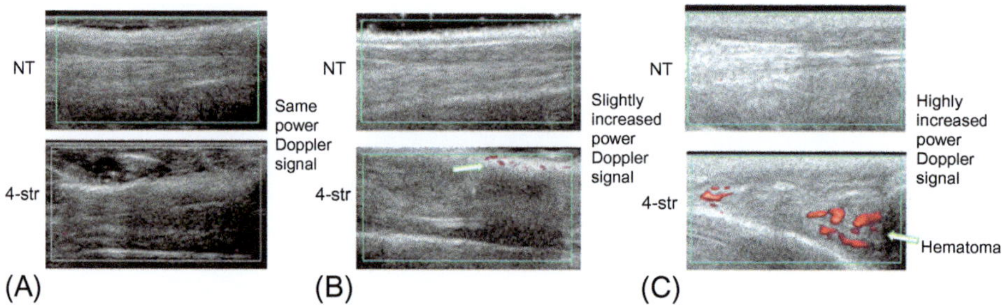


**Fig. 5.5** Power Doppler imaging of rabbit Achilles tendons of four-strand Becker sutured tendons 6 weeks postsurgery (four-str) and the healthy nontreated (NT) contralateral Achilles tendon (AT) for three levels of vascularization. (A) No increased vascularization or perfusion compared with NT, (B) slightly increased perfusion, and (C) formation of a hematoma in the treated tendon.

formation (Meier Buergisser et al., 2014). Conceptually, tendon healing can occur intrinsically, *via* proliferation of EP and endotenon tenoblasts, or extrinsically, by invasion of cells from the surrounding sheath and synovia (Sharma and Maffulli, 2006). Pronounced adhesion formation is associated with a predominance of extrinsic tendon healing (Strickland, 2005), and extrinsic healing is reported to go along with a well-established vascular network (James et al., 2008). With this theoretical background in mind, we compared the extent of adhesion formation determined in histology (Meier Buergisser et al., 2014) with the extent of vascularization and perfusion seen in power Doppler ultrasound images (Fig. 5.5), but only half of all cases were matching the histologically found adhesion degree. We concluded that the extent of vascularization assessed by power Doppler US in our rabbit AT model was not a conclusive indication of the adhesion extent of these tendons—albeit the theoretical background suggesting such a relation. It has to be taken into account, however, that clutter filtering removed low blood velocities, for example, those present in the capillary beds. Moreover, because the volume fraction of blood in tissue containing microvessels is relatively low, which, coupled with the reduced hematocrit in capillaries, significantly reduces the number of red blood cells present, the signal strength in power Doppler was also reduced compromising the theoretical relationship to extrinsic healing and to adhesion.

### *5.2.7 Anatomical studies with US*

US imaging is used also for anatomical studies. For example, Doherty et al. used US imaging and MRI in order to compare the anatomy of rabbit and human AT and found peculiarities about the rabbit AT; in contrast to the human AT, the rabbit exhibits an FDS but has no *flexor digitorum longus* (FDL) (Doherty et al., 2006). This has to be considered when the rabbit AT is used as an animal model intended to mimic human tendons (Buschmann et al., 2013, 2015). Moreover, baseline collection of healthy accessory ligaments of the SDFT in horses was presented, which is essential for comparison of abnormal or injured tendons (Denoix and Busoni, 1999). Such baseline values were also presented for different types of horses including ponies, focusing on the cross-sectional areas (CSAs) of the SDFT and the *deep digital flexor* tendons (DDFT) under the use of a 7.5 MHz transducer (Smith et al., 1994b). Interestingly, the correlation of the CSAs to weight, height, and mid-metacarpal circumference only positively correlated for ponies, but not for horses. Of further interest and of clinical relevance in veterinary medicine was the finding that the ratio of CSAs SDFT/DDFT was constant irrespective of the type of horse nor of its size.

Tendon elasticity can also be assessed by US-sonoelastography. In an in vitro study, where rabbit ATs were analyzed with the baseband speckle-tracking algorithm 10 days postinjury after inducing tendon ischemia by blunt dissection of AT and subsequent suturing, it was found by A-mode 5 MHz circular transducer US of a gelatin-based phantom with the extracted ATs imbedded, which the relative tendon displacements under a compression by a 0.5 kg load gave accurate estimates of strains as a function of relative depth (Kuo et al., 1999). Furthermore, the strains were correlated to the gray-scale findings of B-mode sonography using a 10 MHz transducer. In those images, hyperechogenic regions correlated well to the low-strain values found in the elastographic measurements. Also the thickness and the locality of the low-strain regions (regions of ischemia) were accurately represented in the B-mode sonograms. In addition, histological comparison revealed that high-strain regions corresponded to proliferating Gr, while the low-strain regions were represented by bundles of collagen fibers that were randomly distributed and had no alignment. As such, the authors concluded that compared with the conventional gray-scale B-mode US imaging, the A-mode-based strain distribution inside the tendon tissue may be a better tool for clinical tendon disorders as the elastic modulus reflects the different stages of the regeneration.

Sonoelastography was successfully used in a study applying cellular therapy to rabbit AT defects. Autologous adipose-derived stem cells (ASCs) were applied or not after dissection of the gastrocnemius tendon and filling a gap of 10 mm with either a pure collagen matrix or a collagen matrix seeded with 1 million of ASCs (Gehmert et al., 2012). Real-time color-coded sonoelastography was used to assess the tendon elasticity after 8 weeks using a high-resolution 6–15 MHz matrix linear probe (where tissue with high stiffness was yellow, fibrotic tissue green, scar tissue blue, and normal tendon tissue red), allowing easy examiner-independent analysis of elasticity. As an outcome, the authors reported ASC-supported tendon regeneration leading to a higher elasticity compared with unseeded collagen matrices using a repeated measurement ANOVA.

Recently, contrast enhancement in US imaging is reported based on the injection of an ultrasound contrast agent such as microbubbles consisting of a sulfur hexafluoride gas in a phospholipid shell with a diameter of 2.5 μm (Bertuglia et al., 2014). Such microbubbles are highly echogenic and increase the US backscatter leading to strongly contrasted sonograms. In an equine DDFT model, an intratendinous stab incision was applied and standard and angle contrast US was assessed using a 10 MHz linear probe. Very promising results were obtained, whereas in this acute damage model, surgically induced longitudinal lesions were not identified by standard US (no contrast enhancement), the contrast medium easily diluted within the synovial fluid and diffused into the lesions, delineating very clearly the extension and producing hyperechogenic lines strongly contrasting with the echotexture of the tendon. Moreover, angle contrast-enhanced US further increased this contrast by reducing the echogenicity of the collagen fibers because the microbubbles' echogenicity is not dependent of the angle.

### 5.2.8 Summary

US is a very versatile analytic tool allowing to image tendon-related topics such as anatomy, diseases, lacerations, overuse, tendinopathy, vascularization, and adhesion.

## 5.3 Magnetic resonance

### 5.3.1 Overview

Besides US, magnetic resonance imaging (MR/MRI) is also a widely used noninvasive imaging method to study and evaluate tendons and ligaments (Table 5.3). Frequently, MRI is used for a qualitative in vivo assessment because of its high sensitivity and specificity. MRI is also called nuclear magnetic resonance imaging (NMRI) or magnetic resonance tomography (MRT) and is based on strong magnetic fields and radio waves to generate images. Oscillating magnetic fields (mostly 1.5 T and range 0.2–7 T) are applied in order to excite hydrogen atoms, which emit radio frequency signals. The contrast between different tissues is determined by the rate at which excited atoms (often hydrogen atoms) return to the equilibrium state (relaxation). This return is characterized by two independent processes of relaxation: $T_1$ (spin-lattice) and $T_2$ (spin-spin). To create a $T_1$-weighted image, more magnetization is allowed to recover before measuring the MR signal by changing the repetition time (TR). To create a $T_2$-weighted image, more magnetization is allowed to decay before measuring the MR signal by changing the echo time (TE). As different tissues have different amounts of water (and as such different amounts of hydrogen atoms), differences in relaxation times lead to contrasts between different organs. The advantage of MRI lies in its accurate 3-D imaging; however, in contrast to US, it is bound to a certain locality as the MRI scanners are too heavy to be portable with few reported exceptions (Miltner et al., 2003) (mobile NMR-*mo*bile *u*niversal *s*urface *e*xplorer (MOUSE)). A limitation of MRI lies in the use of water-sensitive pulse sequences; in order to generate differential contrast, MRI requires dephasing caused by water diffusion at a tear site, a site of degeneration or repair. Up to date, MRI has been applied to study animal tendons and ligaments for similar purposes as US, including anatomical studies (Doherty et al., 2006) such as the determination of equine tendon CSAs (Couppe et al., 2014), studies on tendinopathy (Vanel et al., 2012) like tendinitis (Kasashima et al., 2002) caused by potential influence of wrong foot positions (Spriet and Zwingenberger, 2009), studies on lesions (Milner et al., 2012), morphological characterization (Widmer et al., 2000; Baums et al., 2012), biomechanical properties such as viscoelasticity (Wellen et al., 2005), and also vascularization (Stange et al., 2015).

**Table 5.3 Overview on magnetic resonance studies presented in this chapter, grouped by thematic subchapters**

| Aimed T&L | Technical features | Medical findings | Lit. |
|---|---|---|---|
| Rabbit+Hum AT | 1.5 T, $T_1$/PD/ $T_2$/US | Anatomical studies and method comp. | Doherty et al. (2006) |
| Horse T | 1.5/3 T, $T_1$ | Determination of tendon CSA | Couppe et al. (2014) |
| Horse DDFT | 0.27/1 T, 2-/ 3-D $T_1$/$T_2$* | Tendinopathy/ navicular syndrome | Vanel et al. (2012) and Marsh et al. (2012) |
| Horse SDFT/ DDFT | 0.2–0.3 T, 3-D $T_1$/$T_2$/* | Tendinitis | Kasashima et al. (2002), Spriet and Zwingenberger (2009), and Milner et al. (2012) |
| Horse digit T | 1.5 T, $T_1$/PD/ $T_2$/US | Morphology and method comparison | Widmer et al. (2000) |
| Sheep RC | 1.5 T, $T_1$ | Morphology | Baums et al. (2012) |
| Rabbit AT | 2 T, 2-/3-D $T_1$/ $T_2$ | Viscoelasticity | Wellen et al. (2005) |
| Rat PT | USPIO-3 T, $T_2$* | Neovascularization vs biomechanics | Stange et al. (2015) |
| Rabbit RC | 3 T, PD/$T_2$ | AC graft repair | Pan et al. (2015) |
| Horse | 0.3–1.5 T, 3-D-$T_1$/$T_2$* | Lameness study in cadavers | Whitton et al. (1998), Dyson et al. (2005), and Vallance et al. (2012) |
| Horse | 1.5 T, $T_1$/$T_2$* | Comparison with histology | Blunden et al. (2009) |
| Human cadavers | 1.5 T | MR parameter changes under loads | Smith et al. (1994a) |
| Rabbit PT | 3 T/$T_2$* | Cyclical loading | Koff et al. (2014) |
| Rabbit AT | 2 T, $T_1$/$T_2$ | Water diffusion during loading | Han et al. (2000) |
| Human | 1.5 T | Elastography, MRE | Sack et al. (2002), Dresner et al. (2001), and Uffmann et al. (2004) |
| Horse DDFT | 4.7 T, $T_1$/SEM | Contrast-enhanced MRI | Asperio et al. (2000) |

*T&L*, tendons and ligaments; *Hum*, human; *AT*, Achilles tendon; *DDFT*, deep digital flexor tendon; *SDFT*, superficial digital flexor tendons; *RC*, rotator cuff; *PT*, patellar tendon. Technical features, *MRI*, magnetic resonance imaging; *T*, tesla; $T_1$, spin-lattice relaxation process; $T_2$, spin-spin relaxation process; $T_2$*, effective value for the relaxation process, $T_2^* \leq T_2$; *PD*, proton density; *USPIO*, ultrasmall superparamagnetic iron oxide; *CSA*, cross-sectional area, *MRE*, phase-sensitive magnetic resonance elastography; *SEM*, scanning electron microscopy.

## 5.3.2 Investigation of morphological changes after injury

Surgical repair of rotator cuff was monitored by MRI in an in vivo sheep model in order to assess morphological changes after two different suture techniques including a single- or a double-row Mason-Allen stitch (Baums et al., 2012). While biomechanical properties clearly showed that the double-row technique

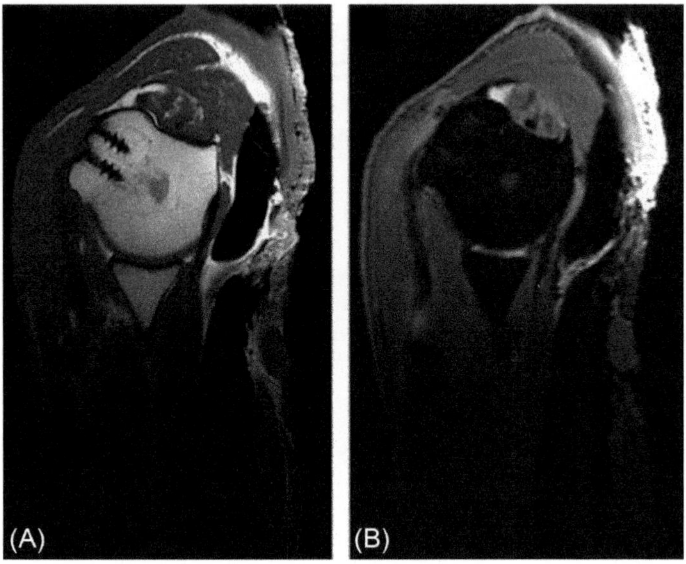


**Fig. 5.6** Postoperative magnetic resonance images of a sheep shoulder, rotator cuff (RC). Fat-suppressed $T_1$-weighted (A) and $T_2$-weighted spin-echo (B) images showing an intact infraspinatus tendon at 26 weeks' follow-up were obtained in the transverse plane with low intensity of the soft tissue presenting type I integrity according to Sugaya et al. (2005). From Baums, M.H., Spahn, G., Buchhorn, G.H., Schultz, W., Hofmann, L., Klinger, H.M., 2012. Biomechanical and magnetic resonance imaging evaluation of a single- and double-row rotator cuff repair in an in vivo sheep model. Arthroscopy 28, 769–777, reprinted from Arthroscopy, Copyright 2012, with permission from Elsevier.

accelerated the mechanical recovery of the corresponding tendons, there were no significant morphological changes in the MRI images that were performed with a 1.5 T under fat-suppressed $T_1$-weighted spin echoes. Specimens were evaluated for evidence of foreign-body reaction to the anchors at the repair site. Moreover, also $T_2$-weighted images were measured in order to see the postoperative cuff integrity (Fig. 5.6).

### *5.3.3 Tissue engineering in tendon repair*

In an in vivo rabbit rotator cuff repair model under the use of decellularized tendon grafts, MRI scanning at 4, 8, and 12 weeks postsurgery revealed hypertrophic scar tissue in the defect group (no graft), while in the graft group, the newly formed tissue was homogenous and regular (Pan et al., 2015). MRI images were acquired with a 3 T unit, and the scanning protocol included the location of the *infraspinatus* tendon using spin-echo $T_2$-weighted sequence (SE/T2WI) and proton density-weighted (PDWI) sequences on transverse view. MRI images demonstrated that the signal intensity of the *infraspinatus* tendon was gradually reduced from a high signal to a low one during recovery, being similar to normal low signal intensity at 12 weeks post operation. Although biomechanical tests revealed that the ultimate tensile load was not statistically significantly different between the graft and the defect groups (at any time point), MRI images in contrast clearly showed the different tissues formed during regeneration, with the homogenous and regular tissue in the graft group opposed to the scar tissue in the defect group.

### 5.3.4 Diagnosis of tendon diseases

Sometimes, there is an evident advantage of imaging a disease by MRI compared with US as has been reported by Widmer et al. The authors mentioned US being of limited value compared with MRI and CT when the navicular syndrome was evaluated because of the poor acoustic window in the palmar aspect of the horses' digits (Widmer et al., 2000). The navicular syndrome is a horse-specific disease including inflammation or degeneration of the navicular bone and its surrounding tissue (as DDFT) ending often up in lameness. MR images were acquired with a 1.5 T scanner, and multiplanar sets of spin echo were obtained and gradient echo and inversion recovery images with $T_1$, $T_2$, or proton density weighting. MR images were related to histological sections including DDFT, articular cartilage, navicular bone, and cysts. In addition, Marsh et al. investigated the influence of corticosteroids and hyaluronan in terms of navicular syndrome by MRI and found that the effect was dependent on the disease process detected on MRI and also the duration of lameness (Marsh et al., 2012).

As lameness in horses is one particular disease gathering a lot of attention in the corresponding community, several MRI studies were performed to accurately image tissue structures related to lameness in cadaver models (Whitton et al., 1998) and in living horses (Dyson et al., 2005; Vallance et al., 2012). Moreover, MRI data acquired in the living horse were successfully linked and correlated to histological data obtained post mortem (Blunden et al., 2009).

### 5.3.5 Investigation of biomechanical properties

Of interest in terms of biomechanical properties of tendons and ligaments are studies that evaluate changes of MR parameters during or after the application of load as reported by Smith et al. where a reduced signal intensity was found upon ligament loading (Smith et al., 1994a). For example, Koff et al. used a 3 T scanner and ultrashort echo imaging (UTE), which allowed displaying contrast within rabbit patellar tendons (PTs) that were manually cyclically loaded (Koff et al., 2014). Such loaded tendons (independently of being fresh or frozen) exhibited significantly shorter $T_2$* values indicating stronger spin-spin interactions, with $T_2$* values showing static and stochastic effects. It was further histologically checked whether the fiber structure and the crimp pattern were not affected by the loading regime, and it was found that reduction in $T_2$* values was caused only by loading and not by any damage of the tendon tissue. It was suggested that the loading led to a recoverable change in tissue organization.

Other MRI studies of loaded tendons focus on the apparent diffusion coefficient (ADC) of water. An increase of ADC of rabbit ATs was measured as the load was increased (Han et al., 2000). Furthermore, it was shown that tensile loading induced water transport from the central to the peripheral region of rabbit AT, indicating that water affects the viscoelastic properties of tendon tissue (Wellen et al., 2005). In this MRI study, Wellen et al. assessed parallel and perpendicular ADC values under the use of a 3-D mapping with a 2 T unit characterizing $T_1$ and $T_2$ relaxation times. The authors emphasized that adequate tendon segmentation in the image processing is crucial with respect to appropriate data interpretation.

In order to assess tissue elasticity, for example, the shear modulus, phase-sensitive magnetic resonance elastography (MRE) has been developed (Sack et al., 2002). The principle behind MRE is the induction of shear waves within the tissue under view by application of HF vibrations in order to induce tissue displacements, which can be correlated to the phase shifts of the MRI signals. MRE has advantages over US elastography due to larger fields of view and easy 3-D analysis. For tendons, only few studies are reported, focusing more on the muscle than on the tendons involved and usually dealing with clinical situations, not animal models (Uffmann et al., 2004; Dresner et al., 2001).

### 5.3.6 Contrast agents in MRI

Contrast-enhanced MRI has been successfully used in order to determine the DDFT in horses (Asperio et al., 2000). Vascular administration of gadopentate dimeglumine (Gd-DTPA/Dimeg) and postinjection MRI was used to evaluate the concentration of water protons in the tendon. The tendons did not have any detectable signal before application of the contrast agent; however, after injecting Gd-DTPA/Dimeg, the internal structure of the tendon was clearly visible, and the structures observed were validated by scanning electron microscopy (SEM).

The extent of (neo)vascularization and angiogenesis in tendons during regeneration can also be visualized and quantified by MRI. In a rat PT model where fibroblasts and microcirculation were partially damaged by liquid nitrogen using a frosted surgical forceps, MRI was performed under the use of a "blood pool" contrast agent; ultrasmall superparamagnetic iron oxide (USPIO) enhanced MR relaxometry (Stange et al., 2015). USPIO is a contrast agent with nanoparticles of 3–5 nm, which reduce the $T_2$*-weighted signal intensity. The relative blood volume that was assessed by $T_2$* mapping was successfully positively correlated to histologically determined microvessel density and vascular endothelial growth factor (VEGF) synthesis. In addition, the authors correlated the corresponding $\Delta R_2$* (which is the relaxation rate change calculated upon $T_2$* data) to the biomechanical properties of the tendons and revealed an inverse proportionality to material properties such as maximum stress and Young's modulus (Fig. 5.7). Tendinopathy goes along with lower material properties predisposing the tendon for recurrent trauma. Such lower material properties are associated with higher levels of hypoxia-inducible factor (HIF-1), VEGF, and matrix metalloproteinase-3 (MMP-3), all of them supporting neoangiogenesis (Krishnan

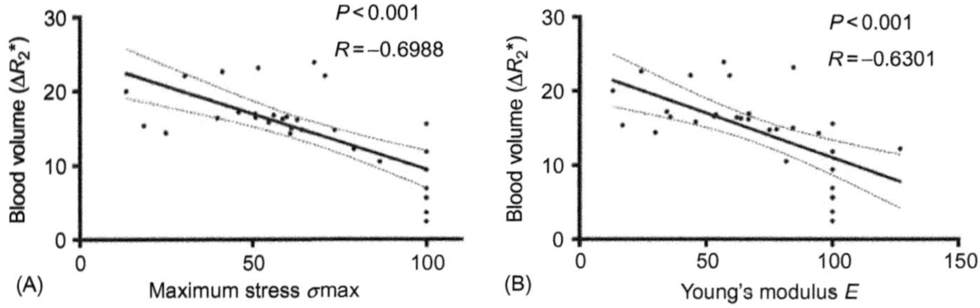


**Fig. 5.7** Statistical analysis correlating magnetic resonance and biomechanical properties in a rat patellar tendon (PT). Pearson's correlation analysis revealed a significant correlation between $\Delta R_2{}^*$ and (A) maximum stress $\sigma_{max}$ and (B) Young's modulus $E$ according to Sahin et al. (2012).
From Stange, R., Sahin, H., Wieskotter, B., Persigehl, T., Ring, J., Bremer, C., Raschke, M.J., Vieth, V., 2015. In vivo monitoring of angiogenesis during tendon repair: a novel MRI-based technique in a rat patellar tendon model. Knee Surg. Sports Traumatol. Arthrosc. 23, 2433–2439, © by Knee Surgery, Sports Traumatology, Arthroscopy, with permission from Springer.

et al., 2007; Sahin et al., 2012). As such, it is not surprising that lower mechanical properties correlated with higher relative blood volume assessed by contrast-enhanced MRI.

## 5.4 Light microscopy, fluorescence microscopy

### 5.4.1 Overview

Light microscopy (LM) and fluorescence microscopy (FM) are imaging techniques that are largely used in research for characterization of tendon and ligament tissues (Table 5.4), ranging from the tendon morphology as such, over tissue-engineered tendon grafts, to analysis of fascicles, fibers, fibrils, microfibrils, collagen macromolecules, and tenocytes (Fig. 5.1). LM, or the less used term optical microscopy, is based on visible light—bright-field (BF) technique—its reflection, and the magnification by lenses in order to examine small items. FM images the fluorescence of a sample, which occurs when the sample has absorbed light or other electromagnetic radiations before. FM involves the application of a beam of light (often ultraviolet, UV) that leads to the excitation of electrons; by "falling" back to their ground state, light is emitted.

Polarized light microscopy (PLM), which is applied when the orientation and organization of tendon fibers are studied, involves—as the name says—illumination of the sample with polarized light. Complex microscopy techniques that also involve polarized light are differential interference contrast microscopy and interference reflection microscopy. Such techniques are mostly used to analyze birefringent (birefractive)

**Table 5.4 Overview on light and fluorescence microscopy studies (LM/FM) presented in this chapter, grouped by thematic subchapters**

| Aimed T&L | Technical features | Medical findings | Lit. |
|---|---|---|---|
| Mouse digit flexors | LM-BF/CM | Anatomy | Wong et al. (2006) |
| Rabbit flexors | LM | Tendon graft fabrication with Tc | Zhang et al. (2009) |
| Nude mice | FM | Synth./coll-scaffold fabr. with Tb and Mst | Xu et al. (2014) |
| Rat PT | LM/TEM | Biomechanics of PT autograft as ACL | Muellner et al. (2000, 2001) |
| Bov. ext/fl/ HumSTT | CM | Fatigue behavior after repetitive loads | Shepherd et al. (2013) |
| Rat tail tendon | MPh-CM/SAXS/ TEM | Viscoelastic alterations due to diabetes | Li et al. (2013) and Fessel et al. (2014) |
| Mouse RC (SSP) | PLM/RS | Biomech./morphol. under immobilization | Schwartz et al. (2013) |
| Bovine flexors | ILM/PLM | Morphology of bundles, helical structure | Vidal (2003) |
| Rat AT/dog T | 2-D-PLM | Morphology of collage alignment | Whittaker and Canham (1991) |
| Rabbit DFTT | LM | Anisotropy assessment | Morgan et al. (2006) |
| Rat AT | PLM | Crimp morphology | Franchi et al. (2007) |
| Mouse tiss./ rat AT | LM/CM | Rebuilding of collagen after electroacup | Almeida et al. (2012) and Langevin et al. (2006) |
| Rabbit FDP | rt-CM | Cell growth and adhesion formation | Branford et al. (2012) |
| Rat tail | CM | Remodeling of ECM | Screen et al. (2004) |
| Rabbit PT | LM | Mechanobiology of cells, cell stretching | Miyazaki et al. (2000) |
| Rat tail | CLM | In situ deflection, cilia loading | Lavagnino et al. (2011, 2013) |
| Rat tail | CLsM | Cell nucleus deformation | Arnoczky et al. (2002) |
| Mouse tail | LM (2P-BF) | In situ cell-cell connection under load | Legerlotz et al. (2014) |
| Rat T/Bovine ext. | ICLsM | Cell morphology under load | Screen et al. (2003) |
| Mouse AT | Endoscopic CFM | Local tissue inhomogeneity under load | Snedeker et al. (2006) |
| Rat tail | CFM | Cell gap junction permeability under l | Maeda et al. (2012) |

*T&L*, tendon and ligaments; *Hum*, human; *PT*, patellar tendon; *STT*, semitendinosus tendon; *RC*, rotator cuff; *SSP*, *supraspinatus* tendon; *AT*, Achilles tendon; *FDP*, flexor digitorum profundus; *ACL*, anterior cruciate ligament. Technical features, *LM*, light microscopy; *FM*, fluorescence microscopy; *BF*, bright field; *CM*, confocal microscopy; *TEM*, transmission electron microscopy; *MPh-CM*, multiphoton confocal microscope; *SAXS*, small-angle X-ray scattering; *ILM*, interference light microscopy; *PLM*, polarized light microscopy; *RS*, Raman spectroscopy; *rt-CM*, real-time confocal microscopy; *CLsM*, confocal laser microscopy; *ICLsM*, inverted confocal laser microscopy; *CFM*, confocal fluorescence microscope; *ECM*, extracellular matrix.

materials—tendons and ligaments are birefringent—where the refractive index depends on the polarization and propagation of the light. In the following paragraphs, application of light, fluorescence, and polarized light microscopic investigations is summarized in a top-down approach from macro- to microstructures and components in order to show the wide application of these techniques in the field of tendon characterization.

More confocal light microscopy (CLM) types were developed with different features (light, fluorescence, and laser).

### 5.4.2 The tendon tissue (engineering) level

Light microscopy and confocal microscopy (CM) have been used in an anatomical study of murine hindpaw tendons (Wong et al., 2006). For this purpose, the two techniques to image several aspects of tendon anatomy were very suitable. Not only were the pulleys, the elastic areas of the digits (Fig. 5.8), and the fibrocartilage regions made visible by different appropriate staining and subsequent analyses by light microscopy (H&E, Miller's elastin staining, and alcian blue staining, respectively) but also the gross anatomy of the vasculature in the hindpaw was assessed successfully by fluorescein isothiocyanate (FITC)-dextran perfusion and fluorescence imaging.

Light microscope analysis is often used for the evaluation of tendon constructs aimed at tendon tissue engineering. For example, intrasynovial tendon grafts were fabricated by first decellularization and then repopulation with intrasynovial tendon cells of the rabbit (Zhang et al., 2009). The attachment of both the EP cells and the endotenon cells to the tendon grafts was observed by light microscopy. Moreover, after fluorescently labeling a subset of EP cells in green and endotenon cells in red, FM was used to determine the preferred place of their attachment to a tissue-engineered tendon construct. Endotenon cells were more likely found in the core of the scaffold material, while EP cells preferred the periphery. Another comprehensive study, including in vitro and in vivo experiments, deals with the effect of mechanical stimulation on the maturation of stem cells derived from rabbit tendons (Xu et al., 2014). The use of fluorescence imaging was chosen in order to track tendon stem cells in vivo that were seeded on a poly(L-lactic acid-*co*-caprolactone) (P(LLA/CL))/collagen scaffold and implanted in the back of nude mice. Of interest is the comparison of in vitro dynamic tendon stem cell cultivation to static cultivation in terms of ultimate stress and Young's modulus of the constructs after 4 and 12 weeks in vivo, showing the obvious positive effect of previous mechanical stimulation during the cultivation period.

For anterior cruciate ligament (ACL) reconstruction, very often, the central third of the PT is transplanted (Kennedy et al., 1980). In animal experiments, however, it has been shown that the PT autografts change into a similar histological appearance as the ACL tissue but with a significant smaller tensile strength (Muellner et al., 2000). Therefore, stress-shielding approaches have been examined. In a rat PT model, the effects of a stress shield represented by a cerclage consisting of Vicryl 2.0 or Prolene 2.0 were evaluated by light microscopy and by transmission electron microscopy (TEM) (Muellner et al., 2001). Light microscopy revealed not only changes in the morphology of the tendon, such as thickness, but also changes in organization of the collagen bundles, which got irregular due to the stress shield applied. Furthermore,

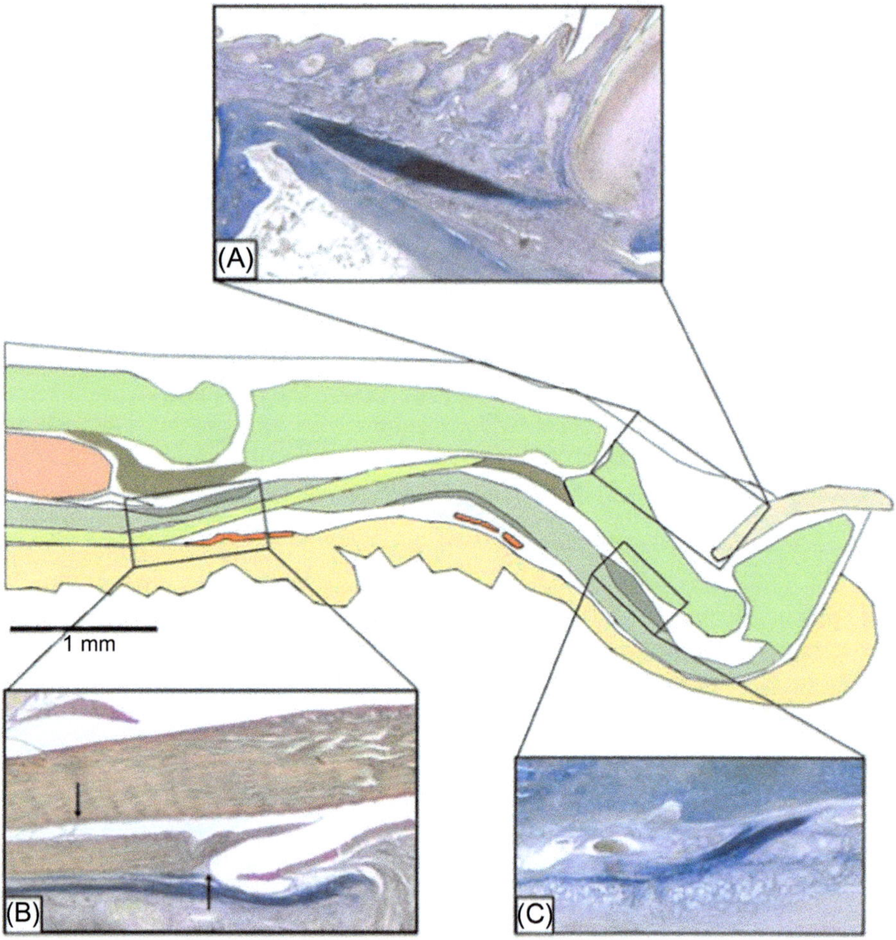


**Fig. 5.8** Elastic areas of the digit of a mouse. Miller's elastin stain highlights elastic fibers in blue. (A) The dorsal elastic ligament of the claw. (B) The proximal extent of the synovial sheath. The plicae border (*arrows*) shows evidence of elastic fibers outside of the sheath but not within the sheath. (C) Elastic fibers seen in a vincular-type attachment close to the dorsal fibrocartilage region of the *DDFT*, deep digital flexor tendon.
From Wong, J., Bennett, W., Ferguson, M.W., Mcgrouther, D.A., 2006. Microscopic and histological examination of the mouse hindpaw digit and flexor tendon arrangement with 3D reconstruction. J. Anat. 209, 533–545, © by Journal of Anatomy, with permission from Wiley.

cellular density, vascularity, and acidic proteoglycans were analyzed and found to be increased. TEM analysis finally showed that the diameter of the collagen fibrils was smaller compared with untreated tendons. The authors concluded from these findings that a stress shield might be good during the first time post operation; however, the stress shield should be continuously reduced, for example, by using a biodegradable material and not inherent Vicryl or Prolene, which are usually applied as cerclage materials.

### 5.4.3 The fascicle level

Shepherd et al. emphasize the fact that tendons of different species and tendons with different usual daily stresses may vary significantly with respect to their fatigue behavior; in other words, repeated microstrains that are below the failure threshold leading to overuse tendinopathy do affect tendons of different species exposed to various stresses in different ways. Therefore, this research group investigated the fatigue behavior (cyclic creep and cyclic stress relaxation) of tendon fascicles of the digital extensor tendon and the digital flexor tendon of the bovine hoof and the human hamstring semitendinosus (Shepherd et al., 2013). While the bovine extensor tendon exhibited the poorest fatigue response, the human hamstring semitendinosus tendon had the best outcome. Nevertheless, all tendons had a similar degree of matrix breakdown, which was analyzed by CM of the fascicles stained with acridine orange (a dye to track tenocytes' nuclei). Obviously, the more energy-storing tendons (human semitendinosus tendon and the bovine flexor tendon) were better able to withstand repetitive loading than the bovine extensor tendon—suggesting that fatigue resistance of tendons is correlated to their daily life loading. Moreover, these differences found among different species emphasize that animal tendon models may indeed not always be a one-to-one reliable source for information intended to be transferred to the human being situation.

Fascicle mechanics in a rat tail tendon model were assessed either without or with methylglyoxal (MGO) treatment (Li et al., 2013). The MGO treatment was intended to form advanced glycation end products (AGEs), which naturally accumulate with age or diabetes and lead to cross-linking of collagen and thereby to altered physical characteristics. While ultimate stress, yield stress, yield strain, and energy to yield of the tendon fascicles were significantly increased after MGO treatment, on the fiber level, the fiber-fiber sliding was significantly reduced, which was assessed by cell nuclei displacements under a multiphoton confocal microscope (MPh-CM). Such findings indicate a loss of tissue viscoelasticity, and the authors emphasize the importance of not only assessing elastic response of age-related changes in tendons but also mainly focusing on the viscoelasticity. In a subsequent study, the widespread assumption was contradicted that tissue stiffening in aging and diabetes is related to the AGE-increased fibril stiffness. Fessel et al. were able to show using synchrotron small-angle X-ray scattering (SAXS) that MGO treatment of rat tail tendons reduced the side-by-side sliding of the collagen molecules within the fibril entity, indicating lateral cross-links between collagen molecules (Fessel et al., 2014). At the same time, no change in maximum fibril modulus was found (being 2.5 GPa $\pm$ MGO treatment). Therefore, AGEs reduce inherent collagen fibril viscoelasticity but not the fibril stiffness—which led to the conclusion that the cause for the observed increased stiffness of tendons with age or diabetes is found in the higher stiffness of hierarchically higher structures such as fibers (Li et al., 2013).

### 5.4.4 The fiber level

In a mice model, load and stiffness of the *supraspinatus* tendon attachment decreased when no muscle loading was applied for 4 and 8 weeks, respectively (Schwartz et al., 2013). In order to elucidate the causes of this reduced biomechanical performance of

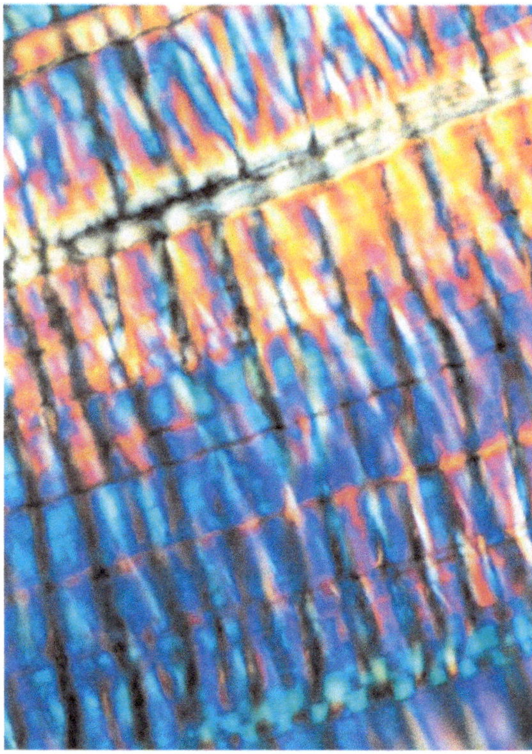

**Fig. 5.9** Birefringence in a 20 μm thick cryo-section of a peeled rat tail tendon, allowing to elucidate the spiral organization of tendon fibers.
From Vidal, B.D.C., 2003. Image analysis of tendon helical superstructure using interference and polarized light microscopy. Micron 34, 423–432, reprinted from Micron, Copyright 2003, with permission from Elsevier.

the tendon-bone enthesis, PLM and Raman spectroscopy (RS) were used to assess the collagen fiber alignment, the mineral characteristics, and the formation of fibrocartilage. As a result, the authors reported a reduced fiber alignment and a reduced crystallinity in the hydroxylapatite, while phosphate was replaced by carbonate to some extent in the crystalline phase. These structural and compositional changes were obviously evoked by the unloading of the muscle and might explain the reduced biomechanical properties of the enthesis.

Tendon helical superstructures such as the spiral organization of the fibers were successfully imaged with interference and PLM (Vidal, 2003). Typical interference color images were obtained (Fig. 5.9). Furthermore, the directional organization was also assessed for fixed rat ATs and canine tendons under the use of 2-D polarized light microscopy (2-D-PLM) at strains in the range of 0–3% (Whittaker and Canham, 1991). As expected, increasing strain revealed increasing alignment of the collagen. In

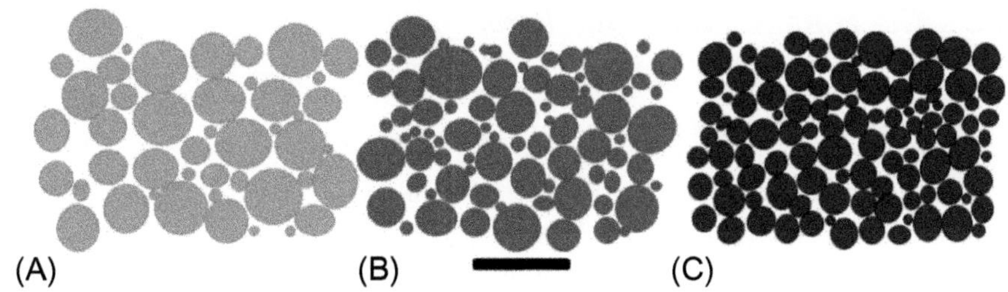


**Fig. 5.10** Rabbit digital tendons, schematic transverse sections of (A) excised tendon, (B) in situ unloaded tendon, and (C) in situ 1 kg loaded and extended tendon, bar = 600 nm. Schemes according to photographs from Morgan, M., Kostyuk, O., Brown, R.A., Mudera, V., 2006. In situ monitoring of tendon structural changes by elastic scattering spectroscopy: correlation with changes in collagen fibril diameter and crimp. Tissue Eng. 12, 1821–1831, original electron micrographs: magnification 35,000 ×.

accordance, while strained fibers exhibited a sharply peaked distribution of orientation, unstrained fibers showed a bimodal fiber orientation.

### 5.4.5 The fibril and microfibril level

Usually, learning books about tendons and ligaments show typical stress-strain curves depicting a toe region with crimped fibers between 0% and 2% of strain (Goodman and Choueka, 2005). This crimped waveform of the fibers, however, is only found when tendons are excised for in vitro mechanical testing, in other words if they are completely free of any tension outside the body. Inside the body, however, under physiological conditions, there is always an in situ tension, which is high enough for the majority of the tendon fibers to be straightened. Morgan et al. used light microscopy in order to assess the anisotropy in digital flexor tendons of fresh rabbit cadavers (Morgan et al., 2006). They determined the backscattered light as a function of wavelength (320–860 nm) at different angles resulting in an anisotropy factor (AF), which is the ratio of backscattered light in the perpendicular plane to the backscattered light in the parallel plane at a specific wavelength. AF at 600 nm was further investigated for different loading regimes of the tendons ($AF_{600}$). Compared with the tensionless specimen, 1 kg loaded flexors exhibited a 68% higher $AF_{600}$. This change in anisotropy went along with a change in the distribution of the fibril diameters that were measured in electron micrographs using TEM (schematic illustration in Fig. 5.10)

As can be expected, the crimp number of fibrils in rat ATs decreases under stretch (Franchi et al., 2007). The crimp morphology was accurately analyzed using PLM in relaxed and stretched tendons, and the angles in the crimps were found to be 148 and 165 degrees, for relaxed and stretched tendons, respectively. Interestingly, under SEM and TEM imaging, these fibrillar crimps were still present, even if the collagen fibril was completely straightened, suggesting that the "fibrillar crimp" is the true structural component of the tendon crimp acting as a shock absorber during stretch.

### 5.4.6 The collagen molecule level

Light microscopy was also used to evaluate the general extracellular matrix and the metachromasy of tendons in a study, where electroacupuncture was applied during the healing in a rat AT model after partial transection (Almeida et al., 2012). Electroacupuncture was hypothesized to be an alternative therapy for tendon injuries as it is reported to have an anti-inflammatory effect and leads to subcutaneous tissue fibroblast cytoskeletal remodeling (Langevin et al., 2006). The authors were able to show that more collagen was rebuilt and the collagen was better organized in the acupuncture group compared with the nonacupuncture group (Almeida et al., 2012).

### 5.4.7 The cell level

Adhesion formation is one of the most occurring postsurgical problems besides rerupture after flexor tendon reconstruction (Elliot and Giesen, 2013). The tendon tissue mechanics of local adhesions are therefore interesting to be studied (van Sterkenburg and van Dijk, 2011). The microstrains and the macrostrains in adhesions of immobilized and mobilized partially lacerated rabbit FDP tendons were assessed; and macromechanics based on tensile testing were related to micromechanics based on real-time confocal microscopy (rt-CM) of fluorescently labeled cell nuclei (Branford et al., 2012). Immobilization was not achieved by a cast, however, by complete transection preventing the rabbit from active motion. Macromechanical results showed that immobilized tendons exhibited adhesions that had a 140% higher stiffness of mobilized tendons and a 160% higher load until failure compared with mobilized tendons. On the micromechanical level, cell nucleus tracking was made with a reference cell nucleus, and local strain was assessed as the displacement change resulting from the applied load. Five different local strain patterns were determined for the cells including stretch, compress, shear, random, and tear pattern, with the local strains for mobilized adhesion being considerably higher than for the immobilized counterparts, in both stretched and compressed regions, showing clearly that tenocytes and tendon fibroblasts were sensitive to their specific mechanical environment. Fibroblasts act through mechanotransduction pathways by remodeling the extracellular matrix (ECM) (Screen et al., 2004). The knowledge gathered by Branford et al. showed the impact of in vivo loading on the adhesion ECM remodeling by the involved tendon fibroblasts. Moreover, this study showed that adhesions of immobilized and mobilized flexors differ on both hierarchical levels in the same way, on the macro- and the micromechanical level. Finally, understanding the mechanisms involved in adhesion formation on the local tissue level is crucial when antiadhesive treatments are developed.

The mechanobiology of single whole cells in tendon tissue can be accurately evaluated with tensile testing systems including a pair of micropipettes gripping the corresponding cell between the tips of the pipettes and a load cell consisting of a pair of stainless steel plates (Miyazaki et al., 2000). The maximum load and elongation at failure were $0.9 \pm 0.2$ μN and $86 \pm 24$ μm, respectively, for rabbit PT fibroblasts. For the detection of the cell stretching (displacement), a light microscope and a CCD camera (charge-coupled device) were used. Other tensional studies of tendon cells include the in situ deflection of the cilia in response to tensile loading (Lavagnino et al., 2011, 2013). The length of the cell cilia was measured with CLM, reflecting the local mechanobiological environment—and thus serving as a biomarker for cell tension (Lavagnino et al., 2011). Moreover, the cilia deflection angles were positively correlated to the applied strain with a coefficient $R^2=0.40$ (Lavagnino et al., 2013), supporting the concept that cell cilia are involved in the mechanotransduction response of tenocytes (see also Chapter 3).

Cell nucleus deformation of tenocytes in tendons under tensile load has been reported by Arnoczky et al. (2002). Using confocal laser microscopy (CLsM), images of individual cells were captured at 0%, 2%, 4%, and 6% tendon strain in a rat tail tendon model stained with acridine orange. Local tissue strain was determined by measuring the distance between the centers of adjacent cell nuclei. As such, nucleus deformation, local tissue strain, and whole-tissue strain were related and compared. Local tissue strain was weakly ($R^2=0.34$) but significantly ($p<0.01$) correlated to the nucleus strain in the $x$-axis, which was suggested to be playing a significant role in the signal transduction of mechanical stimulation.

Cell deformation under load might play a significant role in the mechanical signal transduction in tendons. For example, strain-induced changes in tenocyte shape of mice tendons and the impact of strain on the cell-cell connections in situ were reported (Legerlotz et al., 2014). A study using inverted confocal laser scanning microscope (ICLsM) determined the length/width ratio of tenocytes in order to assess the local strain fields within fascicles (Screen et al., 2003). It was found that increasing applied strain increased local strain with higher length/width ratios of tenocytes in rat tail, rat digital extensor, and bovine extensor tendons. As a main outcome, the authors found that local tissue strains are significantly lower than whole-tissue strains. Another study by Snedeker et al. revealed in addition that the local tissue strains were very inhomogeneous during stretching (Snedeker et al., 2006). Using confocal FM to track fluorescently labeled cells, it was found in a mice AT model by endoscopic measurements that local primary strains in the tissue were ranging from 5% to 55%, while the relative change in the distance between the cells was $20.3\pm3.1\%$. As a conclusion, it was reported that complex relative motions between substructures in the tendon tissue occurred during stretching.

Moreover, the effect of tensile loading on the gap junction between tenocytes in rat tail tendons was assessed by confocal immunofluorescence imaging with calcein Am labeling and subjecting the cells to a loss in fluorescence induced by photobleaching (Maeda et al., 2012). The communication between gap junctions can be influenced by transient events enduring seconds and by longer time periods such as minutes to hours, allowing the exact regulation of their permeability and functionality. While a static tensile load of 1 N for 10 min did not alter gap permeability between the cells, the same load applied for 1 h reduced the gap permeability significantly. This was furthermore accompanied by a suppression of connexin 43, an important gap junction protein. As such, not only gap junctions are transmitting mechanically activated signals but also they are directly regulated by mechanical loading in terms of their permeability.

## 5.5 Electron microscopy

### 5.5.1 Overview

There are two main electron microscopy methods: SEM and TEM. SEM is successfully used to get information about the surface topography of a sample. An electron beam is focused on the sample, and it is scanned over the sample in a raster scan pattern producing an image. The electron beam evokes the emittance of secondary electrons, the amount of which is dependent on the topography—a flat surface emitting less secondary electrons than a tilted one, ending up in a contrasted image. Resolution is below 1 nm, allowing magnifications up to 500,000 ×. In order to prevent nonconductive samples from being charged during the scanning process, they are coated by a thin conductive layer such as gold.

TEM uses also an electron beam focused on a very thin sample. In contrast to SEM, however, the electron beam is transmitted through the specimen, and an image is created by the interaction of the electron beam with the electrons present in the atoms constituting the sample. Resolution of TEM is being optimized, and recent values of 0.05 nm have been achieved, corresponding to a magnification of 50 million times.

Several studies on tendons and ligaments are done with TEM and/or SEM (Table 5.5). As a consequence, these imaging techniques are mostly used for the elucidation of very small structures.

### 5.5.2 Investigation on collagen morphology and structure with TEM

In order to follow the biosynthesis of collagen within the cell, TEM analysis was successfully used by Santander et al. (Gonzalez Santander et al., 1999). They determined the length of collagen fibrils that were excreted into the ECM to be 30–60 nm, with preceding formation of procollagen filaments within the tenocyte. The procollagen filaments made contact in parallel apposition to produce striated spindle-shaped bodies, which got longer by attaching more and more filaments until the 30–60 nm long nascent collagen fibrils were built up and excreted.

Another TEM study elucidated the fact that severed tendons get translucent—accompanied by the fact that the strength gaining is delayed (Ehrlich et al., 2005). They clearly showed that the location of translucency was spiked with frayed, irregular, and thin collagen fibers. As in embryogenesis, tendon fibers grow by the addition of discreet collagen fibril segments (Gonzalez Santander et al., 1999); Ehrlich et al. hypothesized that it might be that during intrinsic wound healing after tendon laceration, collagen fibrils are released from fibers so that these fibrils could get to be reused at the repair site. In other words, not only tenoblasts migrating to the repair site are involved in the production of new fibrils and ECM but also “old” collagen material coming from already before existing fibers could by recycled and attached to give new fibrils and finally regenerated fibers.

The development of the structure of tendon cells and also their arrangement was monitored by SEM and TEM in the rat AT from rats aged 5–90 days (Murata et al., 2000). For example, TEM micrographs of a 90-day-old rat are shown in Fig. 5.11. Very clearly, in

**Table 5.5 Overview on electron microscopy (TEM/SEM) studies presented in this chapter, grouped by thematic subchapters**

| Aimed T&L | Technical features | Medical findings | Lit. |
|---|---|---|---|
| Rabbit AT | TEM | Collagen fibril length/ protocoll. fabric | Gonzalez Santander et al. (1999) |
| Rat AT | TEM | Fibril recycling in healing/translucency | Ehrlich et al. (2005) |
| Rat AT | TEM/SEM | Streak extent of tenocytes | Murata et al. (2000) |
| Rat tail tendon | TEM/SEM | Collagen fibril organization | Squier and Bausch (1984) |
| Rabbit/ dog T | TEM | Components of ECM, elaunin + elastin | Carvalho et al. (1994) |
| Rabbit AT | TEM/AFM | Axial structure of collagen and microfib. | Raspanti et al. (2011) |
| Rabbit AT | TEM | Diameter and density of collagen fibrils | Zhou et al. (2007) |
| Human ACL | TEM | Mechanical weakness, protocoll. levels | Sluss et al. (2001) |
| Rabbit PT/ MCL | TEM | Fibril diameter vs time vs gene expres. | Lo et al. (2004) |
| Mouse AT | TEM | Genetic regulation of biomech. behav. | Rigozzi et al. (2010) |
| Rabbit flexors | SEM | Extrinsic + intrinsic healing in tendon rep | Abrahamsson et al. (1992) and Lundborg et al. (1980) |
| Rabbit FDP | SEM | Tissue eng. for tendon repair | Chen et al. (2014) and Hu et al. (2013) |
| Rabbit SSCT | SEM | Fibril rupture under strain | Morizaki et al. (2012) |
| Hum/div. animals | SEM/ TEM/LM | Morph.+ function of synovial lining cells | Steinberg and Hodde (1990) |
| Rat tail tendon | SEM/TEM | Ultrastructure of tendon sheath | Gotoh et al. (1997) |
| Rat MCL | TEM | Influence of gentamicin application | Vente et al. (1997) |
| Mouse DDFT | SEM | Gliding resistance in intrasynovial T | Hayashi et al. (2013) |
| Rat AT | SEM | Reorg. of collage fibers in healing | Sasaki et al. (2012) |
| Rat, bovine, feline | SEM/ X-ray | Collagen fiber development | Provenzano and Vanderby (2006) |
| Chicken T | SEM/TEM | Force transm. tendon-muscle fibers | Trotter and Baca (1987) |

*T&L*, tendon and ligaments; *AT*, Achilles tendon; *ACL*, anterior cruciate ligament; *PT*, patellar tendon; *MCL*, medial collateral ligament; *FDP*, flexor digitorum profundus; *SSCT*, subsynovial connective tissue; *DDFT*, deep digital flexor tendon. Technical features, *TEM*, transmission electron microscopy; *SEM*, scanning electron microscopy; *LM*, light microscopy; *ECM*, extracellular matrix.

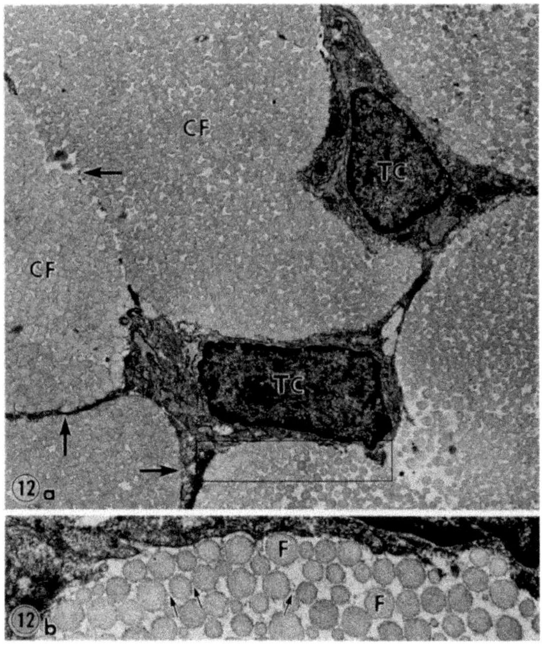


**Fig. 5.11** Rat Achilles tendon (AT), images of tenocytes (TC), collagen fascicle (CF), and fibrils (F) performed by transmission electron microscopy (TEM) at 10,000 × magnification; streaks of dense and flocculent material extend (*arrows*) from the cell surface to the interfascicular clefts (above 12a), and at 27,000 × magnification, connecting filaments (*arrows*) can be seen between fibrils (below 12b).
From Murata, H., Nishizono, H., Miyoshi, M., 2000. Postnatal development of structure and arrangement of tendon cells. A scanning and transmission electron microscope study in the rat calcaneal tendon. Okajimas Folia Anat. Jpn. 77, 77–86, reprinted with permission by Okajimas Folia Anatomica Japonica.

the upper part of the figure, the streaks of the tenocytes extend from the cell surface to the interfascicular space, which are marked by arrows. A similar study had been published 16 years earlier, however, under the use of rat tail tendons (Squier and Bausch, 1984). Therein, the organization of tendon fibroblasts and fibrils was investigated; fibroblasts were regularly arranged in columns and showed a higher association in the longitudinal than in the transverse plane. Thin cytoplasmic sheets transversely formed junctional attachments with similar processes from adjacent cells or from the same cell.

Also, components of the ECM are clearly detectable using TEM. In a study by de Carvalho et al., TEM analysis revealed elaunin fibers in the ECM of tendons wrapped around joints that receive compressive forces besides their actual task of transferring force from muscle to bone (Carvalho et al., 1994). Elaunin is a component of elastic fibers formed after deposition of elastin. It was found that elaunin—among other components—is only present in the pressure-bearing zones of these “wrap around tendons,” implying that components responsible for the elasticity of a tissue also play a role in the attenuation of local compression.

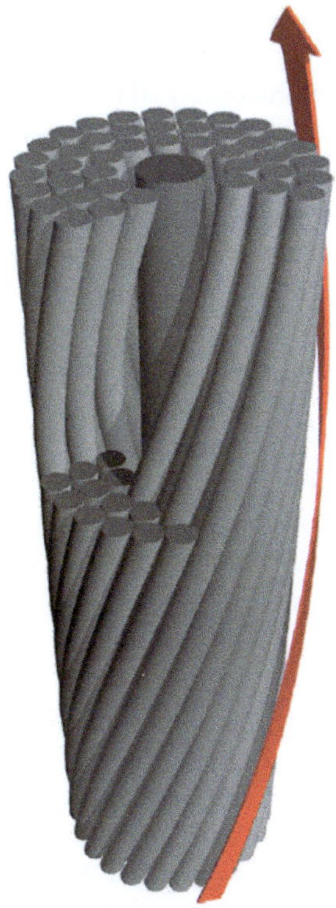

**Fig. 5.12** Tendons and ligaments, proposed architecture of the unimodal collagen fibril and the single-axial structure in the core—surrounded by the helical microfibrils.
Reprinted with permission from Raspanti, M., Reguzzoni, M., Protasoni, M., Martini, D., 2011. Evidence of a discrete axial structure in unimodal collagen fibrils. Biomacromolecules 12, 4344–4347, copyright 2011, American Chemical Society.

Furthermore, a discrete single-axial structure in rabbit AT sheath collagen fibrils was envisioned by TEM after a histochemical reaction intended at revelation of glycogen and polysaccharides (Raspanti et al., 2011). This axial structure that was also confirmed by tapping-mode atomic force microscopy (AFM; for working principle, see Section 5.6 about X-ray), however, was only found in fibrils with helical microfibrils (Fig. 5.12), but not in other tendons whose microfibrils run longitudinal and parallel.

TEM imaging is also suitable to determine the diameter and the density of collagen fibrils. The changing of those parameters as a function of several conditions has been reported. For example, immobilization of rabbit ATs (without any surgical intervention) clearly showed a mechanical weakness of the tendons (Zhou et al., 2007). The authors tried to explain this weakness by histological analyses and determination of the diameter and density of the collagen fibrils by TEM. However, no parameter under view exhibited significant changes due to immobilization. The conclusion was that other readouts are necessary to explain the mechanical weakness, lying in biochemical, gene expression, and functional image marker fields. A gene expression study that tried to correlate procollagen type III to the fibril diameter determined by TEM was performed by Sluss et al. (2001). Mechanical weakness depicted by ultrastructural alterations in the tendon was accompanied by incomplete processing of procollagen III, which was substantially elevated in tendons with smaller diameter fibrils. Moreover, elevated procollagen III was found in grafts with lower loads until failure than with functional ones. Indeed, procollagen III levels may therefore act as a reliable indicator of tendon weakness.

The diameter of collagen fibrils was also monitored as a function of time using TEM, the process of collagen fibrillogenesis being compared in a rabbit PT and medial collateral ligament (MCL) during maturation. For this purpose, TEM imaging was useful and successful, and data were well correlated to gene expression analysis (Lo et al., 2004). During maturation, fibril diameter distribution was peaking in the small-diameter range for early time points, getting more and more spread to a wide range of small, middle, and high diameter range. It was shown that on the mRNA level, collagen I, decorin, and lumican levels were higher in PT than in MCL during maturation during the time studied (up to 52 weeks).

In an attempt to investigate the genetic regulation of tendon biomechanics, architectural baseline values in terms of collagen morphology were assessed with TEM (Rigozzi et al., 2010). For that purpose, two inbred mouse strains, which are widely used in skeletal structure-function studies, were used: C57BL/6J (B6) and C3H/HeJ (C3H). Tendons from C3H mice had larger collagen fibril radius and smaller fibril surface area compared with B6 mice—although the tendons looked similar under macroscopic inspection. Biomechanically, C3H murine tendons exhibited higher elastic moduli. It is speculated that this functional difference could be attributed to the reduced fibril surface area exchange between fibrils and the surrounding proteoglycan-rich matrix where GAG (glycosaminoglycan) chains promote interfibril sliding. However, as B6 mice have a higher GAG content, this stands in contrast with the reported findings. Therefore, another theory of proteoglycan-mediated collagen fibril load-sharing is needed, where more investigations are necessary to come to a conclusive end.

### 5.5.3 *Investigations of surfaces with SEM*

As SEM is used to obtain images of the surface topography, more than 20 years ago, it was often used to get information about the repair mechanism of tendons (Abrahamsson et al., 1992; Lundborg et al., 1980) and more recently about the usefulness of tissue-engineered tendon grafts or augmentations (Chen et al., 2014; Hu et al., 2013). For example, rabbit flexor tendons were peeled in order to get rid of the EP (Abrahamsson et al., 1992). As such, they were implanted subcutaneously in the back of the rabbits after being dissected and sutured. After 5 and 11 weeks, respectively, there was a multiple cell layer of fibroblast-like cells on the surface, indicating intrinsic healing. An older SEM study (Lundborg et al., 1980) revealed that superficial tendon repair in a rabbit flexor tendon had cellular fibroplasia with fibroblasts being of intrinsic local origin, while the surface cells (including also macrophages) contributed to the restoration of the gliding surface were apparently derived from the synovial fluid—thus indicating extrinsic healing. Also, material characterization of the subsynovial connective tissue in the carpal tunnel was assessed by SEM, indicating ruptured fibrils when the specimen was under strain (FDS excursion by 2, 3.5, 5, and 8 mm) (Morizaki et al., 2012).

An impressive and species-comprehensive (rat, rabbit, dwarf goat, sheep, pig, dog, and human) study examined the morphology and functional features of synovial lining cells (SLC) using SEM, TEM, and light microscopy (Steinberg and Hodde, 1990). As a main outcome, the authors defined A-type and B-type cells: while A-type cells were macrophages specialized on the production of hyaluronic acid and phagocytosis of wear and tear products after mechanical loads, B-type cells were fibroblast-like cells concerned with protein synthesis, for example, lubricating glycoprotein (LGP-1). The strength of the study is the compilation of tendon sheath surface images of all the aforementioned species by SEM and TEM images of single cells at the same location determining the size, the width and length, the chromatin, cisternae, mitochondria, vacuoles, lysosomes, microtubule, and filaments. This allowed the direct corroboration of TEM and SEM observations. While A-type cells showed many vacuoles, B-type cells had much endoplasmic reticulum. Generally, the intima of tendon, the tendon sheath, and the synovium had more B-type cells than A-type cells, while in infrapatellar fat pads, A-type cells were predominant.

The ultrastructure of the tendon sheath was also investigated by Gotoh et al. turning out in a schematic representation of the tendon sheath and the fascicles (Gotoh et al., 1997). The influence of gentamicin application was investigated in a rat MCL model (Vente et al., 1997). It was found that the interfibrillar bonds were weakened by this treatment, facilitating the isolation of collagen fibrils. In contrast to other treatments such as phosphate-buffered saline resulting in totally broken fibrils, gentamicin application resulted in both broken and intact fibrils, enabling the analysis of the intact fibrils from healing tendons by TEM.

In a SEM study using lubricin knockout mice and wild-type mice, the role of lubricin was studied in terms of gliding resistance in intrasynovial tendons (Hayashi et al., 2013). It was clearly found that the lubricin knockout group had significantly higher forces to be applied in order to fulfill the 50 degree excursion (flexion/tension cycle). SEM images supported these biomechanical findings; the surface of lubricin knockout tendons was rougher compared with the wild type.

Using rat AT, the reorganization of the collagen fibers during the remodeling phase of tendon healing was nicely imaged by Sasaki et al. (2012). After removing the cells, collagen fiber organization was envisioned by SEM at 4, 7, 14, 28, and 56 days post injury, respectively (Fig. 5.13).

Collagen fiber development was studied by SEM und X-ray in fetal rat, bovine, and feline tendons, as well as in mature rat tendons (Provenzano and Vanderby, 2006). It was found that at early stages of development, the collagen fibrils overlapped and interweaved with the fibrils having apparent ends. Later in the development, fibrils still showed to have ends. However, in mature tendons and ligaments, no fibril ends were found, although extensive microscopic analysis was undertaken including imaging fibrils in individual micrographs and consecutive overlaid micrographs. This result was interpreted in terms of force transmission; fibrils in mature tendons and ligaments are functionally continuous; therefore, force within these tissues was suggested to be transmitted directly through collagen fibrils and not through interfibrillar cross-links such as proteoglycans. Not only was the force transmission within

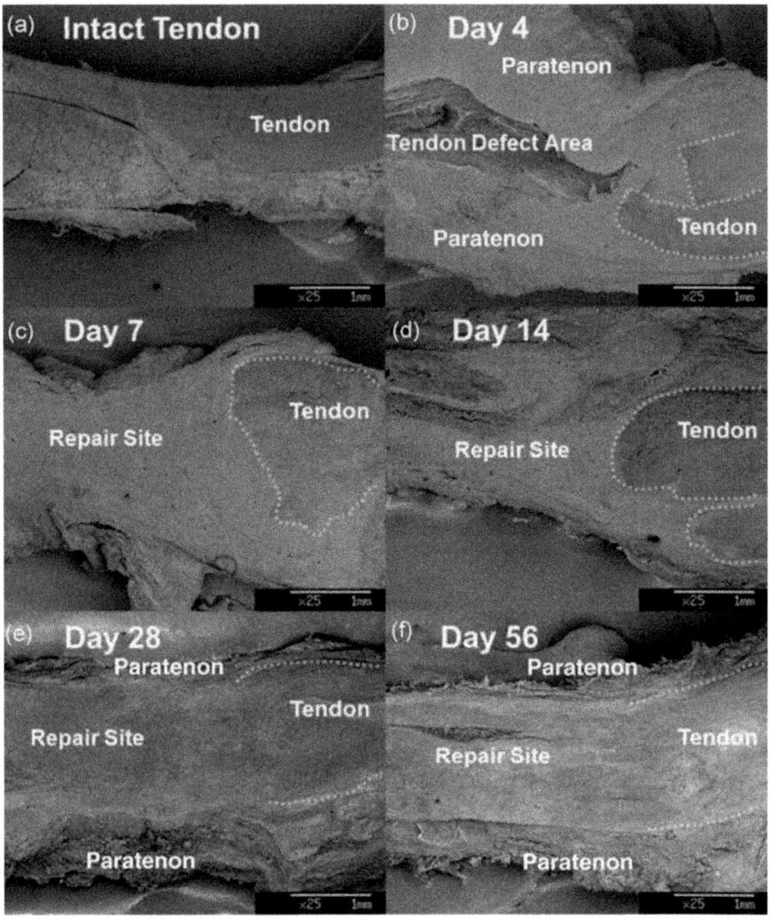


**Fig. 5.13** SEM longitudinal section views of a whole rat Achilles tendon (AT), (A) intact tendon, (B) day 4, (C) day 7, (D) day 14, (E) day 28, and (F) day 56 post injury. The repair site developed on day 7 and grew to become tendon-like tissue by day 56.
From Sasaki, K., Yamamoto, N., Kiyosawa, T., Sekido, M., 2012. The role of collagen arrangement change during tendon healing demonstrated by scanning electron microscopy. J. Electron Microsc. (Tokyo) 61, 327–334, by permission of Oxford University Press.

tendon tissue elucidated by SEM but also at the tendon-muscle junction. Force transmission between tendon tissue and muscle fibers of *anterior and posterior latissimus dorsi* (ALD and PLD, respectively) was studied in a SEM and TEM imaging-based survey in a chicken model (Trotter and Baca, 1987). It was found that the ends of ALD and PLD fibers were structurally different in some respect; however, they were similar with respect to their surface specialization, which was attributed to force transmission. The "fast" fibers of the PLD had 40% more surface area devoted to force transmission compared with the "slow" PLD fibers.

## 5.6 X-ray diffraction, atomic force, and second harmonic generation microscopy

### 5.6.1 Overview

X-ray is mostly used in form of X-ray diffraction (X-ray-D) to analyze the structures of crystals. The atoms in the crystal thereby cause a beam of X-rays to diffract into various directions, which are specific for the crystal under view. Angles and intensities of diffracted X-rays allow the reconstruction of a 3-D image. In the analysis of tendons and ligaments, this analytic tool is rather rarely used; however, a few studies are mentioned—dealing, for example, with the collagen molecule orientation within the tendon tissue (Table 5.6).

### 5.6.2 X-ray for diagnosis after injury

X-ray was applied to measure the distance between two tendon stumps after a rabbit AT tenotomy with or without a suture and with or without a cast (Nystrom and Holmlund, 1983). For accurate X-ray monitoring, two steel markers had to be fixed at the ends. Of all combinations, minimal gap formation was found to be in case of immediately suturing and a cast for at least 1 week.

### 5.6.3 X-ray-D for investigation of structural and molecular changes in function of age and diseases

X-ray-D was used to show age-related and diabetes-related changes in the collagen packing of tendons and fibril diameter, respectively (Naresh and Brodsky, 1992; Birch et al., 1999; Edwards et al., 2005). The molecular packing of the fibrils changed with age, going along with a reported increase in Maillard products (nonenzymatic addition of reducing sugar and dicarbonyl moieties) and the sugar-derived cross-links (Naresh and Brodsky, 1992). Moreover, the changes in the X-ray parameters followed those seen for rat tail tendons subjected to in vitro glycosylation. In an earlier study (Fujimori, 1989), it had been shown that glycation leading to cross-links between collagen molecules had been 10 times as effective using the sugar fructose compared with glucose. Such reports were important to allow further studies gathering insight into the cross-linking by AGEs (Li et al., 2013; Fessel et al., 2014).

### 5.6.4 X-ray-D for investigation of molecular processes at collage level

Another X-ray-D study on rat tail tendon collagen shed light onto the impact of the temperature on the water content of the tendon and on the distribution of the water (Price et al., 1997). X-ray patterns at room temperature were compared with patterns found for frozen tendons (liquid nitrogen, $N_2$). By comparing the corresponding patterns, it was suggested that there were two components of water, one being mobile and

**Table 5.6 Overview on X-ray, X-ray diffraction, AFM, and SHG microscopy studies presented in this chapter, grouped by thematic subchapters**

| Aimed T&L | Technical features | Medical findings | Lit. |
|---|---|---|---|
| Rabbit AT | X-ray | Gap formation after injury | Nystrom and Holmlund (1983) |
| Hum/ equine ext +flex | X-ray-D/ SAXS | Molec. processes vs age and diseases | Naresh and Brodsky (1992) and Fessel et al. (2014) |
| Rat tail | X-ray-D/ XANES-PEEM | Molecular processes in collagen fibrils | Price et al. (1997) and Lam et al. (2012) |
| Rabbit AT | AFM/SHG | Viscoelasticity by nonfibrillic ECM | Kahn et al. (2013) |
| Rat tail T | AFM | Collagen architecture | Baldwin et al. (2014) and Venturoni et al. (2003) |
| Hum PT/rat tail T | AFM | Molecular bonds in tendon biomech. | Svensson et al. (2013) |
| Rat AT | AFM | Ultrastructure/adh. forces in fibrils | Lee et al. (2011, 2013) |
| Rat tail T/ AT | AFM/SEM | Ultrastructure in fibrils/ECM | Raspanti et al. (1997, 2002, 2005) |
| Rat tail T | SHG | Orientation of fibrils | Stoller et al. (2002) |
| Horse SDFT | SHG | Fiber organization after injury | Sivaguru et al. (2010, 2014) |
| Rat AT | SHG | Invest. of fiber alignment after overuse | Abraham et al. (2011) |
| Mouse tail/ AT | SHG/FM | Backscattering effects | Shen et al. (2013) |
| Hum skin/ rat tail T | SHG-cc | Orientation of fibrils | Lo et al. (2005) |
| Chicken T | PS-OCT | Structure of proteoglycans | Yang et al. (2012) |

*T&L*, tendon and ligaments; *Hum*, human; *AT*, Achilles tendon; *PT*, patellar tendon; *SDFT*, superficial digital flexor tendons. Technical features, *X-ray-D*, X-ray diffraction; *SAXS*, small-angle X-ray scattering; *XANES-PEEM*, X-ray absorption near-edge structure and photoemission electron microscopy; *AFM*, atomic force microscopy; *SHG*, second-harmonic generation; *SEM*, scanning electron microscopy; *FM*, fluorescence microscopy; *SHG-cc*, SHG-collar correlation; *OCT*, optical coherence tomography; *PS-OCT*, polarization-sensitive optical coherence tomography; *ECM*, extracellular matrix.

diffusing from the intermolecular to the interfibrillar space upon freezing and one being "bound," which stayed in the intermolecular space. As such, water motion upon freezing resembles water motion upon tendon loading as discussed in the MRI section where the ADC of water was discussed as a function of applied load (Wellen et al.,

2005). Also AFM studies (see below) have been performed to assess the effects of tissue hydration on the structure and mechanics of individual collagen type I fibrils (Kemp et al., 2012).

X-ray absorption near-edge structure and photoemission electron microscopy (XANES-PEEM) was applied to assess the relative positions of carbonyl bonds (C=O) and amino functional groups ($NH_2$) in collagen type I (Lam et al., 2012). As such, the authors were able to show the molecular orientation of the collagen and therefore established a reliable tool to characterize the chemical composition and structural organization at the nanoscale level (Fig. 5.14).

## 5.6.5 Ultrastructural investigations of collagen fibril with AFM

The principle of AFM is scanning the surface of a thin sample by a piezoelectric element, where locally electric charges are evoked under pressure. In a combinatory imaging assessment with AFM and second-harmonic generation (SHG) microscopy, tensile testing results of cyclically loading and unloading of rabbit ATs were interpreted in terms of viscoelastic behavior (Kahn et al., 2013). Based upon fibril diameter assessment and tissue mechanics of the fibrils and the ECM by AFM (Fig. 5.15), it was found that not the fibrils, but the nonfibril ECM consisting of water and proteoglycans, among others, which form together a hydrophilic gel and represent the viscoelastic properties of the tendon. At low strain (<0.1), the interfibrillar material was concluded to be the predominant material acting viscoelastically.

Modulus variation along single rat tail fibrils was assessed by AFM high-resolution modulus (Baldwin et al., 2014). It was shown that subtle changes in molecular

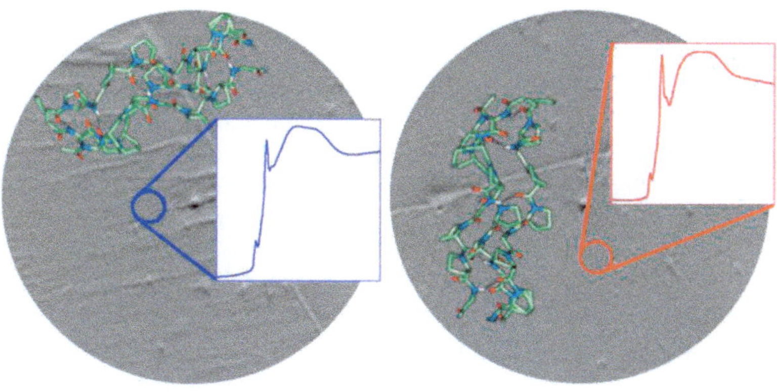

**Fig. 5.14** By changing the position, X-ray absorption near-edge structure and photoemission electron microscopy (XANES-PEEM) allows to assess relative positions of collagen molecules within rat tail tendon tissue.
Reprinted with permission from Lam, R.S., Metzler, R.A., Gilbert, P.U., Beniash, E., 2012. Anisotropy of chemical bonds in collagen molecules studied by X-ray absorption near-edge structure (XANES) spectroscopy. ACS Chem. Biol. 7, 476–480, copyright 2012, American Chemical Society.

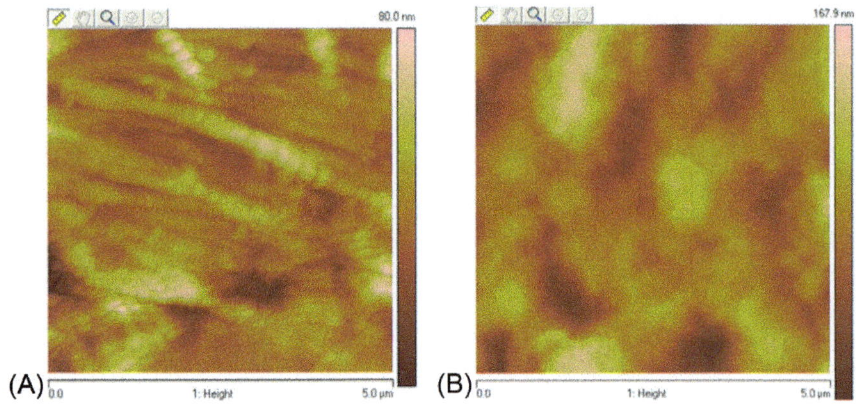


**Fig. 5.15** Atomic force microscopy (AFM) images of rabbit Achilles tendon (AT) of longitudinal (A) and transverse (B) section. AFM allows determining Young's modulus of different structures; for each point on the image, a force-displacement curve is obtained, and the modulus can be calculated locally. In the longitudinal sections, AT exhibited two different Young's moduli, 3.0 GPa (fibrils) and 223 MPa (amorphous structures between the fibrils, ground substance including GAG, proteoglycan, and glycoprotein). In the cross sections, the corresponding values were 2.5 and 372 MPa.
From Kahn, C.J.F., Dumas, D., Arab-Tehrany, E., Marie, V., Tran, N., Wang, X., Cleymand, F., 2013. Structural and mechanical multi-scale characterization of white New-Zealand rabbit Achilles tendon. J. Mech. Behav. Biomed. Mater. 26, 81–89, reprinted from Journal of the Mechanical Behavior of Biomedical Materials, Copyright 2013, with permission from Elsevier.

dynamics and fibrillar architecture occurred, which were caused by elevation of the temperature to 50 or 62°C.

Another AFM study using collagen fibers of rat tail tendons under different conditions of pH and ionic strength elucidated the periodicity of the collagen type I. The typical periodicity is 67 nm, but in the corresponding study, two new periodicities were detected, at 23 nm and around 210 nm (Venturoni et al., 2003).

In order to test the translation of molecular bonds into tendon mechanics, stress-strain curves of individual collagen fibrils extracted from human PT, rat tail tendon (either or not reduced with $NaBH_4$ to reduce cross-links), and rat tail tendons of diabetic fat rats were assessed by AFM (Svensson et al., 2013). The three-phase stress-strain curves showed that there was no alteration whether the cross-links in the rat tail tendons were reduced or not, although it is widely believed that cross-linking in collagen is a major prerequisite for the mechanical strength of the tendon (Fessel and Snedeker, 2009). However, when human PT was compared with rat tail tendon's fracture mechanics, totally different stress-strain curves were determined; rat tail tendon fibrils exhibited a longer plateau in the high-strain region, whereas human PT fibrils had a significant rise in stress and modulus. Therefore, the authors concluded that comparing results across different species—even on the fibril level—has to be considered carefully (Svensson et al., 2013). As for the diabetic rat tail tendon fibrils, they exhibited less strain with similar failure stress as the normal rat tail tendons.

In two subsequent studies, Lee et al. used AFM to elucidate changes in the ultrastructure and adhesion force of collagen fibrils in AT having tendinitis in a collagenase-induced rat model (Lee et al., 2011). As for the adhesion force, it decreased in the tendinitis model compared with noninjured tendons until day 5 post-collagenase application and then increased again (but not to the original level), going along with a decreased diameter of collagen fibrils. While the fibril diameter in healed tendons 9 weeks post collagenase was similar to those of uninjured tendons, the adhesion forces remained different (Lee et al., 2013). AFM as such allowed insight into the healing mechanisms of tendinitis.

Also, structural studies based on AFM were performed quite extensively; not only were the slender filaments on the surface of collagen fibrils determined to represent the side chain of decorin, the most common proteoglycan in the tissue (Raspanti et al., 1997), but also other aspects concerning the ECM such as the interaction of small proteoglycans with collagen fibrils with or without chondroitinase digestion were assessed (Raspanti et al., 2002). The irregular zigzag planar course of the bundles of collagen fibers resulting in crimps was furthermore investigated in the rat AT (Raspanti et al., 2005).

### 5.6.6 Investigations with SHG

The principle of laser SHG analysis is the excitation of electrons by a laser beam of light where the emitted photons with the same frequency (caused by the falling back of the electrons to the ground state of the corresponding atoms) interact with a birefringent material, and upon doing so, the frequency of two such electrons is combined and doubled, while the wavelength attributed to this new photon is halved. As such, imaging of this nonlinear optical process results in nicely contrasted images. Using a 100 fs laser with a rat tail tendon under view, Stoller et al. were able to show that the SHG backscattering was polarization angle-dependent (Stoller et al., 2002). They developed a continuous scanning technique to measure the tendon tissue without damaging the sample. As such, the orientation of the fibrils was investigated and also the $\gamma$ parameter, which is a key ratio of the elements of the nonlinear susceptibility tensor. As the structure and orientation of the fibrils had been examined by many other imaging approaches, for example, SEM and TEM (see above), this biological sample was used to verify the usefulness of laser SHG for the same analytic question.

Other studies using SHG investigated the collagen fiber organization in intact and injured tendons (Sivaguru et al., 2010) or the local mechanical environment of the cells (Sivaguru et al., 2014). In the latter study, the authors claimed that their SHG data could be used for practical suggestions about the load that should be applied postoperatively to injured tendons in order to promote healing. Moreover, early Achilles tendinosis was analyzed by SHG in response to in vivo mechanical loading (Abraham et al., 2011). An association between abnormal tenocyte morphology and changes in the fibrillar collagen matrix of the mechanically loaded rat ATs was found (Fig. 5.16). In tendinosis tendons, the longitudinal alignment of collagen fibers was reduced, which was associated with reduced signal intensity in SHG (Fig. 5.16A). When SHG signal intensity was plotted as a function of position along a line drawn perpendicular to the collagen fiber

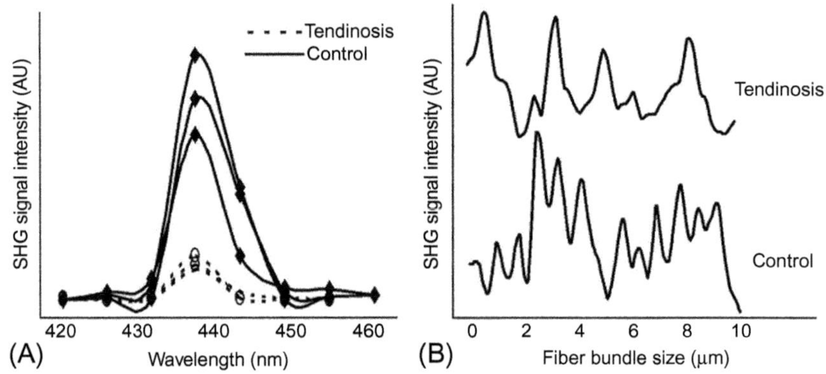


**Fig. 5.16** Analysis of second-harmonic generation (SHG) signal intensity in rat AT from collagen surrounding normal or abnormal tenocytes. (A) Longitudinal alignment of collagen fibers, wavelength. (B) Fiber bundle size.
From Abraham, T., Fong, G., Scott, A., 2011. Second-harmonic generation analysis of early Achilles tendinosis in response to in vivo mechanical loading. BMC Musculoskelet. Disord. 12, 26, open access article.

bundles, the fiber bundle size could be estimated with submicron resolution in normal tendon, but not in tendinosis tendons (Fig. 5.16B).

The backscattering of SHG in mouse tail tendon and mouse AT tissue was determined with confocal multiphoton microscopy in an attempt to optimize and compare signal intensities for different tendon tissues (Shen et al., 2013). Moreover, one aim was to understand the mechanisms behind backscattering SHG. As such, this method paper gave supporting information for the investigation of the generation mechanism of SHG. Another study intended at the optimization of SHG measurement used rat tail tendons and found that the application of a correction collar increased the maximum of the detected multiphoton signal by 57%. The correction collar consisted of a cover glass with a water immersion objective; not only did it improve the multiphoton signal gathered for rat tail tendon but also skin samples were examined where the signal intensity was improved by 36% (Lo et al., 2005).

## 5.6.7 Investigations with OCT

Finally, optical coherence tomography (OCT) was also used to study tendons, in particular the influence of the proteoglycans on the structural stability of the tendons (Yang et al., 2012). OCT captures micrometer-resolution 3-D images by low-coherence interferometry typically employing near-infrared light. In contrast to CM, OCT allows penetration quite deeply into the scattering specimen, however, with a lower resolution than CM.

## 5.7 Summary

Depending on the focus of imaging tendons and ligaments, techniques covering a wide range of resolution are currently available. In this chapter, methods assessing tendon and ligament anatomy and degenerative tissue are presented, among them ultrasound and MRI. Furthermore, basic research dealing with the structure of those tissues demands for higher resolution. For that purpose and to analyze histological sections, light microscopy and FM are imaging methods widely used and easy to handle. Going to higher resolutions, surface characteristics of collagen fibers are nicely imaged by SEM and SHG microscopy, and on the molecular level, tendon cells and ECM composition can be accurately analyzed with TEM. In this chapter, many preclinical and clinical studies using these methods and techniques are reported, and in addition, exemplary research questions solved with these techniques are presented.

## References

Abraham, T., Fong, G., Scott, A., 2011. Second harmonic generation analysis of early Achilles tendinosis in response to in vivo mechanical loading. BMC Musculoskelet. Disord. 12, 26.

Abrahamsson, S.O., Lundborg, G., Lohmander, L.S., 1992. Restoration of the injured flexor tendon surface: a possible role for endotenon cells. A morphological study of the rabbit tendon in vivo. J. Hand Surg. Br. 17, 553–560.

Almeida, M.D.S.D., Aro, A.A.D., Guerra, F.D.R., Vieira, C.P., Vidal, B.D.C., Pimentel, E.R., 2012. Electroacupuncture increases the concentration and organization of collagen in a tendon healing model in rats. Connect. Tissue Res. 53, 542–547.

Arnoczky, S.P., Lavagnino, M., Whallon, J.H., Hoonjan, A., 2002. In situ cell nucleus deformation in tendons under tensile load; a morphological analysis using confocal laser microscopy. J. Orthop. Res. 20, 29–35.

Asperio, R.M., Marzola, P., Sbarbati, A., Osculati, F., Addis, F., 2000. Comparison of results of scanning electron microscopy and magnetic resonance imaging before and after administration of a radiographic contrast agent in the tendon of the deep digital flexor muscle obtained from horse cadavers. Am. J. Vet. Res. 61, 321–325.

Avella, C.S., Ely, E.R., Verheyen, K.L.P., Price, J.S., Wood, J.L.N., Smith, R.K.W., 2009. Ultrasonographic assessment of the superficial digital flexor tendons of National Hunt racehorses in training over two racing seasons. Equine Vet. J. 41, 449–454.

Baldwin, S.J., Quigley, A.S., Clegg, C., Kreplak, L., 2014. Nanomechanical mapping of hydrated rat tail tendon collagen I fibrils. Biophys. J. 107, 1794–1801.

Baums, M.H., Spahn, G., Buchhorn, G.H., Schultz, W., Hofmann, L., Klinger, H.M., 2012. Biomechanical and magnetic resonance imaging evaluation of a single- and double-row rotator cuff repair in an in vivo sheep model. Arthroscopy 28, 769–777.

Bertuglia, A., Mollo, G., Bullone, M., Riccio, B., 2014. Identification of surgically-induced longitudinal lesions of the equine deep digital flexor tendon in the digital flexor tendon sheath using contrast-enhanced ultrasonography: an ex-vivo pilot study. Acta Vet. Scand. 56, 78.

Birch, H.L., Bailey, J.V., Bailey, A.J., Goodship, A.E., 1999. Age-related changes to the molecular and cellular components of equine flexor tendons. Equine Vet. J. 31, 391–396.

Blunden, A., Murray, R., Dyson, S., 2009. Lesions of the deep digital flexor tendon in the digit: a correlative MRI and post mortem study in control and lame horses. Equine Vet. J. 41, 25–33.

Branford, O.A., Lee, D.A., Bader, D.L., Grobbelaar, A.O., 2012. The mechanics of flexor tendon adhesions. J. Hand Surg. Eur. Vol. 37E, 555–563.

Buschmann, J., Meier Buergisser, G., Bonavoglia, E., Neuenschwander, P., Milleret, V., Giovanoli, P., Calcagni, M., 2013. Cellular response of healing tissue to DegraPol tube implantation in rabbit Achilles tendon rupture repair: an in vivo histomorphometric study. J. Tissue Eng. Regen. Med. 7, 413–420.

Buschmann, J., Puippe, G., Meier Buergisser, G., Bonavoglia, E., Giovanoli, P., Calcagni, M., 2014. Correspondence of high-frequency ultrasound and histomorphometry of healing rabbit Achilles tendon tissue. Connect. Tissue Res. 55, 123–131.

Buschmann, J., Calcagni, M., Meier Buergisser, G., Bonavoglia, E., Neuenschwander, P., Milleret, V., Giovanoli, P., 2015. Synthesis, characterization and histomorphometric analysis of cellular response to a new elastic DegraPol® polymer for rabbit Achilles tendon rupture repair. J. Tissue Eng. Regen. Med. 9, 584–594.

Carvalho, H.F.D., Lino Neto, J., Taboga, S.R., 1994. Microfibrils: neglected components of pressure-bearing tendons. Ann. Anat. 176, 155–159.

Chamberlain, C.S., Duenwald-Kuehl, S.E., Okotie, G., Brounts, S.H., Baer, G.S., Vanderby, R., 2013. Temporal healing in rat Achilles tendon: ultrasound correlations. Ann. Biomed. Eng. 41, 477–487.

Chen, S.H., Chen, C.H., Shalumon, K., Chen, J.P., 2014. Preparation and characterization of antiadhesion barrier film from hyaluronic acid-grafted electrospun poly(caprolactone) nanofibrous membranes for prevention of flexor tendon postoperative peritendinous adhesion. Int. J. Nanomedicine 9, 4079–4092.

Couppe, C., Svensson, R.B., Sodring-Elbrond, V., Hansen, P., Kjaer, M., Magnusson, S.P., 2014. Accuracy of MRI technique in measuring tendon cross-sectional area. Clin. Physiol. Funct. Imaging 34, 237–241.

Denoix, J.M., Busoni, V., 1999. Ultrasonographic anatomy of the accessory ligament of the superficial digital flexor tendon in horses. Equine Vet. J. 31, 186–191.

Dillehay, G.L., Deschler, T., Rogers, L.F., Neiman, H.L., Hendrix, R.W., 1984. The ultrasonographic characterization of tendons. Invest. Radiol. 19, 338–341.

Docking, S.I., Daffy, J., Van Schie, H.T., Cook, J.L., 2012. Tendon structure changes after maximal exercise in the Thoroughbred horse: use of ultrasound tissue characterisation to detect in vivo tendon response. Vet. J. 194, 338–342.

Doherty, G.P., Koike, Y., Uhthoff, H.K., Lecompte, M., Trudel, G., 2006. Comparative anatomy of rabbit and human Achilles tendons with magnetic resonance and ultrasound imaging. Comp. Med. 56, 68–74.

Dresner, M.A., Rose, G.H., Rossman, P.J., Muthupillai, R., Manduca, A., Ehman, R.L., 2001. Magnetic resonance elastography of skeletal muscle. J. Magn. Reson. Imaging 13, 269–276.

Dyson, S.J., Murray, R., Schramme, M.C., 2005. Lameness associated with foot pain: results of magnetic resonance imaging in 199 horses (January 2001–December 2003) and response to treatment. Equine Vet. J. 37, 113–121.

Edwards, L.J., Goodship, A.E., Birch, H.L., Patterson-Kane, J.C., 2005. Effect of exercise on age-related changes in collagen fibril diameter distributions in the common digital extensor tendons of young horses. Am. J. Vet. Res. 66, 564–568.

Ehrlich, H.P., Lambert, P.A., Saggers, G.C., Myers, R.L., Hauck, R.M., 2005. Dynamic changes appearing in collagen fibers during intrinsic tendon repair. Ann. Plast. Surg. 54, 201–206.

Elliot, D., Giesen, T., 2013. Primary flexor tendon surgery: the search for a perfect result. Hand Clin. 29, 191–206.

Fessel, G., Snedeker, J.G., 2009. Evidence against proteoglycan mediated collagen fibril load transmission and dynamic viscoelasticity in tendon. Matrix Biol. 28, 503–510.

Fessel, G., Li, Y., Diederich, V., Guizar-Sicairos, M., Schneider, P., Sell, D.R., Monnier, V.M., Snedeker, J.G., 2014. Advanced glycation end-products reduce collagen molecular sliding to affect collagen fibril damage mechanisms but not stiffness. PLoS One 9, e110948.

Franchi, M., Fini, M., Quaranta, M., De Pasquale, V., Raspanti, M., Giavaresi, G., Ottani, V., Ruggeri, A., 2007. Crimp morphology in relaxed and stretched rat Achilles tendon. J. Anat. 210, 1–7.

Fujimori, E., 1989. Cross-linking and fluorescence changes of collagen by glycation and oxidation. Biochim. Biophys. Acta 998, 105–110.

Gehmert, S., Jung, E.M., Kügler, T., Klein, S., Gehmert, S., Zeitler, K., Loibl, M., Prantl, L., 2012. Sonoelastography can be used to monitor the restoration of Achilles tendon elasticity after injury. Ultraschall Med. 33, 581–586.

Gonzalez Santander, R., Plasencia Arriba, M.A., Martinez Cuadrado, G., Lopez Alonso, A., Gonzalez-Santander Martinez, M., Martinez Alonso, F.J., Monteagudo, M., Toledo Lobo, M.V., 1999. Intracellular biogenesis of collagen fibrils in 'activated fibroblasts' of tendo Achillis. An ultrastructural study in the New Zealand rabbit. J. Bone Joint Surg. Br. 81, 522–530.

Goodman, H.J., Choueka, J., 2005. Biomechanics of the flexor tendons. Hand Clin. 21, 129–149.

Gotoh, T., Murashige, N., Yamashita, K., 1997. Ultrastructural observations on the tendon sheath of the rat tail. J. Electron Microsc. (Tokyo) 46, 247–252.

Han, S., Gemmell, S.J., Helmer, K.G., Grigg, P., Wellen, J.W., Hoffman, A.H., Sotak, C.H., 2000. Changes in ADC caused by tensile loading of rabbit Achilles tendon: evidence for water transport. J. Magn. Reson. 144, 217–227.

Hayashi, M., Zhao, C., Thoreson, A.R., Chikenji, T., Jay, G.D., An, K.N., Amadio, P.C., 2013. The effect of lubricin on the gliding resistance of mouse intrasynovial tendon. PLoS One 8, e83836.

Hu, J.Z., Zhou, Y.C., Huang, L.H., Lu, H.B., 2013. Development of biodegradable polycaprolactone film as an internal fixation material to enhance tendon repair: an in vitro study. BMC Musculoskelet. Disord. 14, 246.

Hung, L.K., Fu, S.C., Lee, Y.W., Mok, T.Y., Chan, K.M., 2013. Local vitamin-C injection reduced tendon adhesion in a chicken model of flexor digitorum profundus tendon injury. J. Bone Joint Surg. Am. 95, e41.

James, R., Kesturu, G., Balian, G., Chhabra, A.B., 2008. Tendon: biology, biomechanics, repair, growth factors, and evolving treatment options. J. Hand Surg. 33, 102–112.

Kahn, C.J.F., Dumas, D., Arab-Tehrany, E., Marie, V., Tran, N., Wang, X., Cleymand, F., 2013. Structural and mechanical multi-scale characterization of white New-Zealand rabbit Achilles tendon. J. Mech. Behav. Biomed. Mater. 26, 81–89.

Kasashima, Y., Kuwano, A., Katayama, Y., Taura, Y., Yoshihara, T., 2002. Magnetic resonance imaging application to live horse for diagnosis of tendinitis. J. Vet. Med. Sci. 64, 577–582.

Kemp, A.D., Harding, C.C., Cabral, W.A., Marini, J.C., Wallace, J.M., 2012. Effects of tissue hydration on nanoscale structural morphology and mechanics of individual Type I collagen fibrils in the Brtl mouse model of Osteogenesis Imperfecta. J. Struct. Biol. 180, 428–438.

Kennedy, J.C., Roth, J.H., Mendenhall, H.V., Sanford, J.B., 1980. Presidential address: intraarticular replacement in the anterior cruciate ligament-deficient knee. Am. J. Sports Med. 8, 1–8.

Kobayashi, H., Vanderby, R., 2005. New strain energy function for acoustoelastic analysis of dilatational waves in nearly incompressible, hyper-elastic materials. J. Appl. Mech. 72, 843–851.

Kobayashi, H., Vanderby, R., 2007. Acoustoelastic analysis of reflected waves in nearly incompressible, hyper-elastic materials: forward and inverse problems. J. Acoust. Soc. Am. 121, 879–887.

Koff, M.F., Pownder, S.L., Shah, P.H., Yang, L.W., Potter, H.G., 2014. Ultrashort echo imaging of cyclically loaded rabbit patellar tendon. J. Biomech. 47, 3428–3432.

Krishnan, L., Hoying, J.B., Nguyen, H., Song, H., Weiss, J.A., 2007. Interaction of angiogenic microvessels with the extracellular matrix. Am. J. Physiol. Heart. Circ. Physiol. 293, H3650–H3658.

Kristoffersen, M., Ohberg, L., Johnston, C., Alfredson, H., 2005. Neovascularisation in chronic tendon injuries detected with colour Doppler ultrasound in horse and man: implications for research and treatment. Knee Surg. Sports Traumatol. Arthrosc. 13, 505–508.

Kuo, P.-L., Li, P.-C., Shun, C.-T., Lai, J.-S., 1999. Strain measurements of rabbit Achilles tendons by ultrasound. Ultrasound Med. Biol. 25, 1241–1250.

Lam, R.S., Metzler, R.A., Gilbert, P.U., Beniash, E., 2012. Anisotropy of chemical bonds in collagen molecules studied by X-ray absorption near-edge structure (XANES) spectroscopy. ACS Chem. Biol. 7, 476–480.

Langevin, H.M., Bouffard, N.A., Badger, G.J., Churchill, D.L., Howe, A.K., 2006. Subcutaneous tissue fibroblast cytoskeletal remodeling induced by acupuncture: evidence for a mechanotransduction-based mechanism. J. Cell. Physiol. 207, 767–774.

Lavagnino, M., Arnoczky, S.P., Gardner, K., 2011. In situ deflection of tendon cell-cilia in response to tensile loading: an in vitro study. J. Orthop. Res. 29, 925–930.

Lavagnino, M., Gardner, K., Sedlak, A.M., Arnoczky, S.P., 2013. Tendon cell ciliary length as a biomarker of in situ cytoskeletal tensional homeostasis. Muscles Ligaments Tendons J. 3, 118–121.

Lee, G.J., Choi, S., Chon, J., Yoo, S., Cho, I., Park, H.K., 2011. Changes in collagen fibril pattern and adhesion force with collagenase-induced injury in rat Achilles tendon observed via AFM. J. Nanosci. Nanotechnol. 11, 773–777.

Lee, G.J., Choi, S., Chon, J., Yoo, S.D., Kim, H.S., Park, H.K., Chung, J.H., 2013. Nanostructural and nanomechanical responses of collagen fibrils in the collagenase-induced Achilles tendinitis rat model. J. Nanosci. Nanotechnol. 13, 7279–7286.

Legerlotz, K., Dorn, J., Richter, J., Rausch, M., Leupin, O., 2014. Age-dependent regulation of tendon crimp structure, cell length and gap width with strain. Acta Biomater. 10, 4447–4455.

Li, Y., Fessel, G., Georgiadis, M., Snedeker, J.G., 2013. Advanced glycation end-products diminish tendon collagen fiber sliding. Matrix Biol. 32, 169–177.

Lo, I.K., Marchuk, L.L., Leatherbarrow, K.E., Frank, C.B., Hart, D.A., 2004. Collagen fibrillogenesis and mRNA levels in the maturing rabbit medial collateral ligament and patellar tendon. Connect. Tissue Res. 45, 11–22.

Lo, W., Sun, Y., Lin, S.J., Jee, S.H., Dong, C.Y., 2005. Spherical aberration correction in multiphoton fluorescence imaging using objective correction collar. J. Biomed. Opt. 10, 034006.

Lundborg, G., Hansson, H.A., Rank, F., Rydevik, B., 1980. Superficial repair of severed flexor tendons in synovial environment. An experimental, ultrastructural study on cellular mechanisms. J. Hand Surg. Am. 5, 451–461.

Maeda, E., Ye, S., Wang, W., Bader, D.L., Knight, M.M., Lee, D.A., 2012. Gap junction permeability between tenocytes within tendon fascicles is suppressed by tensile loading. Biomech. Model. Mechanobiol. 11, 439–447.

Maner, M.B., Marsh, M.J., 1981. Ultrasonic findings in a ruptured Achilles tendon. Med. Ultrasound 5, 46–82.

Marsh, C.A., Schneider, R.K., Sampson, S.N., Roberts, G.D., 2012. Response to injection of the navicular bursa with corticosteroid and hyaluronan following high-field magnetic resonance imaging in horses with signs of navicular syndrome: 101 cases (2000–2008). J. Am. Vet. Med. Assoc. 241, 1353–1364.

Martinoli, C., Derchi, L.E., Pastorino, C., Bertolotto, M., Silvestri, E., 1993. Analysis of echotexture of tendons with US. Radiology 186, 839–843.

Meier Buergisser, G., Calcagni, M., Müller, A., Bonavoglia, E., Fessel, G., Snedeker, J.G., Giovanoli, P., Buschmann, J., 2014. Prevention of peritendinous adhesions using an electrospun DegraPol polymer tube: a histological, ultrasonographic, and biomechanical study in rabbits. Biomed. Res. Int. 2014, 656240. 11 pages.

Milner, P.I., Sidwell, S., Talbot, A.M., Clegg, P.D., 2012. Short-term temporal alterations in magnetic resonance signal occur in primary lesions identified in the deep digital flexor tendon of the equine digit. Equine Vet. J. 44, 157–162.

Miltner, O., Schwaiger, A., Schmidt, C., Bucker, A., Kolker, C., Siebert, C.H., Zilkens, K.W., Niethard, F.U., Blumich, B., 2003. Portable NMR-MOUSE: a new method and its evaluation of the Achilles tendon. Z. Orthop. Ihre Grenzgeb. 141, 148–152.

Miyazaki, H., Hasegawa, Y., Hayashi, K., 2000. A newly designed tensile tester for cells and its application to fibroblasts. J. Biomech. 33, 97–104.

Morgan, M., Kostyuk, O., Brown, R.A., Mudera, V., 2006. In situ monitoring of tendon structural changes by elastic scattering spectroscopy: correlation with changes in collagen fibril diameter and crimp. Tissue Eng. 12, 1821–1831.

Morizaki, Y., Vanhees, M., Thoreson, A.R., Larson, D., Zhao, C., An, K.N., Amadio, P.C., 2012. The response of the rabbit subsynovial connective tissue to a stress-relaxation test. J. Orthop. Res. 30, 443–447.

Moshiri, A., Oryan, A., 2011. Structural and functional modulation of early healing of full-thickness superficial digital flexor tendon rupture in rabbits by repeated subcutaneous administration of exogenous human recombinant basic fibroblast growth factor. J. Foot Ankle Surg. 50, 654–662.

Muellner, T., Kwasny, O., Reihsner, R., Loehnert, V., Schabus, R., 2000. Mechanical properties of a rat patellar tendon stress-shielded in situ. Arch. Orthop. Trauma Surg. 120, 70–74.

Muellner, T., Kwasny, O., Loehnert, V., Mallinger, R., Unfried, G., Schabus, R., Plenk Jr., H., 2001. Light and electron microscopic study of stress-shielding effects on rat patellar tendon. Arch. Orthop. Trauma Surg. 121, 561–565.

Murata, H., Nishizono, H., Miyoshi, M., 2000. Postnatal development of structure and arrangement of tendon cells. A scanning and transmission electron microscope study in the rat calcaneal tendon. Okajimas Folia Anat. Jpn. 77, 77–86.

Naresh, M.D., Brodsky, B., 1992. X-ray diffraction studies on human tendon show age-related changes in collagen packing. Biochim. Biophys. Acta 1122, 161–166.

Nystrom, B., Holmlund, D., 1983. Experimental evaluation of immobilization in operative and non-operative treatment of Achilles tendon rupture. A radiographic study in the rabbit. Acta Chir. Scand. 149, 669–673.

Pan, J., Liu, G.-M., Ning, L.-J., Zhang, Y., Luo, J.-C., Huang, F.-G., Qin, T.-W., 2015. Rotator cuff repair using a decellularized tendon slices graft: an in vivo study in a rabbit model. Knee Surg. Sports Traumatol. Arthrosc. 23, 1524–1535.

Peixinho, C.C., Martins, N.S., De Oliveira, L.F., Machado, J.C., 2014. Structural adaptations of rat lateral gastrocnemius muscle-tendon complex to a chronic stretching program and their quantification based on ultrasound biomicroscopy and optical microscopic images. Clin. Biomech. (Bristol, Avon) 29, 57–62.

Price, R.I., Lees, S., Kirschner, D.A., 1997. X-ray diffraction analysis of tendon collagen at ambient and cryogenic temperatures: role of hydration. Int. J. Biol. Macromol. 20, 23–33.

Provenzano, P.P., Vanderby Jr., R., 2006. Collagen fibril morphology and organization: implications for force transmission in ligament and tendon. Matrix Biol. 25, 71–84.

Rantanen, N.W., 1982. The use of diagnostic ultrasound in limb disorders of the horse: a preliminary report. J. Equine Vet. Sci. 2, 62–64.

Rantanen, N.W., Genovese, R.L., Gaines, R., 1983. The use of diagnostic ultrasound to detect structural damage to the soft tissues of the extremities of horses. J. Equine Vet. Sci. 3, 134–135.

Raspanti, M., Alessandrini, A., Ottani, V., Ruggeri, A., 1997. Direct visualization of collagen-bound proteoglycans by tapping-mode atomic force microscopy. J. Struct. Biol. 119, 118–122.

Raspanti, M., Congiu, T., Guizzardi, S., 2002. Structural aspects of the extracellular matrix of the tendon: an atomic force and scanning electron microscopy study. Arch. Histol. Cytol. 65, 37–43.

Raspanti, M., Manelli, A., Franchi, M., Ruggeri, A., 2005. The 3D structure of crimps in the rat Achilles tendon. Matrix Biol. 24, 503–507.

Raspanti, M., Reguzzoni, M., Protasoni, M., Martini, D., 2011. Evidence of a discrete axial structure in unimodal collagen fibrils. Biomacromolecules 12, 4344–4347.

Rigozzi, S., Muller, R., Snedeker, J.G., 2010. Collagen fibril morphology and mechanical properties of the Achilles tendon in two inbred mouse strains. J. Anat. 216, 724–731.

Sack, I., Bernarding, J., Braun, J., 2002. Analysis of wave patterns in MR elastography of skeletal muscle using coupled harmonic oscillator simulations. Magn. Reson. Imaging 20, 95–104.

Sahin, H., Tholema, N., Petersen, W., Raschke, M.J., Stange, R., 2012. Impaired biomechanical properties correlate with neoangiogenesis as well as VEGF and MMP-3 expression during rat patellar tendon healing. J. Orthop. Res. 30, 1952–1957.

Sasaki, K., Yamamoto, N., Kiyosawa, T., Sekido, M., 2012. The role of collagen arrangement change during tendon healing demonstrated by scanning electron microscopy. J. Electron Microsc. (Tokyo) 61, 327–334.

Schie, J.H.T.V., Bakker, E.M., Weeren, P.R.V., 1999. Ultrasonographic evaluation of equine tendons: a quantitative in vitro study of the effects of amplifier gain level, transducer-tilt, and transducer-displacement. Vet. Radiol. Ultrasound 40, 151–160.

Schie, H.T.V., Bakker, E.M., Jonker, A.M., Weeren, P.R.V., 2000. Ultrasonographic tissue characterization of equine superficial digital flexor tendons by means of gray level statistics. Am. J. Vet. Res. 61, 210–219.

Schie, H.T.V., Bakker, E.M., Jonker, A.M., Weeren, P.R.V., 2003. Computerized ultrasonographic tissue characterization of equine superficial digital flexor tendons by means of stability quantification of echo patterns in contiguous transverse ultrasonographic images. Am. J. Vet. Res. 64, 366–375.

Schwartz, A.G., Lipner, J.H., Pasteris, J.D., Genin, G.M., Thomopoulos, S., 2013. Muscle loading is necessary for the formation of a functional tendon enthesis. Bone 55, 44–51.

Scott, A., Khan, K.M., Heer, J., Cook, J.L., Lian, O., Duronio, V., 2005. High strain mechanical loading rapidly induces tendon apoptosis: an ex vivo rat tibialis anterior model. Br. J. Sports Med. 39, e25.

Screen, H.R., Lee, D.A., Bader, D.L., Shelton, J.C., 2003. Development of a technique to determine strains in tendons using the cell nuclei. Biorheology 40, 361–368.

Screen, H.R.C., Bader, D.L., Lee, D.A., Shelton, J.C., 2004. Local strain measurement within tendon. Strain 40, 157–163.

Sharma, P., Maffulli, N., 2006. Biology of tendon injury: healing, modeling and remodeling. J. Musculoskelet. Neuronal Interact. 6, 181–190.

Shen, M., Tian, Y., Chong, S.P., Zhao, J., Zeng, H., Tang, S., 2013. Quantifying the backscattering of second harmonic generation in tissues with confocal multiphoton microscopy. J. Biomed. Opt. 18, 115003.

Shepherd, J.H., Legerlotz, K., Demirci, T., Klemt, C., Riley, G.P., Screen, H.R., 2013. Functionally distinct tendon fascicles exhibit different creep and stress relaxation behaviour. Proc. Inst. Mech. Eng. H 228, 49–59.

Sivaguru, M., Durgam, S., Ambekar, R., Luedtke, D., Fried, G., Stewart, A., Toussaint Jr., K.C., 2010. Quantitative analysis of collagen fiber organization in injured tendons using Fourier transform-second harmonic generation imaging. Opt. Express 18, 24983–24993.

Sivaguru, M., Eichorst, J.P., Durgam, S., Fried, G.A., Stewart, A.A., Stewart, M.C., 2014. Imaging horse tendons using multimodal 2-photon microscopy. Methods 66, 256–267.

Sluss, J.R., Liberti, J.P., Jiranek, W.A., Wayne, J.S., Zuelzer, W.A., 2001. pN collagen type III within tendon grafts used for anterior cruciate ligament reconstruction. J. Orthop. Res. 19, 852–857.

Smith, K.L., Daniels, J.L., Arnoczky, S.P., Dodds, J.A., Cooper, T.G., Gottschalk, A., Shaw, D.A., 1994a. Effect of joint position and ligament tension on the MR signal intensity of the cruciate ligaments of the knee. J. Magn. Reson. Imaging 4, 819–822.

Smith, R.K., Jones, R., Webbon, P.M., 1994b. The cross-sectional areas of normal equine digital flexor tendons determined ultrasonographically. Equine Vet. J. 26, 460–465.

Snedeker, J.G., Pelled, G., Zilberman, Y., Gerhard, F., Muller, R., Gazit, D., 2006. Endoscopic cellular microscopy for in vivo biomechanical assessment of tendon function. J. Biomed. Opt. 11, 8.

Snellenberg, W.V., Wiley, J.P., Brunet, G., 2007. Achilles tendon pain intensity and level of neovascularization in athletes as determined by color Doppler ultrasound. Scand. J. Med. Sci. Sports 17, 530–534.

Spriet, M., Zwingenberger, A., 2009. Influence of the position of the foot on MRI signal in the deep digital flexor tendon and collateral ligaments of the distal interphalangeal joint in the standing horse. Equine Vet. J. 41, 498–503.

Squier, C.A., Bausch, W.H., 1984. Three-dimensional organization of fibroblasts and collagen fibrils in rat tail tendon. Cell Tissue Res. 238, 319–327.

Stange, R., Sahin, H., Wieskotter, B., Persigehl, T., Ring, J., Bremer, C., Raschke, M.J., Vieth, V., 2015. In vivo monitoring of angiogenesis during tendon repair: a novel MRI-based technique in a rat patellar tendon model. Knee Surg. Sports Traumatol. Arthrosc 23, 2433–2439.

Steinberg, P.J., Hodde, K.C., 1990. The morphology of synovial lining of various structures in several species as observed with scanning electron microscopy. Scanning Microsc. 4, 987–1019 (discussion 1019-20).

Stoller, P., Kim, B.M., Rubenchik, A.M., Reiser, K.M., Da Silva, L.B., 2002. Polarization-dependent optical second-harmonic imaging of a rat-tail tendon. J. Biomed. Opt. 7, 205–214.

Strickland, J.W., 2005. The scientific basis for advances in flexor tendon surgery. J. Hand Ther. 18, 94–110.

Sugaya, H., Maeda, K., Matsuki, K., Moriishi, J., 2005. Functional and structural outcome after arthroscopic full-thickness rotator cuff repair: single-row versus dual-row fixation. Arthroscopy J. Arthrosc. Relat. Surg. 21, 1307–1316.

Svensson, R.B., Mulder, H., Kovanen, V., Magnusson, S.P., 2013. Fracture mechanics of collagen fibrils: influence of natural cross-links. Biophys. J. 104, 2476–2484.
Trotter, J.A., Baca, J.M., 1987. A stereological comparison of the muscle-tendon junctions of fast and slow fibers in the chicken. Anat. Rec. 218, 256–266.
Uffmann, K., Maderwald, S., Ajaj, W., Galban, C.G., Mateiescu, S., Quick, H.H., Ladd, M.E., 2004. In vivo elasticity measurements of extremity skeletal muscle with MR elastography. NMR Biomed. 17, 181–190.
Vallance, S.A., Bell, R.J., Spriet, M., Kass, P.H., Puchalski, S.M., 2012. Comparisons of computed tomography, contrast-enhanced computed tomography and standing low-field magnetic resonance imaging in horses with lameness localised to the foot. Part 2: lesion identification. Equine Vet. J. 44, 149–156.
van Sterkenburg, M.N., van Dijk, C.N., 2011. Mid-portion Achilles tendinopathy: why painful? An evidence-based philosophy. Knee Surg. Sports Traumatol. Arthrosc. 19, 1367–1375.
Vanel, M., Olive, J., Gold, S., Mitchell, R.D., Walker, L., 2012. Clinical significance and prognosis of deep digital flexor tendinopathy assessed over time using MRI. Vet. Radiol. Ultrasound 53, 621–627.
Vente, J.E.D., Lester, G.E., Trotter, J.A., Dahners, L.E., 1997. Isolation of intact collagen fibrils from healing ligament. J. Electron Microsc. (Tokyo) 46, 353–356.
Venturoni, M., Gutsmann, T., Fantner, G.E., Kindt, J.H., Hansma, P.K., 2003. Investigations into the polymorphism of rat tail tendon fibrils using atomic force microscopy. Biochem. Biophys. Res. Commun. 303, 508–513.
Vidal, B.D.C., 2003. Image analysis of tendon helical superstructure using interference and polarized light microscopy. Micron 34, 423–432.
Wang, P.-H., Luh, J.-J., Chen, W.-S., Li, M.-L., 2011. In vivo photoacoustic micro-imaging of microvascular changes for Achilles tendon injury on a mouse model. Biomed. Opt. Express 2, 1462–1469.
Wellen, J., Helmer, K.G., Grigg, P., Sotak, C.H., 2005. Spatial characterization of T1 and T2 relaxation times and the water apparent diffusion coefficient in rabbit Achilles tendon subjected to tensile loading. Magn. Reson. Med. 53, 535–544.
Whittaker, P., Canham, P.B., 1991. Demonstration of quantitative fabric analysis of tendon collagen using two-dimensional polarized light microscopy. Matrix 11, 56–62.
Whitton, R.C., Buckley, C., Donovan, T., Wales, A.D., Dennis, R., 1998. The diagnosis of lameness associated with distal limb pathology in a horse: a comparison of radiography, computed tomography and magnetic resonance imaging. Vet. J. 155, 223–229.
Widmer, W.R., Buckwalter, K.A., Fessler, J.F., Hill, M.A., Vansickle, D.C., Ivancevich, S., 2000. Use of radiography, computed tomography and magnetic resonance imaging for evaluation of navicular syndrome in the horse. Vet. Radiol. Ultrasound 41, 108–116.
Wong, J., Bennett, W., Ferguson, M.W., Mcgrouther, D.A., 2006. Microscopic and histological examination of the mouse hindpaw digit and flexor tendon arrangement with 3D reconstruction. J. Anat. 209, 533–545.
Wong, J.K., Alyouha, S., Kadler, K.E., Ferguson, M.W., Mcgrouther, D.A., 2010. The cell biology of suturing tendons. Matrix Biol. 29, 525–536.
Xu, Y., Dong, S., Zhou, Q., Mo, X., Song, L., Hou, T., Wu, J., Li, S., Li, Y., Li, P., Gan, Y., Xu, J., 2014. The effect of mechanical stimulation on the maturation of TDSCs-poly(L-lactide-co-e-caprolactone)/collagen scaffold constructs for tendon tissue engineering. Biomaterials 35, 2760–2772.
Yang, X., Coleman, D.P., Pugh, N.D., Nokes, L.D.M., 2011. A novel 3-D power Doppler ultrasound approach to the quantification of Achilles tendon neovascularity. Ultrasound Med. Biol. 37, 1046–1055.

Yang, Y., Rupani, A., Bagnaninchi, P., Wimpenny, I., Weightman, A., 2012. Study of optical properties and proteoglycan content of tendons by polarization sensitive optical coherence tomography. J. Biomed. Opt. 17, 081417.

Yeh, C.K., Chen, J.J., Li, M.L., Luh, J.J., Chen, J.J., 2009. In vivo imaging of blood flow in the mouse Achilles tendon using high-frequency ultrasound. Ultrasonics 49, 226–230.

Zhang, A.Y., Bates, S.J., Morrow, E., Pham, H., Pham, B., Chang, J., 2009. Tissue-engineered intrasynovial tendons: optimization of acellularization and seeding. J. Rehabil. Res. Dev. 46, 489–498.

Zhou, J., Koike, Y., Uhthoff, H.K., Trudel, G., 2007. Quantitative histology and ultrastructure fail to explain weakness of immobilized rabbit Achilles' tendons. Arch. Phys. Med. Rehabil. 88, 1177–1184.

# Part Two

# Repair and regeneration of tendons and ligaments

# Autograft, allograft, and xenograft scaffolds for tendon and ligament repair: Materials and biomechanics

**6**

## Abstract

In order to regenerate lacerated tendons and ligaments, one viable option is to transplant autologous tendon or ligament tissue (autograft). The problem going along with this approach is primarily limited availability. Therefore, also allogeneic tendon and ligament (T&L) tissue is in use, however, the tissue has to be decellularized before transplanting it to the defect site in order to avoid tissue rejection. Such decellularized T&L grafts can also be reseeded by autologous cells in order to facilitate later biointegration. These two processing steps, however, prolong the time window to have a graft ready. Finally, also xenografts are in use, again decellularized to avoid rejection by the host. In the following chapter, auto-, allo-, and xenografts are discussed and possible processing options such as in vitro preseeding, static versus dynamic cultivation, and incorporation of growth factors such as basal fibroblast growth factor, platelet-derived growth factor, or epidermal growth factor among others are presented.

**Keywords:** Tendon grafts, Natural replacement materials, Commercial available natural products, Cell seeding, Mechanical stimulation.

## Abbreviations

| | |
|---|---|
| **AB** | antibiotic (antimicrobial treatment) |
| **ac** | acellular (tissue/graft/scaffold) (elements in tissue engineering) |
| **ACL** | anterior cruciate ligament (tendons and ligaments) |
| **AcT** | acellular tendon (graft for tissue engineering) |
| **ASC** | adipose-derived stem cell (cells for tissue engineering, cell therapy) |
| **AT** | Achilles tendon (tendons and ligaments) |
| **bFGF** | basal fibroblast growth factor (protein in living cells) |
| **BMP** | bone morphogenetic protein (element in tissues) |
| **C** | cells for cell seeding (elements in tissue engineering) |
| **cDNA** | complementary deoxyribonucleic acid (element of living cells) |
| **Chit/HA, Chit ± HA** | chitosan/hyaluronan acid (natural polymers for tissue engineering) |
| **cy** | cycles (mechanical treatment in tissue engineering) |
| **dyn** | dynamic stimulation (mechanical treatment in tissue engineering) |

**Biomechanics of Tendons and Ligaments. http://dx.doi.org/10.1016/B978-0-08-100489-0.00006-5**

| | |
|---|---|
| **ECM** | extra cellular matrix (matrix in living cells) |
| **EGF** | epidermal growth factor (protein in living cells) |
| **F/T** | freezing/thawing procedure (mechanical treatment in tissue engineering) |
| **Fb** | fibroblasts (cells in connective tissue) |
| **FDA** | Food and Drug Administration (American agency) |
| **FDP** | *flexor digitorum profundus* (tendons and ligaments) |
| **FDS** | *flexor digitorum superficialis* (tendons and ligaments) |
| **FHL** | *flexor hallucis longus* (tendons and ligaments) |
| **Flex** | flexors (tendons and ligaments) |
| **FPL** | *flexor pollicis longus* (tendons and ligaments) |
| **FS** | fibrin sealant (agent for tissue agglutination) |
| **H&E** | hematoxylin and eosin (histological staining technique) |
| **hASCs** | human adipose stem cells (stem cells in fatty tissue) |
| **HDM** | human dermal matrix (elements in tissue engineering) |
| **Hum** | human (specie classification) |
| **HUV** | human umbilical vein (tissue) |
| **IGF (-1)** | insulin-like growth factor (protein in living cells) |
| **IV** | in vitro (laboratory experiment) |
| **kDa** | kilo dalton (unity for protein size) |
| **MCL** | medial collateral ligament (tendons and ligaments) |
| **MSC** | mesenchymal stem cells (cell type) |
| **MSt** | mechanical stimulation (mechanical treatment in tissue engineering) |
| **MSTE** | mechanical stimulation for tissue engineering (mechanical treatment in tissue engineering) |
| **na** | not available (when data are not available) |
| **nat** | natural/native (tendons and ligaments) |
| **nf** | nanofiber (structure of tissues) |
| **Ø** | mean, average of values (statistical element) |
| **P(LLA/CL)** | synthetic polymer (element in tissue engineering) |
| **PAA** | peracetic acid (chemical agents for tissue engineering) |
| **PB** | *peroneus brevis* (tendons and ligaments) |
| **PDGF** | platelet-derived growth factor (protein in living cells) |
| **PDGF-BB** | platelet-derived growth factor subunit B (protein in living cells) |
| **PDS-c** | polydioxanone cord (elements in tissue engineering) |
| **PDS-t** | polydioxanone thread (elements in tissue engineering) |
| **PEEK** | polyether ether ketone (elements in tissue engineering) |
| **PGA** | poly(glycolide) acid (synthetic polymer for tissue engineering) |
| **PLT** | posterior cruciate ligament (tendons and ligaments) |
| **PS** | picrosirius red (histological staining technique) |
| **PT** | patellar tendon (tendons and ligaments) |
| $R^2$ | correlation coefficient (element of statistics) |
| **Rab** | rabbit (animal model) |
| **RC** | rotator cuff (tendons and ligaments) |
| **RGD** | arg–gly–asp (tri-peptide, cell binding protein) |
| **scor** | scoring (mechanical treatment in tissue engineering) |
| **SD** | standard deviation (statistical element) |

| | |
|---|---|
| **SDS** | sodium dodecyl sulfate (chemical agents for tissue engineering) |
| **SEM** | standard error of mean (statistical element) |
| **SIS** | small intestinal submucosa (tissue) |
| **STT** | semitendinosus tendon (tendons and ligaments) |
| **Sut** | suture (surgical treatment) |
| **T** | tendon (tendons and ligaments) |
| **T&L** | tendon and ligaments (tendons and ligaments) |
| **Tc** | tenocytes (cells in tendons and ligaments) |
| **TCP** | tricalcium phosphate (calcium salt) |
| **TGF-β1** | transforming growth factor beta 1 (protein in living cells) |
| **VEGF** | vascular endothelial growth factor (protein in living cells) |
| **w** | week (time) |

## 6.1 Introduction

Transplantation of autologous tissue to a defect site is one common approach to regenerate lacerations or traumatized tendons and ligaments; however, donor site morbidity and limited availability often hamper or restrict such proceedings in daily surgery. Autograft failure in e.g., ACL (anterior cruciate ligament) reconstruction is reported to be primarily caused by harvesting of not sufficiently strong graft material, followed by inaccurate anatomical bone tunnel placement, inappropriate tension of the bone-tendon-bone autograft and graft impingement (Vališ et al., 2014). The performance of autograft tendons may also be limited by the fact that not the same tendon is transplanted but another one with slightly different characteristics regarding biomechanics, cell density, and gliding ability among others. For example, Karabekmez and Zhao (2012) were able to show in a canine model that *peroneus longus autografts* used for *flexor digitorum profundus* (FDP) reconstruction was not as successful as *flexor digitorum profundi* obtained from other dogs used as allografts after decellularization. The allografts performed better with respect to gliding capability (less peritendinous adhesion formation) which is one crucial characteristic of flexor tendons.

Not only may full tendons or ligaments be replaced by autologous tissue or allografts, but also partial replacement or augmentation of the lacerated material is a common approach. Moreover, xenografts after decellularization avoiding immune rejection, with or without repopulation by autologous cells, are tested for their suitability in tendon and ligament (T&L) reconstructions. Furthermore, material properties are investigated to be optimized by pretreatments in vitro such as adding growth factors, appropriate cultivation of the grafts including not only static but also dynamic cell-graft cultivation in bioreactors. In this chapter, an overview of different approaches using T&L grafts and augmentations (of auto-, allo-, and xenobiotic origin) will be addressed. Moreover, biological tissues such as small intestinal submucosa (SIS) or human umbilical veins (HUVs) which may serve as successful tendon augmentations are discussed with respect to their preparations as well as their biomechanical performance in vitro and in vivo.

# 6.2 Tendon grafts

Achilles tendon (AT) ruptures and their operative treatment have gained attention for decades (Kellam et al., 1985). Not only operative treatment approaches, but also conservative approaches are in use and their outcomes with respect to biomechanics and structure of the regenerated tendons are investigated (Krapf et al., 2012).

### *6.2.2 Tendon grafts in animal models and human cadaver experiments*

In a rabbit model, Ilhami et al. (2004) were able to demonstrate that the outcome of AT defects reconstructed by augmentation of *plantaris* tendon autografts highly depends on the location of the reinforcing tendon graft. An intratendinous placement of the *plantaris* tendon where it is passed through the AT rupture from the proximal end to the distal end after transection from its insertion, leads to a twofold maximum load 6 weeks postoperatively compared to a simple peritendinous augmentation of the transected AT by the *plantaris* tendon (Table 6.1c). AT augmentation by an intratendinous *plantaris* tendon was also studied in human cadavers (Gebauer et al., 2007); the impact on ultimate load, however, was much more expressed by another parameter than the location of the reinforcement. Variation of the suture technique (Bunnel vs Kessler) and the suture material (cord vs polydioxanone thread) led to a twofold increase in ultimate load (Bunnel), while the Kessler technique only increased it by a factor of 1.3 (baseline: corresponding thread sutures). Moreover, when an additional augmentation by an autologous plantaris graft was used and combined with a Bunnel technique and a polydioxanone thread, the highest stability of all combinations was obtained. However, compared to the values obtained by the suture technique and the suture material, the augmentation by a *plantaris* tendon as a single contribution increased the ultimate load even less (factors of only 1.1–1.2; Table 6.1c).

AT reconstructions using *peroneus brevis* (PB) or *flexor hallucis longus* (FHL) augmentation was studied in a human cadaver model (Sebastian et al., 2007). Biomechanical outcomes for the transferred segments of PB as well as FHL were satisfactory because failure loads were accompanied by AT stump failure, while the transferred segments remained intact. Moreover, stiffness measured for both, PB and FHL reconstructions, respectively, were similar (Table 6.1c).

For ACL reconstructions, *semitendinosus* tendon (STT) seems to be an ideal autograft (Goradia et al., 2000a; Kondo et al., 2012). It has been shown by Goradia et al. that a STT autograft transformed into a histologically very similar tendon as the ACL, after one year. For that purpose, the authors performed an ACL reconstruction in one of the hind limbs of eleven sheep and used a doubled STT autograft which was attached to the femur with an endoscopic button and to the tibia with sutures tied around a screw. At different time points postsurgery, histological analysis revealed a successful adaptation of the tendon structure compared to the natural ACL. No biomechanics were assessed in this study; nevertheless the conclusion was that a STT

**Table 6.1 Properties of reconstructed tendons with auto-, allo-, and xenograft augmentation or reconstruction**

| Graft | Origin | Specifications | Ultimate stress (MPa) | Elastic modulus (MPa) | Ref. |
|---|---|---|---|---|---|
| (a) Tissue engineered grafts: in vitro mechanical properties | | | | | |
| Flexors (fore+hind paw) | Rabbit | Fresh tendons | 62–66 | 628–630 | Chong et al. (2009) |
| | | Frozen tendons | 58–69 | 628–788 | |
| | | AcT | 55–67 | 628–711 | |
| | | AcT+Tc | 49–50 | 534–553 | |
| FDP (hind paw) | Rabbit | Native FDP | 73±15 | 1166±281 | Saber et al. (2010) |
| | | Frozen tendon | 59±13 | 852±139 | |
| | | AcT, frozen | 41±9 | 735±116 | |
| | | AcT, static | 46±7 | 700±105 | |
| | | AcT+dyn | 41±7 | 704±159 | |
| | | AcT+Tc | 36±6 | 632±86 | |
| | | AcT+Tc+dyn | 71±14 | 1091±169 | |
| | | Native FDP (fresh) | 64±12 | 980±150 | Angelidis et al. (2010) |
| | | AcT+ASC, static | 48±20 | 716±140 | |
| | | AcT+ASC+dyn | 67±22 | 907±140 | |
| | | AcT+Fb, static | 45±7 | 680±80 | |
| | | AcT+Fb+dyn | 68±6 | 730±100 | |
| Flexors | Human | AcT | 26±14 | 653±354 | Woon et al. (2011b) |
| | | AcT+PAA+Fb | 37±13 | 914±267 | |
| FDP | Human | Native FDP | 55±19 | 520±200 | Pridgen et al. (2011) |
| | | AcT (SDS) | 52±7 | 510±100 | |
| FDP, FDS, FPL | Human | AcT+Fb | 34±11 | 583±175 | Woon et al. (2011a) |
| | | AcT+Fb+dyn | 60±28 | 1070±430 | |
| | | AcT+ASC | 37±10 | 722±259 | Woon et al. (2012) |
| | | AcT+scoring+ASC | 47±11 | 903±390 | |
| PT | Human | Native PT | 12±4 | 36±9 | Lomas et al. (2004) |
| | | PT+PAA disinfection | 12±3 | 38±14 | |

*Continued*

**Table 6.1 Continued**

| Model | Target T&L | Augmented/ reconstr. by graft | Specifications | Healing time (weeks) | Ultimate stress (MPa) | Elastic modulus (MPa) | Ref. |
|---|---|---|---|---|---|---|---|
| (b) Mechanical properties of reconstructed tendon and ligaments with auto-, allo-, and xenografts | | | | | | | |
| Rabbit | ACL | In situ frozen/ thawed | Native ACL | | 98±18 | 680±170[a] | Nagumo et al. (2005) |
| | | | FS only | 12 | 48±8 | 340±130[a] | |
| | | | FS+TGF-β1 | | 72±10 | 560±150[a] | |
| | | | FS+EGF | | 45±6 | 350±100[a] | |
| | | | FS+PDGF-BB | | 48±10 | 350±190[a] | |
| | | | Native ACL | | 76±10 | 464±66[a] | Sakai et al. (2002) |
| | | | F/T only | 6 | 58±5 | 350±64[a] | |
| | | | FS only | | 53±10 | 360±99[a] | |
| | | | FS+TGF-β1/ EGF (ng)[b] | | 62±9 | 387±90[a] | |
| | | | FS+TGF-β1/ EGF (μg)[c] | | 51±12 | 424±115[a] | |
| | | | F/T only | 12 | 36±14 | 254±70[a] | |
| | | | FS only | | 44±6 | 320±148[a] | |
| | | | FS+TGF-β1/ EGF (ng)[b] | | 62±16 | 445±180[a] | |
| | | | FS+TGF-β1/ EGF (μg)[c] | | 37±18 | 314±115[a] | |
| | Flexors | Flexors (autolog., hind paw) | Sut+AcT+Tc, static | 4 | 36±11 | 505±148 | Thorfinn et al. (2012) |
| | | | Sut+AcT+Tc +MSt | | 46±15 | 699±197 | |

| Model | Target tendon or ligament | Augmented/ reconstr. by graft | Specifications | Healing time (weeks) | Ultimate load (N) | Stiffness (N/mm) | Ref. |
|---|---|---|---|---|---|---|---|
| (c) Structural properties of reconstructed tendon and ligaments with auto-, allo-, and xenografts | | | | | | | |
| hCAD | AT | PLT (autograft, intratendin. augment.) | Bunnel tech.+PDS-t | 0 | 139 ± 30 | | Gebauer et al. (2007) |
| | | | Bunnel tech.+PDS-c | | 291 ± 55 | | |
| | | | Bunnel tech.+PDS-t+PLT | | 326 ± 125 | | |
| | | | Kessler tech.+PDS-t | | 137 ± 37 | | |
| | | | Kessler tech.+PDS-c | | 180 ± 41 | | |
| | | | Kessler tech.+PDS-t+PLT | | 211 ± 60 | | |
| | | PB, FHL | Native PB | 0 | 333 ± 137 | | Sebastian et al. (2007) |
| | | | AT reconstructed with PB | | 349 ± 125 | 17 ± 6 | |
| | | | Native FHL | | 511 ± 164 | | |
| | | | AT reconstructed with FHL | | 242 ± 82 | 14 ± 4 | |
| | PCL | AT | PCL reconstr. with AT, short | 0 | | 154 ± 42 | Defrate et al. (2004) |
| | | | PCL reconstr. with AT, med. | | | 116 ± 37 | |
| | | | PCL reconstr. with AT, long | | | 88 ± 36 | |
| Rabbit | AT | PLT (autograft) | Native AT | | 208 ± 35 | | Ilhami et al. (2004) |
| | | | Peritendinous augment | 6 | 83 ± 8 | | |
| | | | Intratendinous placement | | 160 ± 31 | | |

*Continued*

**Table 6.1 Continued**

| Model | Target tendon or ligament | Augmented/ reconstr. by graft | Specifications | Healing time (weeks) | Ultimate load (N) | Stiffness (N/mm) | Ref. |
|---|---|---|---|---|---|---|---|
| Goat | ACL | PT | Native ACL | | 3151 ± 407 | 480 ± 42 | Badylak et al. (1999) |
| | | | ACL reconstr. with PT | 0 | 253 ± 55 | 51 ± 12 | |
| | | | | 13 | 611 ± 138 | 147 ± 54 | |
| | | | | 52 | 879 ± 145 | 217 ± 64 | |
| Sheep | ACL | STT (autograft) | Native ACL | | 100% | 100% | Goradia et al. (2000a) |
| | | | ACL reconstruction with doubled STT (tendon-to-bone healing) | 0 | 15% | | |
| | | | | 4 | | 18% | |
| | | | | 8 | | 39% | |
| | | | | 12 | 19% | | |
| | | | | 24 | 28% | 52% | |
| | | | | 52 | 40% | 82% | |
| | | | Native ACL | | 780 ± 60 | 210 ± 60 | Yoshikawa et al. (2006) |
| | | | ACL recontr. with STT | 12 | 250 ± 80 | 80 ± 12 | |
| | | | ACL recontr. with STT + VEGF | | 220 ± 100 | 40 ± 30 | |
| | | | Native ACL | 0 | 1490 ± 500 | 270 ± 120 | Kondo et al. (2012) |
| | | | ACL recontr. with STT | | 400 ± 10 | 40 ± 10 | |
| | | | Native ACL | 12 | 1350 ± 350 | 360 ± 70 | |
| | | | ACL recontr. with STT | | 300 ± 150 | 70 ± 20 | |
| | | | Native ACL | 24 | 1650 ± 300 | 340 ± 30 | |
| | | | ACL recontr. with STT | | 450 ± 200 | 120 ± 30 | |
| | | | Native ACL | 52 | 1500 ± 300 | 320 ± 150 | |
| | | | ACL recontr. with STT | | 700 ± 300 | 230 ± 120 | |

| | | | | | | | |
|---|---|---|---|---|---|---|---|
| Dog | FDP | Fibrin/heparin | Sut only | 3 | 72 ± 14 | 15 ± 2 | Thomopoulos et al. (2010) |
| | | | Sut + 500 ng bFGF | 3 | 82 ± 18 | 15 ± 5 | |
| | | | Sut only | 3 | 76 ± 19 | 16 ± 2 | |
| | | | Sut + 1000 ng bFGF | 3 | 74 ± 19 | 15 ± 3 | |
| Rat | Subcut. implantation | Flexor T (human) | Sut + AcT | 4 | 0.78 ± 0.69 | | Schmitt et al. (2013) |
| | | | Sut + AcT + hASC | 4 | 0.71 ± 0.46 | | |

[a] Tangent modulus.
[b] 4 ng + 100 ng.
[c] 2 μg + 50 μg.

*ACL*, anterior cruciate ligament; *AcT*, acellular tendon; *ASC*, adipose stem cells; *AT*, Achilles tendon; *bFGF*, basal fibroblast growth factor; *dyn*, dynamic stimulation; *EGF*, epidermal growth factor; *F/T*, freezing/thawing procedure; *Fb*, fibroblasts; *FDP*, flexor digitorum profundus; *FDS*, flexor digitorum superficialis; *FHL*, flexor hallucis longus; *FPL*, flexor pollicis longus; *FS*, fibrin sealant; *hASC*, human adipose stem cells; *hCAD*, human cadaver; MSt, mechanical stimulation; *Ø ± SD*, mean ± standard deviation; *PAA*, peracetic acid; *SDS*, sodium dodecyl sulfate; *PB*, peroneus brevis; *PDGF-BB*, platelet-derived growth factor subunit B; *PDS-c*, polydioxanone cord; *PDS-t*, polydioxanone thread; *PLT*, posterior cruciate ligament; *PT*, patellar tendon; *scor*, scoring; *STT*, semitendinosus tendon; *T&L*, tendons and ligaments; *Tc*, tenocytes; *TGF-β1*, transforming growth factor beta 1; *VEGF*, vascular endothelial growth factor.

autograft performed as good as the usually used patellar tendon (PT) autograft (Goradia et al., 2000b). Moreover, in a very similar study in sheep 12 years later, Kondo et al. (2012) were able to demonstrate that while histological adaptation to similar patterns observed in ACL occurred, the biomechanical performance of a doubled STT autografts was only half of the maximum load and only around 70% of the natural stiffness achieved after 52 weeks (Table 6.1c). This is not a bad outcome; however, patients receiving a STT autograft for ACL reconstruction should be aware of the grafts' lower biomechanical characteristics and therefore not be vigorously active during the first year postoperation. Otherwise, the STT autograft is very useful in terms of ACL reconstruction (Table 6.1c).

Often, the worse biomechanical performance of such tendon autografts is considered to be overcome by the addition of growth factors—in order to accelerate the healing process and theoretically improve strength at early time points postoperation. However, one has to be careful with respect to growth factor application to the STT graft. For example, Yoshikava et al. soaked a sheep *semitendinosus* autograft in a VEGF (vascular endothelial growth factor) solution. As they hypothesized that VEGF promoted cell infiltration from the blood system required for tendon healing, the angiogenic factor was believed to accelerate the healing process and thereby enhance biomechanics. However, they found the opposite. The treatment by VEGF reduced the ultimate load to 88% at 12 weeks postoperation compared to no-VEGF *semitendinosus* autografts, and the linear stiffness was even reduced to 50% (Yoshikawa et al., 2006). Such negative biomechanical impacts have to be taken into account in case clinical VEGF application is considered (Table 6.1c).

Also exogenous application of basic fibroblast growth factor (bFGF) is not suggested with regards to tendon healing and biomechanics; despite substantial acceleration of cell proliferation, no improvements in biomechanical or functional properties were found in a canine flexor tendon model (Thomopoulos et al., 2010). The authors used a fibrin matrix and a heparin-based delivery system for bFGF providing a controlled release of the growth factor during three weeks. Such implants at two different doses of loaded bFGF were placed between the tendon stumps of flexor tendons in dogs. Three weeks postsurgery, the adhesion extent of the flexor tendons to the surrounding tissue was significantly higher in the bFGF-treated dogs and the biomechanical properties such as the load until failure and the stiffness were similar to the controls—but not improved (Table 6.1c). Thus, administration of bFGF is not suggested in terms of accelerating intrasynovial tendon healing such as the flexor tendons.

However, growth factors other than VEGF or bFGF were beneficially applied to in situ frozen-thawed rabbit ACL. The model of in situ frozen ACL in the rabbit model had been developed by Sakai et al. (2002) in order to mimic frozen-thawed tendon autografts or allografts that are anatomically equal but acellular (Jackson et al., 1991). While transforming growth factor beta 1 (TGF-β1) significantly inhibited some of the material deterioration that occurs in the in situ frozen–thawed ACL, epidermal growth factor (EGF) or platelet derived growth factor (PDGF) did not have such an effect (Nagumo et al., 2005). Moreover, TGF-β1 performed better with respect to biomechanics such as tangent modulus (70% higher compared to EGF or PDGF-BB applications) and ultimate stress—called also tensile strength—(40% higher compared to EGF or PDGF-BB applications) of the ACL 12 weeks postoperation (Table 6.1b). The combination of TGF-β1 which stimulates fibroblast (Fb) proliferation, collagen, and noncollagenous protein synthesis with EGF known to accelerate Fb proliferation in vitro led also to an enhanced biomechanical performance in terms of ultimate stress and tangent modulus, but only if applied in low doses such as 4 ng TGF-β1 and 100 ng EGF compared to 2 μg TGF-β1 and 50 ng EGF, respectively (Sakai et al., 2002).

Based on the abovementioned studies where TGF-β1 had been shown to be beneficial with respect to biomechanical performance of reconstructed tendons, it is not surprising that TGF-β1 has been investigated in terms of tendon healing. Because problems in accurate delivery of proteins to the site of tendon injury are not yet fully solved, gene transfer is one way to overcome such limitations. For example, TGF-β1 transduction of a muscle autograft for AT reconstruction in rats significantly increased ultimate load and stiffness compared to muscle autograft control (Table 6.2b; Majewski et al., 2012). The authors used adenovirus to deliver TGF-β1 cDNA (complementary deoxyribonucleic acid) and found that stiffness of AT was significantly increased by TGF-β1 delivery 1 week postoperation compared to muscle autografts without transduction. At later time points postoperation (2, 4, and 8 weeks, respectively), stiffness was still higher compared to nontreated grafts, however, no statistical difference was obtained anymore.

Although no auto- or allograft study, but still interesting with respect to growth factor application, insulin-like growth factor (IGF-1), transduced rat tendon Fbs were shown to enhance ultimate load values of a cell-seeded polymer in vitro, while PDGF had only half of the ultimate loads that were found for IGF-1 (Dines et al., 2007). In addition, slow release of bFGF from dextran nanoparticles in an electrospun PLLA (poly-L-lactic acid) mesh for rat AT reconstruction did not improve biomechanics after 3 weeks in vivo (see Chapter 8; Liu et al., 2013).

### 6.2.3 *In vitro mechanical properties of tendon grafts*

A concisely examined approach in flexor tendon reconstruction is tissue engineering based on decellularization of flexor tendons. In vitro experiments using decellularized rabbit fore and hind paw flexor tendons showed that ultimate stress as well as elastic modulus did not change significantly compared to fresh as well as frozen tendons (Table 6.1a; Chong et al., 2009). Moreover, when the decellularized flexor tendons were seeded with rabbit tenocytes for 4 days and cultivated in a static culture, there was no significant difference in biomechanical properties, rather, a trend towards lower ultimate stress and lower elastic modulus was determined (Fig. 6.1). In Fig. 6.1, the linear modulus ($y$-axis) of the corresponding grafts is plotted against their ultimate stress values ($x$-axis). As such, the range of biomechanical main properties of the grafts and the native rabbit flexor tendons can be envisioned and compared at once. Although the seeding with tenocytes led to slightly less strong and less elastic tendon grafts compared to Fb-seeded or even adipose-derived stem cell (ASC)-seeded constructs (magnification, Fig. 6.1 right), such an approach of first decellularization and

**Table 6.2 Properties of reconstructed tendons with natural graft augmentation or reconstruction**

| Model | Target tendon or ligament | Augmented/ reconstr. by graft | Specifications | Healing time (weeks) | Ultimate stress (MPa) | Elastic modulus (MPa) | Ref. |
|---|---|---|---|---|---|---|---|
| (a) Mechanical properties | | | | | | | |
| ***Small intestinal submucosa (SIS)*** | | | | | | | |
| IV | T&L | SIS (pig) | Ac SIS+Tc | – | | 9±3 | Citation in Androjna et al. (2007) |
| | | | Ac SIS+Tc+dyn | | | 21±3 | |
| | Abdomen | | Ac SIS | | 9–15 | | Zhang et al. (2011) |
| | Bladder | | Native SIS graft | – | | 2200±500 | Kropp et al. (1996) |
| | Oesophag. | | Ac tubular SIS | – | 600–1650 | 1500–4800 | Syed et al. (2014) |
| Rabbit | MCL | SIS (pig) | Native MCL | | 76±14 | 936±284[a] | Musahl et al. (2004) |
| | | | Sut only | 12 | 11±4 | 149±77[a] | |
| | | | Sut+SIS | | 16±4 | 280±132[a] | |
| | | | Native MCL | | 84±19 | 1142±412 | Liang et al. (2006) |
| | | | Sut only | 26 | 24±9 | 291±123 | |
| | | | Sut+SIS | | 36±11 | 387±99 | |
| Rat | RC acute/ chronic repair | SIS (pig) | Sut only acute repair | 8+16 | | 9±6 | Perry et al. (2007) |
| | | | Sut+SIS acute repair | | | 9±4 | |
| | | | Sut+SIS chronic repair | | | 15±5 | |
| ***Fascia lata (ECM)*** | | | | | | | |
| IV | T&L (RC) | Fascia lata | AlloPatch ECM (fascia lata) | – | | 304±52 | Aurora et al. (2007) |
| | | | Human fresh Fascia lata | – | | 389±189 | Derwin et al. (2008) |
| | | | AlloPatch+AB | | | 532±106 | |
| | | | Fascia lata ac ECM+AB | | | 440±326 | |

| | | | | | | | |
|---|---|---|---|---|---|---|---|
| ***Human umbilical vein (HUV)*** | | | | | | | |
| IV | T&L | HUV | Human native tendons | – | 13–125 | | Abousleiman et al. (2008) |
| | | | Native HUV | | 1.6±1.0 | | |
| | | | HUV cellular | | 0.6±0.1 | 4.5±2.0 | |
| | | | HUV acellular | | 0.9±0.1 | 3.5±0.7 | |
| | | | HUV ac+MSC (rat) 1 w | | 0.9±0.2 | 6.3±0.5 | |
| | | | HUV ac+MSC (rat) 2 w | | 1.7±0.2 | 9.1±1.8 | |
| | | | HUV ac+MSC (static) 2 w | – | 1.7±0.3 | 9±2 | Abousleiman et al. (2009) |
| | | | HUV ac+MSC (perfusion) 2 w | | 1.5±0.8 | 7±3 | |
| | | | HUV ac+MSC+MSTE 2 w | | 2.7±0.7 | 10±2 | |
| | | HUV+var. seeding rates | HUV ac | – | 0.9±0.1 | | Issa et al. (2011) |
| | | | HUV ac+MSC (static) | | 1.5±0.5 | | |
| | | | HUV ac+MSC+MSTE 1 w (3 Mio cells/mL) | | 2.7±0.8 | | |
| ***Silk+chitin*** | | | | | | | |
| Rabbit | AT | Silk nanofiber photoact., electrospun | IV: silk preannealed | | 1.1±0.4 | 24±13 | Ni et al. (2012) |
| | | | post-an./photoactiv. | | 0.9±0.7 | 30±21 | |
| | | | Native rabbit contralateral AT | | 9.9±1.5 | 19±3 | |

*Continued*

**Table 6.2 Continued**

| Model | Target tendon or ligament | Augmented/ reconstr. by graft | Specifications | Healing time (weeks) | Ultimate stress (MPa) | Elastic modulus (MPa) | Ref. |
|---|---|---|---|---|---|---|---|
| | | | Standard sut | 2 | 4.3 ± 0.8 | 7 ± 1 | |
| | | | Single st. sut + silk-nf | | 1.5 ± 0.3 | 4 ± 1 | |
| | | | Stand. sut + silk-nf | | 4.5 ± 1.5 | 20 ± 5 | |
| | | | Standard sut | 4 | 5.9 ± 1.0 | 14 ± 3 | |
| | | | Single st. sut + silk-nf | | 4.2 ± 1.0 | 8 ± 3 | |
| | | | Single sut + silk-nf | | na | na | |
| | RC | Chitosan-HA fi | Native rabbit infraspinatus T | | 54 ± *13* | 583 ± *93*[a] | Funakoshi et al. (2005) |
| | | | Sut only | 12 | 6 ± *1* | 37 ± *5*[a] | |
| | | | Sut + chit/HA | 4 | 5 ± *2* | 59 ± *7*[a] | |
| | | | Sut + chit/HA | 12 | 7 ± *1* | 60 ± *7*[a] | |
| | | | Sut + chit/HA + Fb = Tb (PT) | 4 | 5 ± *1* | 55 ± *6*[a] | |
| | | | Sut + chit/HA + Fb | 12 | 9 ± *1* | 89 ± *7*[a] | |
| ***Dermis*** | | | | | | | |
| IV | RC | HDM | ac HDM | – | 9 ± 2 | | Fini et al. (2012) |
| | | | Cellularized control | | 7 ± 4 | | |
| ***Veins*** | | | | | | | |
| IV | Flexors | Veins | Vein from pigs, diameter 3–4 mm | – | 3 ± 1 | | Rodger et al. (2015) |

| Animal model | Target tendon or ligament | Augmented/ reconstr. by graft | Specifications | Healing time (weeks) | Ultimate load (N) | Stiffness (N/mm) | Ref. |
|---|---|---|---|---|---|---|---|
| (b) Structural properties | | | | | | | |
| ***Small intestinal submucosa (SIS)*** | | | | | | | |
| In vitro | T&L | SIS (pig) | Ac SIS | – | $0.8 \pm 0.1^b$ | $1.1 \pm 0.1$ | Androjna et al. (2007) |
| | | | Ac SIS + dyn | | $1.3 \pm 0.2^b$ | $1.4 \pm 0.3$ | |
| | | | Ac SIS + Tc | | $0.7 \pm 0.1^b$ | $1.0 \pm 0.1$ | |
| | | | Ac SIS + Tc + dyn | | $1.6 \pm 0.1^b$ | $1.7 \pm 0.1$ | |
| | Abdomen | SIS (pig) | Ac SIS | – | 23–30 | | Zhang et al. (2011) |
| Dog | Infraspin. | SIS (pig) | Native infraspinatus | | $187 \pm 31$ | | Dejardin et al. (2001) |
| | | | Sut only | 0 | $68 \pm 21$ | | |
| | | | Sut + SIS | | $31 \pm 5$ | | |
| | | | Sut only | 13.5 | $93 \pm 20$ | | |
| | | | Sut + SIS | | $78 \pm 33$ | | |
| | | | Sut only | 27 | $91 \pm 12$ | | |
| | | | Sut + SIS | | $85 \pm 24$ | | |
| Rabbit | MCL | SIS (pig) | Native MCL | | $332 \pm 51$ | $90 \pm 15$ | Musahl et al. (2004) |
| | | | Sut only | 6 | $103 \pm 52$ | $35 \pm 9$ | |
| | | | Sut + SIS | | $108 \pm 27$ | $42 \pm 13$ | |
| | | | Sut only | 12 | $66 \pm 31$ | $29 \pm 9$ | |
| | | | Sut + SIS | | $117 \pm 35$ | $46 \pm 13$ | |
| | PT | SIS (pig) | Native PT | | 200–262 | 128–150 | Karaoglu et al. (2008) |
| | | | Sut only | 12 | $46 \pm 25$ | $25 \pm 13$ | |
| | | | Sut + SIS | | $78 \pm 39$ | $39 \pm 16$ | |

*Continued*

**Table 6.2 Continued**

| Animal model | Target tendon or ligament | Augmented/ reconstr. by graft | Specifications | Healing time (weeks) | Ultimate load (N) | Stiffness (N/mm) | Ref. |
|---|---|---|---|---|---|---|---|
| Goat | ACL | SIS (pig) | Native ACL | | 3151±407 | 480±42 | Badylak et al. (1999) |
| | | | ACL reconstr. with SIS | 0 | 721±68 | 50±4 | |
| | | | | 13 | 293±254 | 92±73 | |
| | | | | 52 | 706±544 | 180±113 | |
| Sheep | Infraspinat. | SIS (pig) | Native infraspinatus | | 5233±736 | 545±47 | Schlegel et al. (2006) |
| | | | Sut only | 12 | 985±459 | 154±63 | |
| | | | Sut+SIS | | 1252±402 | 215±44 | |
| Rat | RC Suprasp. | SIS (pig) | Native supraspinatus | | 29–42 | 24–64 | Zalavras et al. (2006) |
| | | | Sut only | 6 | 15±6 | 10–22 | |
| | | | Sut+SIS | | 18±8 | 8–40 | |
| | | | Sut only | 16 | 12±5 | 17±9 | |
| | | | Sut+SIS | | 23±9 | 18±8 | |
| | RC acute/chronic repair | SIS (pig) | Sut only acute repair | 8+16 | | 34±16 | Perry et al. (2007) |
| | | | Sut+SIS acute repair | | | 24±9 | |
| | | | Sut+SIS chronic repair | | | 37±15 | |
| ***Fascia lata (ECM)*** | | | | | | | |
| IV | T&L | Fascia lata | AlloPatch ECM | – | | 98±16 | Aurora et al. (2007) |
| | | | Human fresh Fascia lata | – | | 85±15 | Derwin et al. (2008) |
| | | | AlloPatch+AB | | | 98±16 | |
| | | | Fascia lata ac ECM +AB | | | 71±39 | |

| | | | | | | | |
|---|---|---|---|---|---|---|---|
| ***Muscle*** | | | | | | | |
| Rat | AT | Muscle flap + BMP-12 cDNA del. | Native AT | | 60–80 | 36–86 | Majewski et al. (2008) |
| | | | Sut only | 1 | ~40 | ~18 | |
| | | | Sut+muscle | | ~35 | ~15 | |
| | | | Sut+muscle+BMP-12 | | ~55 | ~27 | |
| | | | Sut only | 2 | ~70 | ~32 | |
| | | | Sut+muscle | | ~75 | ~33 | |
| | | | Sut+muscle+BMP-12 | | ~85 | ~58 | |
| | | | Sut only | 4 | ~85 | ~37 | |
| | | | Sut+muscle | | ~90 | ~36 | |
| | | | Sut+muscle+BMP-12 cDNA del. | | ~100 | ~63 | |
| | | | Sut only | 8 | ~90 | ~53 | |
| | | | Sut+muscle | | ~90 | ~58 | |
| | | | Sut+muscle+BMP-12 cDNA del. | | ~100 | ~53 | |
| | | Muscle (autograft) +TGF-β1 cDNA | Native AT | | 60–80 | 30–85 | Majewski et al. (2012) |
| | | | Sut only | 1 | ~40 | ~18 | |
| | | | Sut+muscle | | ~36 | ~17 | |
| | | | Sut+muscle+TGF-β1 | | ~67 | ~35 | |
| | | | Sut only | 2 | ~66 | ~32 | |
| | | | Sut+muscle | | ~72 | ~36 | |
| | | | Sut+muscle+TGF-β1 | | ~68 | ~57 | |
| | | | Sut only | 4 | ~90 | ~38 | |
| | | | Sut+muscle | | ~93 | ~37 | |
| | | | Sut+muscle+TGF-β1 | | ~92 | ~45 | |
| | | | Sut only | 8 | ~93 | ~52 | |
| | | | Sut+muscle | | ~95 | ~57 | |
| | | | Sut+muscle+TGF-β1 | | ~108 | ~68 | |

*Continued*

Table 6.2 **Continued**

| Animal model | Target tendon or ligament | Augmented/ reconstr. by graft | Specifications | Healing time (weeks) | Ultimate load (N) | Stiffness (N/mm) | Ref. |
|---|---|---|---|---|---|---|---|
| ***Silk + chitin*** | | | | | | | |
| IV | ACL | Silk fiber and caloric extraction | Native silk | – | 11 ± 1 | 1.8 ± 0.1 | Altman et al. (2002) |
| | | | Silk single fiber, 37 °C | | 10 ± 1 | 1.7 ± 0.2 | |
| | | | Silk single fiber, 90 °C | | 9 ± 1 | 1.3 ± 0.2 | |
| | | Silk (div. arch.) | Human ACL | | 2160 ± 157 | 242 ± 28 | Woo et al. (1991) |
| | | | Silk matrix | | 2337 ± 72 | 354 ± 26 | |
| | | | Parallel silk matrix | | 2214 | 1470 | |
| | | | Native silk yarn | – | 9.4 ± 0.3 | 2.0 ± 0.1 | Li and Snedeker (2013) |
| | | | Silk sericin extracted dry | | 7.3 ± 0.4 | 1.4 ± 0.2 | |
| | | | Silk sericin extracted wet | | 6.0 ± 0.3 | 1.0 ± 0.2 | |
| | | | Div. arch: straight dry | | 1600 ± 100 | 1280 ± 100 | |
| | | | Div. arch: straight wet | | 1450 ± 100 | 800 ± 60 | |
| | | | Div. arch: wired dry | | 1900 ± 150 | 550 ± 50 | |
| | | | Div. arch: wired wet | | 1550 ± 80 | 290 ± 30 | |
| | | | Div. arch: braided dry | | 1870 ± 50 | 530 ± 50 | |
| | | | Div. arch: braided wet | | 1650 ± 50 | 240 ± 20 | |
| | | | Silk wired | | 1550 ± 150 | 285 ± 25 | |
| | | | Silk wired + MSt (250 cy) | | 900 ± 200 | 425 ± 30 | |
| | | | Silk wired + MSt (100,000 cy) | | 500 ± 100 | 490 ± 20 | |
| | | | Silk wired + Fb + MSt (100,000 cy) | | 480 ± 50 | 485 ± 15 | |

| | | | | | | | |
|---|---|---|---|---|---|---|---|
| | | | Silk braided | | 1600 ± 100 | 250 ± 30 | |
| | | | Silk braided + MSt (250 cy) | | 800 ± 70 | 520 ± 70 | |
| | | | Silk braided + MSt (100,000 cy) | | 400 ± 100 | 550 ± 50 | |
| | | | Silk braided + Fb + MSt (100,000 cy) | | 395 | 560 | |
| Pig | ACL | Silk/TCP/PEEK anchor | IV: Silk graft | | 800 ± 200 | 285 ± 20 | Li et al. (2014) |
| | | | Native pig ACL | 13 | 1384 ± 181 | 192 ± 22 | |
| | | | | 26 | 1749 ± 284 | 259 ± 15 | |
| | | | Sut + silk graft spiry | 23 | 311 ± 103 | 148 ± 19 | |
| | | | | 26 | 566 ± 29 | 183 ± 10 | |
| IV | RC | Chitosan/HA fibers | Dry scaffold | – | 67 ± *2* | 19 ± *1* | Funakoshi et al. (2005) |
| | | | Wet scaffold after incubation 4 w | | 16 ± *2* | 7 ± *1* | |
| Rabbit | RC Infraspin. | Chitin (unwoven fabric) | IV: chitin dry/wet | 12 | 9–12 | | Funakoshi et al. (2006) |
| | | | Sut only | | 13 ± 7 | 6 ± 3 | |
| | | | Sut + chitin fabric | | 135 ± 28 | 47 ± 14 | |
| | MCL | Chitosan/HA fibers | Native rabbit MCL | – | 106 ± 28 | 93 ± 27 | Irie et al. (2011) |
| | | | Gap only | 3 | 13 ± 4 | 7 ± 3 | |
| | | | Gap only | 6 | 63 ± 14 | 29 ± 6 | |
| | | | Gap only | 12 | 90 ± 25 | 37 ± 10 | |
| | | | Sut + chit + HA | 3 | 14 ± 4 | 1 ± 0.4 | |
| | | | Sut + chit + HA | 6 | 27 ± 17 | 11 ± 15 | |
| | | | Sut + chit + HA | 12 | 83 ± 20 | 31 ± 7 | |
| | | | Sut + chit + HA + Fb = Tb (AT) | 3 | 33 ± 17 | 9 ± 6 | |
| | | | Sut + chit + HA + Fb | 6 | 64 ± 6 | 23 ± 6 | |
| | | | Sut + chit + HA + Fb | 12 | 125 ± 28 | 32 ± 9 | |

*Continued*

**Table 6.2** Continued

| Animal model | Target tendon or ligament | Augmented/ reconstr. by graft | Specifications | Healing time (weeks) | Ultimate load (N) | Stiffness (N/mm) | Ref. |
|---|---|---|---|---|---|---|---|
| ***Dermis*** | | | | | | | |
| IV | RC | HDM | ac HDM | – | 47 ± 11 | 7 ± 1 | Fini et al. (2012) |
| | | | Cellularized control | | 43 ± 18 | 5 ± 1 | |
| Dog | RC | Allografts vs GraftJacket | Native dog infraspinatus | 26 | | 100 ± 30 | Smith et al. (2012) |
| | Infraspin. | | Allograft repair | | | 71 ± 26 | |
| | | | GraftJacket repair | | | 58 ± 23 | |

[a] Tangent modulus.
[b] Load at 8% strain.
*AB*, antibiotika soaking; *ac SIS*, acellular SIS; *ac*, acellular; *ac*, acellularised scaffold; *ACL*, anterior cruciate ligament; *AT*, Achilles tendon; *BMP*, bone morphogenetic proteins; *cDNA*, complementary deoxyribonucleic acid; *cy*, cycles; *dyn*, dynamic stimulation; *ECM*, extra cellular matrix; *Fb*, fibroblasts; *HDM*, human dermal matrix; *HUV*, human umbilical vein; *IV*, in vitro; *MCL*, medial collateral ligament; *MSC*, mesenchymal stem cells; *MSt*, mechanical stimulation; *MSTE*, mechanical stimulation for tissue engineering; *na*, not available; *nf*, nanofiber; *Ø ± SD*, mean ± standard deviation; *Ø ± SEM*, mean ± *standard error of mean*; *T&L*, tendons and ligaments; *PEEK*, polyether ether ketone; *PT*, patellar tendon; *RC*, rotator cuff; *SIS*, small intestine submucosa; *Sut*, suture; *Tb*, tenoblast; *Tc*, tenocytes; *TCP*, tricalcium phosphate; *TGF*, transforming growth factor.

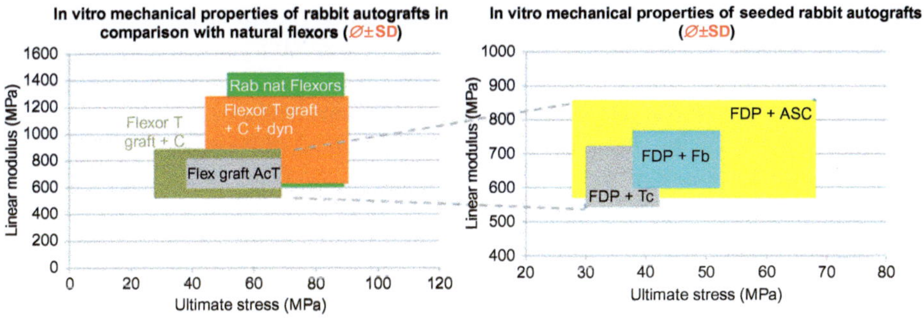


**Fig. 6.1** Autografts for rabbit flexors: in vitro biomechanical correlation of autografts, in comparison with the corresponding in vivo natural tendons. Key: *AcT*, acellular tendon; *ASC*, adipose-derived stem cells; *C*, cells for cell seeding; *dyn*, dynamic mechanical stimulation; *Fb*, fibroblasts; *FDP*, *flexor digitorum profundus*; *Flex*, flexors; *nat*, natural; *Rab*, rabbit; *T*, tendon; *Tc*, tenocytes. Data: $\varnothing \pm SD$, mean ± standard deviation.
Data from Chong, A.K.S., Riboh, J., Smith, R.L., Lindsey, D.P., Pham, H.M., Chang, J., 2009. Flexor tendon tissue engineering: acellularized and reseeded tendon constructs. Plast. Reconstr. Surg. 123, 1759–1766; Saber, S., Zhang, A.Y., Ki, S.H., Lindsey, D.P., Smith, R.L., Riboh, J., Pham, H., Chang, J., 2010. Flexor tendon tissue engineering: bioreactor cyclic strain increases construct strength. Tissue Eng. Part A 16, 2085–2090; Angelidis, I.K., Thorfinn, J., Connolly, I.D., Lindsey, D., Pham, H.M., Chang, J., 2010. Tissue engineering of flexor tendons: the effect of a tissue bioreactor on adipoderived stem cell-seeded and fibroblast-seeded tendon constructs. J. Hand Surg. 35, 1466–1472.

then reseeding with a patient's own cells (bone-marrow stromal cells, (b)MSCs, or ASCs, rather than tenocytes), may find its way into clinical application. A tissue bank of decellularized human flexor tendons is needed and the corresponding biopsies for (stem) cell extraction so that they can be reseeded upon the grafts. In case stem cells are used for reseeding the graft, potential unwanted differentiation may occur such as ossification in the tendon grafts as found for PGA (poly(glycolide) acid) grafts seeded with MSCs (Pietschmann et al., 2013).

In contrast to the repopulated grafts in static culture, cyclic loading of reseeded hind paw flexor tendons in a bioreactor significantly improved the biomechanical properties compared to cell-free tendons (Table 6.1a; Saber et al., 2010). Exposed to a 1 cycle/min load at alternating 1 h periods of mechanical loading (for 5 days) had an impact of a twofold increase in ultimate stress and 1.7-fold increase in elastic modulus. As expected, the same cyclic loading had only a minor impact on biomechanics when the decellularized flexor tendons were not reseeded because loading affects to a much greater extent the mechanobiology of the cells compared to the mechanical properties of the decellularized graft material, especially when only short periods of loading are applied. In addition, the positive effect of dynamically cultivating reseeded tendons was not only shown in vitro; reseeded constructs under dynamic loading were also shown to have 1.3 times higher ultimate stress and a 1.4 times higher elastic modulus when implanted for 4 weeks in vivo to bridge a zone II defect in a rabbit model compared to unloaded grafts that were implanted as controls

(Table 6.1b; Thorfinn et al., 2012). Because Saber et al. (2010) used exactly the same protocol for the in vitro dynamic stimulation during the cultivation period of reseeded constructs as Thorfinn et al. did before implanting the grafts into the rabbit model, relative increases caused by dynamic stimulation can be quantitatively compared. While the increase in ultimate stress is expressed 1.5 times more in vitro than in vivo, the increase in elastic modulus is expressed to a smaller degree; only 1.2 times more in vitro than in vivo—implying that biomechanical parameters such as modulus can be more accurately predicted in terms of in vivo-outcome from in vitro-data than ultimate strength (see also chapter on in vitro and in vivo comparisons).

Dynamic loading has also been reported to be beneficial with respect to biomechanical properties when decellularized human *flexor pollicis longus*, FDP, and *flexor digitorum superficialis* (FDS) were reseeded with Fbs and stimulated for 5 days in a bioreactor. These experiments resulted in a 1.6-fold higher elastic modulus and 1.8-fold higher ultimate stress compared to nonstretched tendons, respectively (Woon et al., 2011a; Table 6.1a and Fig. 6.2). In the same study it was also reported that while

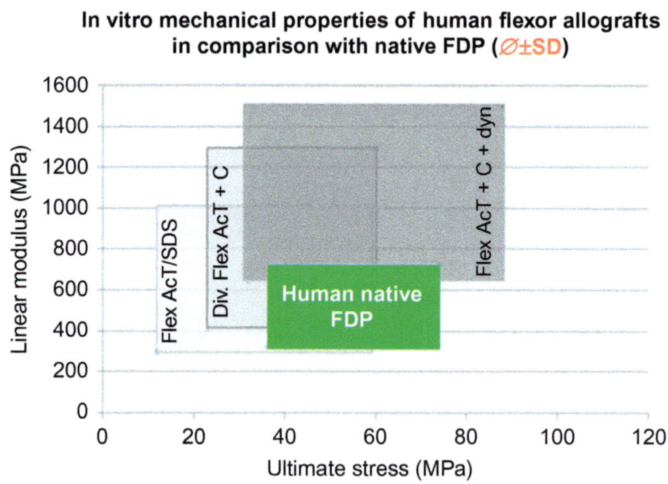


**Fig. 6.2** Allografts for human flexor tendons: correlation of in vitro human allografts in comparison with native FDP. Key: *AcT*, acellular tendon; *C*, cells for cell seeding (adipose-derived stem cells/fibroblasts); *dyn*, dynamic mechanical stimulation; *FDP*, flexor digitorum profundus; *Flex*, flexor tendons; *SDS*, sodium dodecyl sulfate. Data: *Ø* ± *SD*, mean ± standard deviation.

Data from Pridgen, B.C., Woon, C.Y.L., Kim, M., Thorfinn, J., Lindsey, D., Pham, H., Chang, J., 2011. Flexor tendon tissue engineering: acellularization of human flexor tendons with preservation of biomechanical properties and biocompatibility. Tissue Eng. Part C Methods 17, 819–828; Woon, C.Y.L., Kraus, A., Raghavan, S.S., Pridgen, B.C., Megerle, K., Pham, H., Chang, J., 2011a. Three-dimensional-construct bioreactor conditioning in human tendon tissue engineering. Tissue Eng. Part A, 17, 2561–2572; Woon, C.Y.L., Pridgen, B.C., Kraus, A., Bari, S., Pham, H., Chang, J., 2011b. Optimization of human tendon tissue engineering: peracetic acid oxidation for enhanced reseeding of acellularized intrasynovial tendon. *Plast. Reconstr. Surg.* 127, 1107–1117.

the conditioning time had a significant effect on the biomechanical properties of the reseeded grafts (increasing ultimate stress and elastic moduli with time), the magnitude of the applied load did not affect biomechanics significantly; conditioning duration was varied with end-points after 3, 5, and 8 days, while the applied load was ranging from 0.625, 1.25 up to 2.5 N. However, interestingly with respect to the assessment of biomechanical properties of tissue engineered grafts, it was reported that the ultimate stress as well as the elastic modulus were significantly lowered when the material testing was delayed for 2 days after the dynamic loading had been finished in the bioreactor compared to materials that were tested only 2 h after completion; while the ultimate stress was reduced by 37%, the elastic modulus was reduced by 52%. Such information is very important regarding the fact that reseeding of decellularized constructs improved biomechanics in the same range (+60% and +80%, for elastic modulus and ultimate stress, respectively).

Also other studies emphasize the role of dynamic stimulation of cell-seeded constructs fabricated from previously decellularized tendons. For example, decellularized human flexor tendons reseeded with human adipose stem cells (hASCs) without loading or stretching in a bioreactor did not lead to significant improvement of biomechanical properties after subcutaneous implantation in a rat model 4 weeks postsurgery, although stem cells had been shown to bio-integrate into the allograft which was determined by green fluorescent-protein labeling of the hASCs (Schmitt et al., 2013). Besides, ultimate loads were lower than 1 N, which is remarkably low compared to natural human flexor tendons (Table 2.1 in Chapter 2 and Table 6.1); moreover, comparison of ultimate loads 2 and 4 weeks after implantation revealed that ultimate loads at later time points were even lower by approximately 30% which might be explained by the locality of implantation. There is no load applied to such tendon constructs when they are in the back of the mice (subcutaneous) and biomechanical functionality is lost over time.

In the same research group, Angelidis et al. examined hind paw rabbit flexor tendons that were reseeded by ASCs or by Fb with or without cyclic loading in a bioreactor (Angelidis et al., 2010). As found for the Tc-seeded (tenocytes) tendons (Saber et al., 2010), an increase in biomechanical characteristics was noted, however, compared to the Tc-seeded tendons, to a lower extent; 1.4- and 1.5-fold increase in ultimate stress and 1.3- and 1.1-fold increase in elastic modulus for the ASC- and Fb-seeded tendons, respectively (Angelidis et al., 2010; Table 6.1a). In addition, decellularized flexor tendons from human cadavers were compared to fresh human flexor tendons. With respect to biomechanics, ultimate stress and elastic modulus were not altered significantly by the SDS-induced decellularization process (Pridgen et al., 2011). In order to improve cell seeding on decellularized human flexor tendons, a 5% peracetic acid (PAA) treatment for 4 h was suggested by Woon et al., 2011b, because they found that the matrix porosity of the decellularized flexor tendon was significantly increased by oxidation, offering higher surface area for cell attachment. Moreover, these authors also showed that scoring gently 10 times with a 15-blade on the decellularized flexor tendon even improved the reseeding (Woon et al., 2012). Besides, it had been shown earlier that a PAA treatment of tendons has no adverse effects with respect to cytotoxicity or inflammation, nor to biomechanical properties such as ultimate stress or elastic modulus; although PAA disinfection increased the ultimate strain of human ACLs (Lomas et al., 2004; Table 6.1a).

## 6.3 Other tissues of biological origin

For soft-tissue repair, natural extracellular matrices (ECMs) have been procured, processed, and examined for their capability of reinforcing the repair site; marketed natural products that have an ECM rich in collagen include SIS (Restore, DePuy Orthopaedics, Warsaw, Indiana; CuffPatch, Arthrotek, Warsaw, Indiana) or dermis (GraftJacket, Wright Medical Technology, Arlington, Tennessee; TissueMend, Stryker Orthopedics, Mahwah, New Jersey; TEI Biosciences, Boston, Massachusetts; Zimmer Collagen Repair Patch, Zimmer, Warsaw, Indiana). The problem with such natural products when applied in T&L rupture repair is their low stiffness compared to the corresponding healthy organ, with approximately 10 times lower stiffnesses (Derwin et al., 2006). However, other collagen-rich natural scaffolds such as *fascia lata* do have similar biomechanical properties as tendons and ligaments and are therefore valuable substitutes with respect to reinforcements at the repair site (Butler et al., 1984). Nevertheless, strategies in terms of improving biomechanical properties of such biological scaffold materials have been undertaken widely and tested in different animal models. These strategies include cell seeding, static and dynamic cultivation in vitro before implantation, and genetic modification.

### *6.3.1 Small intestinal submucosa*

SIS is a biodegradable and absorbable scaffold that has been used for a long time in tissue reconstruction inducing remodeling in arterial and venous vascular grafts (Badylak et al., 1989; Lantz et al., 1992), bladder repair (Zhang and Liao, 2014), and tendon (Badylak et al., 1995) and ligament reconstruction (Aiken et al., 1994). The use of a single layer of porcine SIS (Cook Biotech Inc., Bloomington, Indiana), for bridging a gap in the MCL (medial collateral ligament) of rabbits has been shown to be beneficial with respect to collagen fiber alignment, collagen fiber diameter, and the ratio of collagen V/I 12 weeks postsurgery (Woo et al., 2006) and even more expressed 26 weeks postsurgery (Liang et al., 2006). Moreover, SIS-treated rabbit MCL had better biomechanical properties compared to nonSIS-treated MCL when a gap of 6 mm was created (Fig. 6.3). Twelve weeks postoperation, ultimate stress, elastic modulus, ultimate load, and stiffness were 1.6–1.9 times higher, when a strip of SIS was sutured onto the ends of the MCL (Table 6.2a and b and Fig. 6.4; Musahl et al., 2004). Compared to natural healthy rabbit MCLs, the application of SIS in the repair of lacerated MCL lead to similar elastic moduli, however, with respect to ultimate stress, the biomechanical outcome was still five times lower (Fig. 6.4). When the biomechanical properties of SIS grafts acquired in vitro are compared to the 12-week postsurgery performance in vivo, the ultimate stress and the elastic moduli are very similar.

It was also shown that in vitro mechanical cyclically loading of cell-free SIS grafts in a bioreactor was increasing loads at 8% strain by a factor of 1.7 (Table 6.2b); even more so when SIS were seeded with tenocytes for 6 h and then made to undergo mechanical stimulation for 2 weeks in a bioreactor (Androjna et al., 2007). Again,

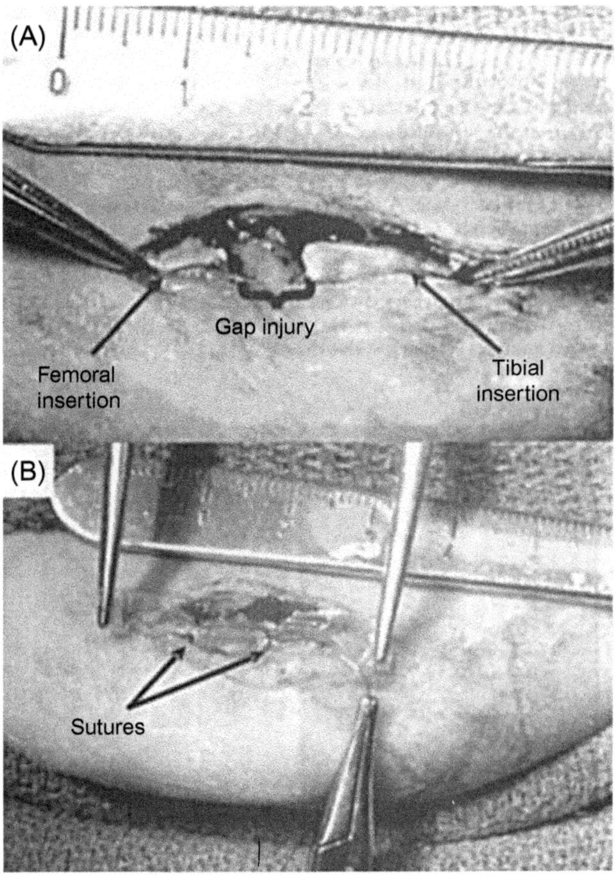


**Fig. 6.3** Rabbit medial collateral ligament (MCL) with 6 mm wide gap (A) and a strip of SIS sutured on top of the gap in the MCL (B).
Based on Musahl, V., Abramowitch, S.D., Gilbert, T.W., Tsuda, E., Wang, J.H.C., Badylak, S.F., Woo, S.L.Y., 2004. The use of porcine small intestinal submucosa to enhance the healing of the medial collateral ligament—a functional tissue engineering study in rabbits. J. Orthop. Res. 22, 214–220, © by Journal of Orthopaedic Research, with permission from Wiley.

the dynamic stimulation of the cells seeded on the SIS grafts leads to cell tensioning related to an increase in ECM production; not only tenocytes, but also Fbs are known to act through mechanotransduction pathways by remodeling the ECM (Screen et al., 2004).

Porcine SIS was also used for rotator cuff (RC) augmentation in a sheep model. Twelve weeks postoperation, ultimate failure loads were 1.3 times higher, however, no statistically significant difference was obtained when the results were compared to nonaugmented tendons, although the number of animals involved was $n = 13$ for augmented and nonaugmented tendons (Schlegel et al., 2006). In contrast, a significant increase in stiffness by a factor of 1.4 was found, which is clinically relevant because

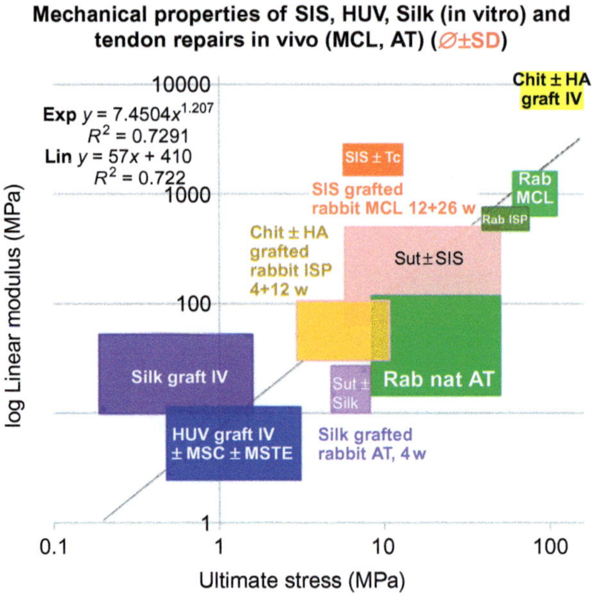


**Fig. 6.4** Natural grafts for tendon repairs: correlations of mechanical properties of small intestine submucosa (SIS), human umbilical vein (HUV), silk and chitosan/hyaluronan acid hybrid polymer fiber (Chit ± HA) in vitro and ex vivo after tendon repair in rabbit (Rab) models. *Key:* Tendons: *AT*, Achilles tendon; *MCL*, medial collateral ligament; *sut*, suture only; *w*, week. Mechanical stimulation: *MSTE*, mechanical stimulation for tissue engineering. *sut*, suture; *nat*, native; *IV*, in vitro. Data: *Ø ± SD*, mean ± standard deviation.
Data from Androjna, C., Spragg, R.K., Derwin, K.A., 2007. Mechanical conditioning of cell-seeded small intestine submucosa: a potential tissue-engineering strategy for tendon repair. Tissue Eng. 13, 233–243; Kropp, B.P., Sawyer, B.D., Shannon, H.E., Rippy, M.K., Badylak, S.F., Adams, M.C., Keating, M.A., Rink, R. C., Thor, K.B., 1996. Characterization of small intestinal submucosa regenerated canine detrusor: assessment of reinnervation, in vitro compliance and contractility. J. Urol. 156, 599–607; Musahl, V., Abramowitch, S.D., Gilbert, T.W., Tsuda, E., Wang, J.H.C., Badylak, S.F., Woo, S.L.Y., 2004. The use of porcine small intestinal submucosa to enhance the healing of the medial collateral ligament—a functional tissue engineering study in rabbits. J. Orthop. Res. 22, 214–220; Woo, S.L., Takakura, Y., Liang, R., Jia, F., Moon, D.K., 2006. Treatment with bioscaffold enhances the fibril morphology and the collagen composition of healing medial collateral ligament in rabbits. Tissue Eng. 12, 159–166; Abousleiman, R.I., Reyes, Y., McFetridge, P., Sikavitsas, V., 2008. The human umbilical vein: a novel scaffold for musculoskeletal soft tissue regeneration. Artif. Organs 32, 735–742; Abousleiman, R.I., Reyes, Y., McFetridge, P., Sikavitsas, V., 2009. Tendon tissue engineering using cell-seeded umbilical veins cultured in a mechanical stimulator. Tissue Eng. Part A 15, 787–795; Ni, M., Rui, Y.F., Tan, Q., Liu, Y., Xu, L.L., Chan, K.M., Wang, Y., Li, G., 2013. Engineered scaffold-free tendon tissue produced by tendon-derived stem cells. Biomaterials 34, 2024–2037; Ni, T., Senthil-Kumar, P., Dubbin, K., Aznar-Cervantes, S.D., Datta, N., Randolph, M.A., Cenis, J.L., Rutledge, G.C., Kochevar, I.E., Redmond, R.W., 2012. A photoactivated nanofiber graft material for augmented Achilles tendon repair. Lasers Surg. Med. 44, 645–652; Meier Buergisser, G., Calcagni, M., Muller, A., Bonavoglia, E., Fessel, G., Snedeker, J.G., Giovanoli, P., Buschmann, J., 2014. Prevention of peritendinous adhesions using an electrospun DegraPol polymer tube: a histological, ultrasonographic, and biomechanical study in rabbits. BioMed Res. Int. 2014, 656240; Funakoshi, T., Majima, T., Iwasaki, N., Suenaga, N., Sawaguchi, N., Shimode, K., Minami, A., Harada, K., Nishimura, S.-I., 2005a. Application of tissue engineering techniques for rotator cuff regeneration using a chitosan-based hyaluronan hybrid fiber scaffold. J. Biomed. Mater. Res. A 33, 1193–1201; Funakoshi, T., Majima, T., Iwasaki, N., Yamane, S., Masuko, T., Minami, A., Harada, K., Tamura, H., Tokura, S., Nishimura, S.-I. 2005b. Novel chitosan-based hyaluronan hybrid polymer fibers as a scaffold in ligament tissue engineering. J. Biomed. Mater. Res. A 74A, 338–346.

stiffness is the biomechanical parameter representing the tissue response to destructive loads seen in early rehabilitation (Table 6.2b). Moreover, RC repair by SIS were compared for an acute and a chronic rat model (after 8 weeks of an untreated defect; Table 6.2a and b); while acute repairs with SIS were similar to acute repairs without SIS with respect to elastic modulus and stiffness, chronic repairs with SIS significantly increased the modulus compared to chronic repairs without SIS—whereas stiffness was unaffected (Perry et al., 2007).

In a rabbit PT model, porcine SIS augmentation was shown not only to increase biomechanical characteristics beneficially (1.7 times higher failure loads and 1.6 times higher stiffness), but also a welcome antiadhesive effect of such an SIS barrier placed between PT and fat pad (Karaoglu et al., 2008). Moreover, in a goat ACL model, augmentation with ECM-SIS from genetically modified pigs was shown to increase ultimate load by a factor of 2.6 and stiffness by a factor of 2.2 (Fisher et al., 2012).

### 6.3.2 Fascia lata

*Fascia lata* is the deep fascia of the thigh. It originates from the *tensor fascia lata* muscle and is inserted into the Gerdy's tubercle at the knee. In a study by Derwin et al. (2008), human *fascia lata* was characterized in terms of its regional variability of biophysical (thickness) and mechanical properties (stiffness). Moreover, protocols were established in order to reduce antigenicity (decrease the DNA content of the graft) as much as possible by decellularization and/or antibiotic soaking. While minimal variability in mechanical and geometric properties were found, biomechanical properties changed upon antibiotic soaking (19% higher stiffness; 33% higher linear modulus) and upon decellularization in combination with antibiotic soaking (5% lower stiffness, not significant; 17% higher linear modulus) (Table 6.2a and b). Hence, having similar biomechanical properties as tendons and ligaments (Butler et al., 1984) and undergoing only a slight variation in biomechanics upon decellularization and/or antibiotic treatment, *fascia lata* was concluded to be an appropriate auto- or allograft for T&L rupture repair. Not surprisingly, *fascia lata* has been used recently in the clinical setting as AT reinforcement (Soons et al., 2015; Fig. 6.5), PT (De Baere et al., 2014) or quadriceps reinforcement (Helito et al., 2014). Besides, when *fascia lata* is used as a reinforcing graft, it has to be considered that the biomechanical properties are age-dependent (Steinke et al., 2012; Hammer et al., 2012), with *fascia lata* of young donors having lower elastic modulus and lower stiffness compared to old donors. Stiffening with increasing age is reported to be related to a decrease of elastin, an increase of collagen-crosslinks, and loss of body water (Hammer et al., 2012).

### 6.3.3 Muscle

A rather unconventional approach has been pursued by Majewski et al. using a genetically modified muscle flap aimed at transferring BMP-12 (bone morphogenetic proteins) (Majewski et al., 2008) or TGF-β1 (Majewski et al., 2012) on the healing of rat ATs. Rat muscle biopsies were transduced with recombinant adenovirus encoding

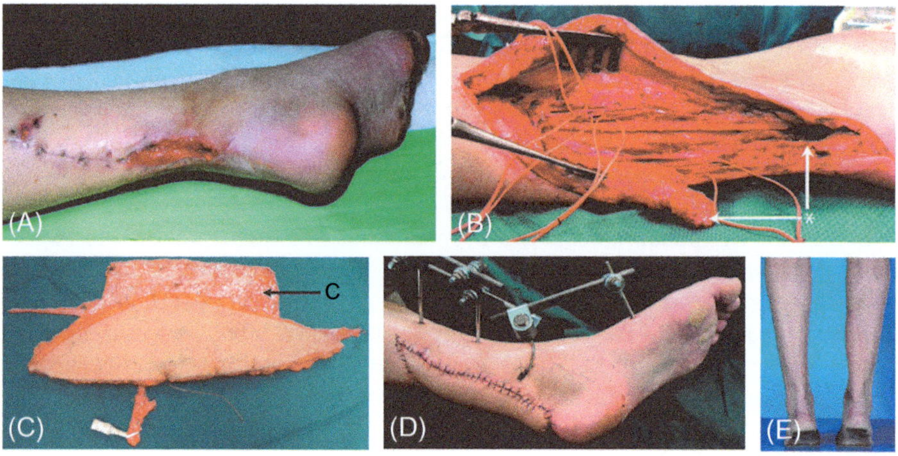


**Fig. 6.5** *Fascia lata* for Achilles tendon reinforcement: photographs of a one-stage operation showing (A) combined Achilles tendon and overlying soft-tissue defect of left lower leg, (B) after radical debridement of necrotic and infected tissues (* proximal and distal ends of the Achilles tendon), (C) free anterolateral thigh flap with vascularized tensor *fascia lata* graft, (D) postoperative result with the external fixator in place and (E) esthetic result 5 years after reconstruction showing tiptoe position.
Reproduced with permission from Soons, J., Rakhorst, H.A., Ruettermann, M., Luijsterburg, A.J.M., Bos, P.K., Zöphel, O.T., 2015. Reconstruction of defects involving the Achilles tendon and local soft tissues: a quick solution for a lingering problem. Bone Joint J. 97-B, 215–220, © of the British Editorial Society of Bone and Joint Surgery, Figures 1a, 1b, 1c, 1d and 1e).

BMP-12 or TGF-β1 and grafted onto surgically transected ATs in recipient animals. The maximum failure load of the healing tendons was significantly increased in the BMP-12 group 1 week postsurgery, compared to controls, and the tendon stiffness was significantly higher at 1, 2, and 4 weeks postoperatively (Majewski et al., 2008). As for the TGF-β1 gene transfer to rat ATs, ultimate failure loads and stiffness were significantly higher 1 week postsurgery, however, no significant differences were observed for later time points (2, 4, and 8 weeks, respectively; Majewski et al., 2012). As such, increased stability at early time points postoperatively was achieved by genetically modified muscle flaps, which is a crucial and beneficial result, because early active mobilization is claimed to be very important for proper tendon healing (Kuwata et al., 2007; Silfverskiold and May, 1995), having the prerequisite of improved strength at the repair site at early time points.

### 6.3.4 Umbilical vein

The umbilical vein contains Wharton's jelly, a gelatinous substance made largely from mucopolysaccharides protecting the blood vessels inside. Moreover, it contains smooth muscle cells that are poorly differentiated and a few tiny myofilaments. Ligament tissue engineering using decellularized HUV—after removal of Wharton's jelly—as a scaffold material has been examined biomechanically when (b)MSCs were

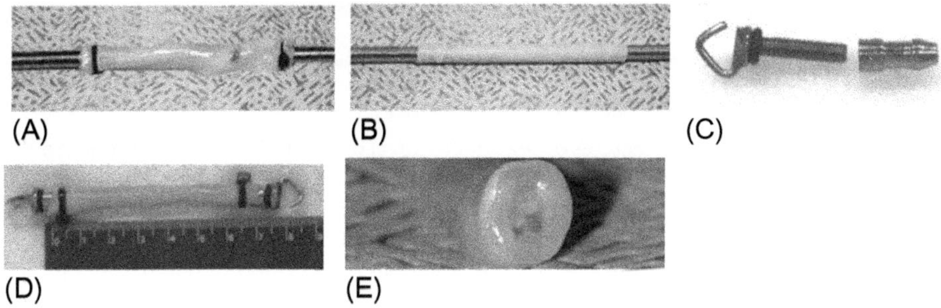


**Fig. 6.6** Human umbilical vein (HUV): a novel scaffold for musculoskeletal soft-tissue regeneration. (A) An 8.5-cm section of umbilical cord before processing; (B) HUV after automatic dissection; (C) novel design of an adapter to facilitate handling and seeding of HUV; (D) HUV seeded with collagen and MSCs (mesenchymal stem cells), and cultured for 1 week; and (E) cross section of seeded HUV 1 week postculture.
Based on Abousleiman, R.I., Reyes, Y., McFetridge, P., Sikavitsas, V., 2008. The human umbilical vein: a novel scaffold for musculoskeletal soft tissue regeneration. Artif. Organs 32, 735–742, © by Artificial Organs, with permission from Wiley.

seeded and cultivated for 1 or 2 weeks in vitro and compared to scaffolds without cells (Abousleiman et al., 2008; Fig. 6.6). While after 2 weeks, ultimate stress was increased 1.8-fold compared to the decellularized HUV, the elastic modulus was increased by a factor of 2.6 (Table 6.2a).

Given the differentiation capacity of the MSCs, the authors concluded, that MSC-seeded HUVs are potential candidates for ligament reconstruction (Abousleiman et al., 2008). Although cell seeding densities may affect tensile properties (Issa et al., 2011), optimally seeded HUVs exhibit still at least two orders of magnitude lower ultimate stress values compared to natural tendons; making it difficult to withstand loads in vivo during early active motion in the rehabilitation period. Although perfusion of seeded HUVs in a bioreactor and additional cyclic loading lead to a further 1.6-fold increase in ultimate load and a 10% increase in elastic modulus, absolute values are still low compared to values obtained for natural tendons and ligaments (Abousleiman et al., 2009; Table 6.2a). Therefore, it is not surprising that research has so far refrained from using HUV as a tendon graft in animal models in vivo. No studies dealing with this topic are reported up-to-date.

## 6.3.5 Silk

Silk is a Food and Drug Administration (FDA)-approved natural material with outstanding mechanical properties (Omenetto and Kaplan, 2010). After the removal of sericin, silk fibroin is a biocompatible and biodegradable material whose physical and mechanical properties can be easily manipulated through structural readjustments (Hofmann et al., 2013; Kundu et al., 2013). It can be derived from domesticated silkworm cocoons of *Bombyx mori*, often referred to mulberry silk, which contains at least two major fibroin proteins, light and heavy chains, 25 and 325 kDa, respectively

Fig. 6.7 Three different architectures of silk after removal of sericin: (A) wired, (B) braided, and (C) straight; the wired structure was found to approximate best the biomechanics of ACL. Based on Li, X., Snedeker, J.G., 2013. Wired silk architectures provide a biomimetic ACL tissue engineering scaffold. J. Mech. Behav. Biomed. Mater. 22, 30–40, reprinted from Journal of the Mechanical Behavior of Biomedical Materials, Copyright 2013, with permission from Elsevier.

(Altman et al., 2003). On the other hand, it can also be derived from wild species of silkworms (nonmulberry silk; Musson et al., 2015). Silk fibroin provides an excellent combination of strong yet adjustable mechanical properties (failure stresses up to 4.8 GPa), remarkable toughness and elasticity (up to 35% failure strain), and a large degree of biological stability (Shen et al., 2014)—rendering it an ideal material for T&L repair.

For example, an electrospun silk mesh was wrapped around a transected and sutured rabbit AT and photochemically sealed (light activation of the photoactive dye Rose Bengal) in order to enhance the strength and to reduce peritendinous adhesions (Ni et al., 2012). While the elastic modulus was 2.8 times higher 2 weeks and 1.4 times higher 4 weeks postsurgery in the silk-augmented specimen compared to the just conventionally sutured ones, the ultimate stress was very similar (Table 6.2a). Moreover, different architectures of silk such as fibers, bundles, yarns, and cords that were wired, braided, or straight, were compared with respect to their biomechanical properties after cyclic loading up to 100,000 cycles (Li and Snedeker, 2013; Fig. 6.7). Aimed at ACL reconstruction, these different silk scaffolds (the antigenic protein sericin had been removed by sodium carbonate at 95°C) were also seeded with human foreskin Fbs and compared to their nonseeded equivalents. As a main outcome, a wired structure best approximated human ACL (Table 6.2b). Moreover, cyclic loading decreased the strength and increased the stiffness of the silk ligament construct—however, these two parameters remained unaffected by Fb seeding (Li and Snedeker, 2013).

Furthermore, nonmulberry silk scaffolds that contain the cell binding RGD tripeptide (Wang et al., 2006) were examined for their use in tendon tissue engineering; in an in vitro study, Musson et al. (2015) assessed the cytocompatibility and the immunogenicity of a knitted nonmulberry silk fibroin scaffold, Spidrex. It was shown that Spidrex promoted a more tenocytic morphology and supported tenocyte cell growth which was attributed to the fact that nonmulberry silks contain the cell binding RGD (arg–gly–asp), which renders Spidrex more tenocyte compatible compared to mulberry silks. Although no biomechanical tests were performed, it was concluded that the nonmulberry silk fibroin Spidrex might be a valuable candidate for the improvement of clinical outcomes by promoting and enhancing tendon repair.

### *6.3.6 Chitosan*

Chitosan is a polycationic polysaccharide, a copolymer of *N*-acetyl-glucosamine and *N*-glucosamine. It is generally obtained by alkaline deacetylation of chitin, which is the primary component in fungi cell walls and also the exoskeleton of shrimp, crab, and lobster.

A braided chitosan-hyaluronan composite scaffold was used for an *infraspinatus* defect in a rabbit model either seeded or not with autologous Fbs from rabbit PT (Funakoshi et al., 2005). Compared to nonaugmented repairs, scaffold-augmented repairs were 16% higher in ultimate stress and even 50% higher when cell seeded and 62% and 140%, respectively, higher in modulus, going along with an enhanced production of collagen I (Funakoshi et al., 2005; Table 6.2a and b). Although the biocompatibility and the biodegradability of this scaffold remained unclear, as other reports dealt with inflammatory reactions towards chitosan, the regenerating tendon tissue did not show significant inflammation. Therefore, in the same research group, this composite material was also used in a MCL reconstruction in the rabbit model (Irie et al., 2011). The biomechanical outcomes indicated an increase in ultimate failure load (39% higher in the cell-seeded scaffold compared to nonaugmented specimen), however, a slight decrease in stiffness was observed (14% lower stiffness compared to nonaugmentation) (Irie et al., 2011; Table 6.2b).

## 6.3.7 Dermis

Dermis has been widely explored as an ECM graft material to mechanically reinforce tendon repair sites, including studies with human dermis (Adams et al., 2006), porcine dermis (Nicholson et al., 2007), and bovine dermis (Valentin et al., 2006). Although studies in animal models using these materials have been reported to be successful for augmentation in RC repair, clinical reports were less enthusiastic; surgical failures and graft rejection were reported, when for example porcine dermis was used as a bridging construct in RC tears (Soler et al., 2007).

For that reason, Shea et al. (2010) examined different dermis grafts and seeded them with human tenocytes, in order to be closer to the clinical setting and to overcome the apparent dichotomy between laboratory animal studies and clinical outcomes. They found that among several commercially available dermis-based grafts *GraftJacket* (Wright Medical Technology, Arlington, Tennessee), a freeze-dried human dermis, had a significantly higher number of attached human tenocytes compared to *Conexa* (Tornier, Edina, Minnesota), which is a porcine dermis that is treated by the removal of α-galactose residues in order to minimize immunorejection, or compared to *Zimmer Collagen Repair Patch* (Zimmer, Warsaw, Indiana) being a porcine dermis with chemically cross-linked collagen to increase its strength. However, all three dermis grafts behaved very well with respect to cell proliferation. Their performance was significantly better compared to tenocyte proliferation on other commercially available ECM grafts such as human fascia lata (Musculoskeletal Transplant Foundation, Edison, New Jersey), OrthADAPT (Pegasus Biologics, Irvine, California) which is an equine pericardium, or even a synthetic mesh such as a woven biodegradable polyurethane (SportMesh, Biomet Sports Medicine, Warsaw, Indiana) that were included in the study.

However, when GraftJacket judged to be a beneficial graft for tendon repair augmentation in the in vitro study mentioned above (Shea et al., 2010) was used in a canine model, its performance turned out to be poor (Smith et al., 2012). GraftJacket

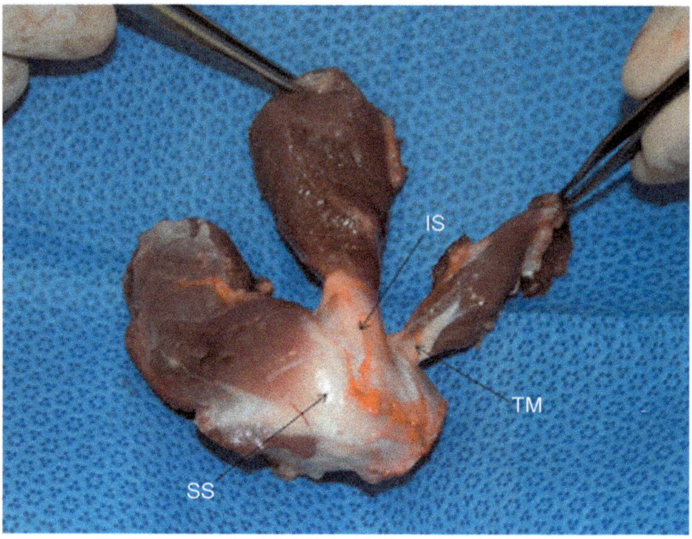


**Fig. 6.8** Canine rotator cuff tissue harvests for grafts (*IS*, infraspinatus; *SS*, supraspinatus; *TM*, teres minor).
Based on Smith, M.J., Cook, J.L., Kuroki, K., Jayabalan, P.S., Cook, C.R., Pfeiffer, F.M., Waters, N.P., 2012. Comparison of a novel bone-tendon allograft with a human dermis–derived patch for repair of chronic large rotator cuff tears using a canine model. Arthroscopy, 28, 169–177, reprinted from Arthroscopy: The Journal of Arthroscopic & Related Surgery, Copyright 2012, with permission from Elsevier.

was used for the repair of a transected *infraspinatus tendon* with a 3-week delay until surgery in order to generate a chronic RC tear model. It acted as an augmentation patch besides a conventional suture bridge. As a result, lymphoplasmacytic infiltrates along the periphery of the graft, tendon degeneration, hypocellularity, and lack of any definitive tendon-bone integration were determined (Smith et al., 2012). Moreover, compared to bone-tendon allografts of RC tissue which was investigated as another option of surgical treatment of RC tears (Fig. 6.8), the biomechanical performance of the GraftJacket patch was minor; 6 month postsurgery, the stiffness was reduced to around 60% compared to 70% (allograft) in comparison with the nontreated control (Table 6.2b). It was therefore concluded that the bone–tendon allograft is preferred over the dermis patch for this kind of repair.

An important feature when dermis is used as an augmentation patch for tendon repair is its performance in a suture pull-out test to guarantee time 0 mechanical stability at the repair site. Therefore Fini et al. (2012) examined a human dermal matrix which was physico-chemically decellularized either seeded or not with rat tenocytes. While decellularized dermis had failure loads and strengths comparable to the cellularized controls, the stiffness of decellularized dermis was significantly higher (Table 6.2a and b). Compared to natural infraspinatus stiffness ($100 \pm 30$ N/mm) (Smith et al., 2012), however, the stiffness of these decellularized dermis was lower by three orders of magnitudes ($6.7 \pm 1.3$ N/m) (Fini et al., 2012).

## 6.4 Summary

Large tendon or ligament defects (critical size defects) demand for tissue grafts in order to bridge and connect the tendon or ligament stumps. Autologous grafts would offer the best option for such treatments; however, limited availability, donor site morbidity, and a second incision to obtain the graft are disadvantageous. In this chapter, other options are presented, covering the suitability of allografts and xenografts and the necessary processing steps to achieve these implants; importantly, to avoid immune rejection, allo- and xenografts have to be decellularized prior to implantation. Moreover, decellularization techniques do vary the characteristics (for example mechanics) of the grafts and therefore different techniques are discussed and compared. Furthermore, such decellularized grafts can be reseeded by cells and made bioactive by adding selected growth factors. By doing so, T&L healing is considered to be improved and accelerated.

## References


Abousleiman, R.I., Reyes, Y., McFetridge, P., Sikavitsas, V., 2008. The human umbilical vein: a novel scaffold for musculoskeletal soft tissue regeneration. Artif. Organs 32, 735–742.

Abousleiman, R.I., Reyes, Y., McFetridge, P., Sikavitsas, V., 2009. Tendon tissue engineering using cell-seeded umbilical veins cultured in a mechanical stimulator. Tissue Eng. Part A 15, 787–795.

Adams, J.E., Zobitz, M.E., Reach Jr., J.S., An, K.-N.A., Steinmann, S.P., 2006. Rotator cuff repair using an acellular dermal matrix graft: an in vivo study in a canine model. Arthroscopy 22, 700–709.

Aiken, S.W., Badylak, S.F., Toombs, J.P., Shelbourne, K.D., Hiles, M.C., Lantz, G.C., Vansickle, D., 1994. Small-intestinal submucosa as an intraarticular ligamentous graft material—a pilot-study in dogs. Vet. Comp. Orthop. Traumatol. 7, 124–128.

Altman, G.H., Horan, R.L., Lu, H.H., Moreau, J., Martin, I., Richmond, J.C., Kaplan, D.L., 2002. Silk matrix for tissue engineered anterior cruciate ligaments. Biomaterials 23, 4131–4141.

Altman, G.H., Diaz, F., Jakuba, C., Calabro, T., Horan, R.L., Chen, J.S., Lu, H., Richmond, J., Kaplan, D.L., 2003. Silk-based biomaterials. Biomaterials 24, 401–416.

Androjna, C., Spragg, R.K., Derwin, K.A., 2007. Mechanical conditioning of cell-seeded small intestine submucosa: a potential tissue-engineering strategy for tendon repair. Tissue Eng. 13, 233–243.

Angelidis, I.K., Thorfinn, J., Connolly, I.D., Lindsey, D., Pham, H.M., Chang, J., 2010. Tissue engineering of flexor tendons: the effect of a tissue bioreactor on adipoderived stem cell-seeded and fibroblast-seeded tendon constructs. J. Hand Surg. 35, 1466–1472.

Aurora, A., Mccarron, J., Iannotti, J.P., Derwin, K., 2007. Commercially available extracellular matrix materials for rotator cuff repairs: state of the art and future trends. J. Shoulder Elb. Surg. 16, S171–S178.

Badylak, S.F., Lantz, G.C., Coffey, A., Geddes, L.A., 1989. Small intestinal submucosa as a large diameter vascular graft in the dog. J. Surg. Res. 47, 74–80.

Badylak, S.F., Tullius, R., Kokini, K., Shelbourne, K.D., Klootwyk, T., Voytik, S.L., Kraine, M.R., Simmons, C., 1995. The use of xenogeneic small-intestinal submucosa as a biomaterial for Achilles-tendon repair in a dog-model. J. Biomed. Mater. Res. 29, 977–985.

Badylak, S., Arnoczky, S., Plouhar, P., Haut, R., Mendenhall, V., Clarke, R., Horvath, C., 1999. Naturally occurring extracellular matrix as a scaffold for musculoskeletal repair. Clin. Orthop. Relat. Res. Suppl. 367, S333–S343.

Butler, D.L., Grood, E.S., Noyes, F.R., Zernicke, R.F., Brackett, K., 1984. Effects of structure and strain-measurement technique on the material properties of young human tendons and fascia. J. Biomech. 17, 579–596.

Chong, A.K.S., Riboh, J., Smith, R.L., Lindsey, D.P., Pham, H.M., Chang, J., 2009. Flexor tendon tissue engineering: acellularized and reseeded tendon constructs. Plast. Reconstr. Surg. 123, 1759–1766.

De Baere, T., De Muylder, J., Deltour, A., 2014. An unusual knee trauma: combined rupture of medial collateral ligament and patellar tendon. Case Rep. Orthop. 2014, 657296.

Defrate, L.E., Van Der Ven, A., Gill, T.J., Li, G., 2004. The effect of length on the structural properties of an Achilles tendon graft as used in posterior cruciate ligament reconstruction. Am. J. Sports Med. 32, 993–997.

Dejardin, L.M., Arnoczky, S.P., Ewers, B.J., Haut, R.C., Clarke, R.B., 2001. Tissue-engineered rotator cuff tendon using porcine small intestine submucosa. Histologic and mechanical evaluation in dogs. Am. J. Sports Med. 29, 175–184.

Derwin, K.A., Baker, A.R., Spragg, R.K., Leigh, D.R., Iannotti, J.P., 2006. Commercial extracellular matrix scaffolds for rotator cuff tendon repair. Biomechanical, biochemical, and cellular properties. J. Bone Joint Surg. (Am. Vol.) 88, 2665–2672.

Derwin, K.A., Baker, A.R., Spragg, R.K., Leigh, D.R., Farhat, W., Iannotti, J.P., 2008. Regional variability, processing methods, and biophysical properties of human fascia lata extracellular matrix. J. Biomed. Mater. Res. A 84, 500–507.

Dines, J.S., Grande, D.A., Dines, D.M., 2007. Tissue engineering and rotator cuff tendon healing. J. Shoulder Elb. Surg. 16, S204–S207.

Fini, M., Bondioli, E., Castagna, A., Torricelli, P., Giavaresi, G., Rotini, R., Marinelli, A., Guerra, E., Orlandi, C., Carboni, A., Aiti, A., Benedettini, E., Giardino, R., Melandri, D., 2012. Decellularized human dermis to treat massive rotator cuff tears: in vitro evaluations. Connect. Tissue Res. 53, 298–306.

Fisher, M.B., Liang, R., Jung, H.-J., Kim, K.E., Zamarra, G., Almarza, A.J., Mcmahon, P.J., Woo, S.L.Y., 2012. Potential of healing a transected anterior cruciate ligament with genetically modified extracellular matrix bioscaffolds in a goat model. Knee Surg. Sports Traumatol. Arthrosc. 20, 1357–1365.

Funakoshi, T., Majima, T., Iwasaki, N., Suenaga, N., Sawaguchi, N., Shimode, K., Minami, A., Harada, K., Nishimura, S.-I., 2005. Application of tissue engineering techniques for rotator cuff regeneration using a chitosan-based hyaluronan hybrid fiber scaffold. Am. J. Sports Med. 33, 1193–1201.

Funakoshi, T., Majima, T., Suenaga, N., Iwasaki, N., Yamane, S., Minami, A., 2006. Rotator cuff regeneration using chitin fabric as an acellular matrix. J. Shoulder Elb. Surg. 15, 112–118.

Gebauer, M., Beil, F.T., Beckmann, J., Sarvary, A.M., Ueblacker, P., Ruecker, A.H., Holste, J., Meenen, N.M., 2007. Mechanical evaluation of different techniques for Achilles tendon repair. Arch. Orthop. Trauma Surg. 127, 795–799.

Goradia, V.K., Rochat, M.C., Grana, W.A., Rohrer, M.D., Prasad, H.S., 2000a. Tendon-to-bone healing of a semitendinosus tendon autograft used for ACL reconstruction in a sheep model. Am. J. Knee Surg. 13, 143–151.

Goradia, V.K., Rochat, M.C., Kida, M., Grana, W.A., 2000b. Natural history of a hamstring tendon autograft used for anterior cruciate ligament reconstruction in a sheep model. Am. J. Sports Med. 28, 40–46.

Hammer, N., Lingslebe, U., Aust, G., Milani, T.L., Hädrich, C., Steinke, H., 2012. Ultimate stress and age-dependent deformation characteristics of the iliotibial tract. J. Mech. Behav. Biomed. Mater. 16, 81–86.

Helito, C.P., Bonadio, M.B., Angelini, F.J., Passarelli Tirico, L.E., Pecora, J.R., Canianho, G.L., Demange, M.K., 2014. Reinforcement with fascia lata as an alternative in the repair of chronic quadriceps tendon injuries. Ups. J. Med. Sci. 119, 354–356.

Hofmann, S., Hilbe, M., Fajardo, R.J., Hagenmueller, H., Nuss, K., Arras, M., Mueller, R., Von Rechenberg, B., Kaplan, D.L., Merkle, H.P., Meinel, L., 2013. Remodeling of tissue-engineered bone structures in vivo. Eur. J. Pharm. Biopharm. 85, 119–129.

Ilhami, K., Gokhan, M., Ulukan, I., Eray, B.M., Levent, A., Cigdem, T., 2004. Biomechanical and histologic comparison of Achilles tendon ruptures reinforced with intratendinous and peritendinous plantaris tendon grafts in rabbits: an experimental study. Arch. Orthop. Trauma Surg. 124, 608–613.

Irie, T., Majima, T., Sawaguchi, N., Funakoshi, T., Nishimura, S., Minami, A., 2011. Biomechanical and histologic evaluation of tissue engineered ligaments using chitosan and hyaluronan hybrid polymer fibers: a rabbit medial collateral ligament reconstruction model. J. Biomed. Mater. Res. A 97A, 111–117.

Issa, R.I., Engebretson, B., Rustom, L., McFetridge, P.S., Sikavitsas, V.I., 2011. The effect of cell seeding density on the cellular and mechanical properties of a mechanostimulated tissue-engineered tendon. Tissue Eng. Part A 17, 1479–1487.

Jackson, D.W., Grood, E.S., Cohn, B.T., Arnoczky, S.P., Simon, T.M., Cummings, J.F., 1991. The effects of insitu freezing on the anterior cruciate ligament—an experimental-study in goats. J. Bone Joint Surg. (Am. Vol.) 73A, 201–213.

Karabekmez, F., Zhao, C., 2012. Surface treatment of flexor tendon autograft and allograft decreases adhesion without an effect of graft cellularity: a pilot study. Clin. Orthop. Relat. Res. 470, 2522–2527.

Karaoglu, S., Fisher, M.B., Woo, S.L.Y., Fu, Y.-C., Liang, R., Abramowitch, S.D., 2008. Use of a bioscaffold to improve healing of a patellar tendon defect after graft harvest for ACL reconstruction: a study in rabbits. J. Orthop. Res. 26, 255–263.

Kellam, J.F., Hunter, G.A., Mcelwain, J.P., 1985. Review of the operative treatment of Achilles tendon rupture. Clin. Orthop. Relat. Res. 201, 80–83.

Kondo, E., Yasuda, K., Katsura, T., Hayashi, R., Kotani, Y., Tohyama, H., 2012. Biomechanical and histological evaluations of the doubled semitendinosus tendon autograft after anterior cruciate ligament reconstruction in sheep. Am. J. Sports Med. 40, 315–324.

Krapf, D., Kaipel, M., Majewski, M., 2012. Structural and biomechanical characteristics after early mobilization in an Achilles tendon rupture model: operative versus nonoperative treatment. Orthopedics 35, E1383–E1388.

Kropp, B.P., Sawyer, B.D., Shannon, H.E., Rippy, M.K., Badylak, S.F., Adams, M.C., Keating, M.A., Rink, R.C., Thor, K.B., 1996. Characterization of small intestinal submucosa regenerated canine detrusor: assessment of reinnervation, in vitro compliance and contractility. J. Urol. 156, 599–607.

Kundu, B., Rajkhowa, R., Kundu, S.C., Wang, X., 2013. Silk fibroin biomaterials for tissue regenerations. Adv. Drug Deliv. Rev. 65, 457–470.

Kuwata, S., Mori, R., Yotsumoto, T., Uchio, Y., 2007. Flexor tendon repair using the two-strand side-locking loop technique to tolerate aggressive active mobilization immediately after surgery. Clin. Biomech. 22, 1083–1087.

Lantz, G.C., Badylak, S.F., Coffey, A.C., Geddes, L.A., Sandusky, G.E., 1992. Small intestinal submucosa as a superior vena cava graft in the dog. J. Surg. Res. 53, 175–181.

Li, X., Snedeker, J.G., 2013. Wired silk architectures provide a biomimetic ACL tissue engineering scaffold. J. Mech. Behav. Biomed. Mater. 22, 30–40.

Li, X., He, J., Bian, W., Li, Z., Zhang, W., Li, D., Snedeker, J.G., 2014. A novel silk-based artificial ligament and tricalcium phosphate/polyether ether ketone anchor for anterior cruciate ligament reconstruction—safety and efficacy in a porcine model. Acta Biomater. 10, 3696–3704.

Liang, R., Woo, S.L.Y., Takakura, Y., Moon, D.K., Jia, F., Abramowitch, S.D., 2006. Long-term effects of porcine small intestine submucosa on the healing of medial collateral ligament: a functional tissue engineering study. J. Orthop. Res. 24, 811–819.

Liu, S., Qin, M., Hu, C., Wu, F., Cui, W., Jin, T., Fan, C., 2013. Tendon healing and anti-adhesion properties of electrospun fibrous membranes containing bFGF loaded nanoparticles. Biomaterials 34, 4690–4701.

Lomas, R.J., Jennings, L.M., Fisher, J., Kearney, J.N., 2004. Effects of a peracetic acid disinfection protocol on the biocompatibility and biomechanical properties of human patellar tendon allografts. Cell Tissue Bank. 5, 149–160.

Majewski, M., Betz, O., Ochsner, P.E., Liu, F., Porter, R.M., Evans, C.H., 2008. Ex vivo adenoviral transfer of bone morphogenetic protein 12 (BMP-12) cDNA improves Achilles tendon healing in a rat model. Gene Ther. 15, 1139–1146.

Majewski, M., Porter, R.M., Betz, O.B., Betz, V.M., Clahsen, H., Fluckiger, R., Evans, C.H., 2012. Improvement of tendon repair using muscle grafts transduced with TGF-beta1 cDNA. Eur. Cell. Mater. 23, 94–101. discussion 101–2.

Musahl, V., Abramowitch, S.D., Gilbert, T.W., Tsuda, E., Wang, J.H.C., Badylak, S.F., Woo, S.L.Y., 2004. The use of porcine small intestinal submucosa to enhance the healing of the medial collateral ligament—a functional tissue engineering study in rabbits. J. Orthop. Res. 22, 214–220.

Musson, D.S., Naot, D., Chhana, A., Matthews, B.G., McIntosh, J.D., Lin, S.T.C., Choi, A.J., Callon, K.E., Dunbar, R., Lesage, S.C., Coleman, B., Cornish, J., 2015. In vitro evaluation of a novel non-mulberry silk scaffold for use in tendon regeneration. Tissue Eng. Part A 21, 1539–1551.

Nagumo, A., Yasuda, K., Numazaki, H., Azuma, H., Tanabe, Y., Kikuchi, S.-I., Harata, S., Tohyama, H., 2005. Effects of separate application of three growth factors (TGF-beta1, EGF, and PDGF-BB) on mechanical properties of the in situ frozen-thawed anterior cruciate ligament. Clin. Biomech. (Bristol, Avon) 20, 283–290.

Ni, T., Senthil-Kumar, P., Dubbin, K., Aznar-Cervantes, S.D., Datta, N., Randolph, M.A., Cenis, J.L., Rutledge, G.C., Kochevar, I.E., Redmond, R.W., 2012. A photoactivated nanofiber graft material for augmented Achilles tendon repair. Lasers Surg. Med. 44, 645–652.

Nicholson, G.P., Breur, G.J., Van Sickle, D., Yao, J.Q., Kim, J., Blanchard, C.R., 2007. Evaluation of a cross-linked acellular porcine dermal patch for rotator cuff repair augmentation in an ovine model. J. Shoulder Elb. Surg. 16, 184S–190S.

Omenetto, F.G., Kaplan, D.L., 2010. New opportunities for an ancient material. Science 329, 528–531.

Perry, S.M., Gupta, R.R., Van Kleunen, J., Ramsey, M.L., Soslowsky, L.J., Glaser, D.L., 2007. Use of small intestine submucosa in a rat model of acute and chronic rotator cuff tear. J. Shoulder Elb. Surg. 16, S179–S183.

Pietschmann, M.F., Frankewycz, B., Schmitz, P., Docheva, D., Sievers, B., Jansson, V., Schieker, M., Mueller, P.E., 2013. Comparison of tenocytes and mesenchymal stem cells

seeded on biodegradable scaffolds in a full-size tendon defect model. J. Mater. Sci. Mater. Med. 24, 211–220.
Pridgen, B.C., Woon, C.Y.L., Kim, M., Thorfinn, J., Lindsey, D., Pham, H., Chang, J., 2011. Flexor tendon tissue engineering: acellularization of human flexor tendons with preservation of biomechanical properties and biocompatibility. Tissue Eng. Part C Methods 17, 819–828.
Rodger, M.P., Theobald, P., Giddins, G., 2015. Vein grafts to augment flexor tendon repairs: a biomechanical study on strength and gap resistance. J. Hand Surg. Eur. Vol. 40, 695–699.
Saber, S., Zhang, A.Y., Ki, S.H., Lindsey, D.P., Smith, R.L., Riboh, J., Pham, H., Chang, J., 2010. Flexor tendon tissue engineering: bioreactor cyclic strain increases construct strength. Tissue Eng. Part A 16, 2085–2090.
Sakai, T., Yasuda, K., Tohyama, H., Azuma, H., Nagumo, A., Majima, T., Frank, C.B., 2002. Effects of combined administration of transforming growth factor-beta 1 and epidermal growth factor on properties of the in situ frozen anterior cruciate ligament in rabbits. J. Orthop. Res. 20, 1345–1351.
Schlegel, T.F., Hawkins, R., Lewis, C.W., Motta, T., Turner, A.S., 2006. The effects of augmentation with swine small intestine submucosa on tendon healing under tension: histologic and mechanical evaluations in sheep. Am. J. Sports Med. 34, 275–280.
Schmitt, T., Fox, P.M., Woon, C.Y., Farnebo, S.J., Bronstein, J.A., Behn, A., Pham, H., Chang, J., 2013. Human flexor tendon tissue engineering: in vivo effects of stem cell reseeding. Plast. Reconstr. Surg. 132, 567e–576e.
Screen, H.R.C., Bader, D.L., Lee, D.A., Shelton, J.C., 2004. Local strain measurement within tendon. Strain 40, 157–163.
Sebastian, H., Datta, B., Maffulli, N., Neil, M., Walsh, W.R., 2007. Mechanical properties of reconstructed Achilles tendon with transfer of peroneus brevis or flexor hallucis longus tendon. J. Foot Ankle Surg. 46, 424–428.
Shea, K.P., McCarthy, M.B., Ledgard, F., Arciero, C., Chowaniec, D., Mazzocca, A.D., 2010. Human tendon cell response to 7 commercially available extracellular matrix materials: an in vitro study. Arthroscopy 26, 1181–1188.
Shen, W., Chen, X., Hu, Y., Yin, Z., Zhu, T., Hu, J., Chen, J., Zheng, Z., Zhang, W., Ran, J., Heng, B.C., Ji, J., Chen, W., Ouyang, H.-W., 2014. Long-term effects of knitted silk-collagen sponge scaffold on anterior cruciate ligament reconstruction and osteoarthritis prevention. Biomaterials 35, 8154–8163.
Silfverskiold, K.L., May, E.J., 1995. Early active mobilization after tendon transfers using mesh reiforced suture techniques. J. Hand Surg. (Br. Eur. Vol.) 20B, 291–300.
Smith, M.J., Cook, J.L., Kuroki, K., Jayabalan, P.S., Cook, C.R., Pfeiffer, F.M., Waters, N.P., 2012. Comparison of a novel bone-tendon allograft with a human dermis-derived patch for repair of chronic large rotator cuff tears using a canine model. Arthroscopy 28, 169–177.
Soler, J.A., Gidwani, S., Curtis, M.J., 2007. Early complications from the use of porcine dermal collagen implants (Permacol (TM)) as bridging constructs in the repair of massive rotator cuff tears—a report of 4 cases. Acta Orthop. Belg. 73, 432–436.
Soons, J., Rakhorst, H.A., Ruettermann, M., Luijsterburg, A.J.M., Bos, P.K., Zöphel, O.T., 2015. Reconstruction of defects involving the Achilles tendon and local soft tissues: a quick solution for a lingering problem. Bone Joint J. 97-B, 215–220.
Steinke, H., Lingslebe, U., Böhme, J., Slowik, V., Shim, V., Hädrich, C., Hammer, N., 2012. Deformation behavior of the iliotibial tract under different states of fixation. Med. Eng. Phys. 34, 1221–1227.
Syed, O., Walters, N.J., Day, R.M., Kim, H.-W., Knowles, J.C., 2014. Evaluation of decellularization protocols for production of tubular small intestine submucosa scaffolds for use in oesophageal tissue engineering. Acta Biomater. 10, 5043–5054.

Thomopoulos, S., Kim, M., Das, R., Silva, M.J., Sakiyama-Elbert, S., Amiel, D., Gelberman, R.H., 2010. The effects of exogenous basic fibroblast growth factor on intrasynovial flexor tendon healing in a canine model. J. Bone Joint Surg. 92 (13), 2285–2293.

Thorfinn, J., Angelidis, I.K., Gigliello, L., Pham, H.M., Lindsey, D., Chang, J., 2012. Bioreactor optimization of tissue engineered rabbit flexor tendons in vivo. J. Hand Surg. Eur. Vol. 37, 109–114.

Valentin, J.E., Badylak, J.S., McCabe, G.P., Badylak, S.F., 2006. Extracellular matrix bioscaffolds for orthopaedic applications—a comparative histologic study. J. Bone Joint Surg. (Am. Vol.) 88A, 2673–2686.

Vališ, P., Sklenský, J., Repko, M., Rouchal, M., Novák, J., Otaševič, T., 2014. Most frequent causes of autologous graft failure in anterior cruciate ligament replacement. Acta Chir. Orthop. Traumatol. Cech. 81, 371–379.

Wang, Y., Kim, H.-J., Vunjak-Novakovic, G., Kaplan, D.L., 2006. Stem cell-based tissue engineering with silk biomaterials. Biomaterials 27, 6064–6082.

Woo, S.L., Hollis, J.M., Adams, D.L., Lyon, R.M., Takai, S., 1991. Tensile properties of the human femur-anterior cruciate ligament-tibia complex. The effects of specimen age and orientation. Am. J. Sports Med. 19, 217–225.

Woo, S.L., Takakura, Y., Liang, R., Jia, F., Moon, D.K., 2006. Treatment with bioscaffold enhances the fibril morphology and the collagen composition of healing medial collateral ligament in rabbits. Tissue Eng. 12, 159–166.

Woon, C.Y.L., Kraus, A., Raghavan, S.S., Pridgen, B.C., Megerle, K., Pham, H., Chang, J., 2011a. Three-dimensional-construct bioreactor conditioning in human tendon tissue engineering. Tissue Eng. Part A 17, 2561–2572.

Woon, C.Y.L., Pridgen, B.C., Kraus, A., Bari, S., Pham, H., Chang, J., 2011b. Optimization of human tendon tissue engineering: peracetic acid oxidation for enhanced reseeding of acellularized intrasynovial tendon. Plast. Reconstr. Surg. 127, 1107–1117.

Woon, C.Y.L., Farnebo, S., Schmitt, T., Kraus, A., Megerle, K., Pham, H., Yan, X., Gambhir, S.S., Chang, J., 2012. Human flexor tendon tissue engineering: revitalization of biostatic allograft scaffolds. Tissue Eng. Part A 18, 2406–2417.

Yoshikawa, T., Tohyama, H., Katsura, T., Kondo, E., Kotani, Y., Matsumoto, H., Toyama, Y., Yasuda, K., 2006. Effects of local administration of vascular endothelial growth factor on mechanical characteristics of the semitendinosus tendon graft after anterior cruciate ligament reconstruction in sheep. Am. J. Sports Med. 34, 1918–1925.

Zalavras, C.G., Gardocki, R., Huang, E., Stevanovic, M., Hedman, T., Tibone, J., 2006. Reconstruction of large rotator cuff tendon defects with porcine small intestinal submucosa in an animal model. J. Shoulder Elb. Surg. 15, 224–231.

Zhang, F., Liao, L., 2014. Tissue engineered cystoplasty augmentation for treatment of neurogenic bladder using small intestinal submucosa: an exploratory study. J. Urol. 192, 544–551.

Zhang, J., Wang, G.Y., Xiao, Y.P., Fan, L.Y., Wang, Q., 2011. The biomechanical behavior and host response to porcine-derived small intestine submucosa, pericardium and dermal matrix acellular grafts in a rat abdominal defect model. Biomaterials 32, 7086–7095.

# Collagen for tendon and ligament repair: Preparations and biomechanics

7

## Abstract

Collagen is the main component of native tendons and ligaments. Tissue-engineered constructs as well as tendon and ligament (T&L) augmentations are therefore often based on collagen; collagen sponges, gels, films, or ribbons among others are used and different processing methods are currently investigated including cross-linking or extrusion. Biomechanical assessment of differently fabricated collagen-based scaffold materials is reported for in vitro studies, either cell-free or seeded with cells and either statically or dynamically treated. In addition, many studies report biomechanical outcomes of collagen scaffolds after implantation to close a full defect as well as augmentation devices in vivo. In this chapter, an overview of different strategies using collagen as a means for T&L reconstruction is presented and discussed.

**Keywords:** Cross-linking, GraftJacket, Extrusion, Collagen gel, Collagen sponge.

## Abbreviations

| | |
|---|---|
| **2D** | two dimensional |
| **3D** | three dimensional |
| **3D-ex** | three-dimensional extruded collagen (element in tissue engineering) |
| **ACL** | anterior cruciate ligament (tendons and ligaments) |
| **AT** | Achilles tendon (tendons and ligaments) |
| **C** | cells for cell seeding (elements in tissue engineering) |
| **cd-HA** | carbodiimide-derivatized hyaluronic acid (element in tissue engineering) |
| **CL** | cross-linked collagen (chemical, physical, or biological treatment in tissue engineering) |
| **Coll** | collagen (element of living beings) |
| **Coll MCA** | collagen matrices commercially available (elements in tissue engineering) |
| **cy** | cycles (mechanical treatment in tissue engineering) |
| **d** | days (duration) |
| **der** | dermis (element of living beings and tissue for tissue engineering) |
| **div. tr.** | diverse treatments (elements in tissue engineering) |
| **DTEC** | direct transitional electrical current (treatment in tissue engineering) |
| **ECM** | extra cellular matrix (matrix in living cells) |
| **EDC** | 1-ethyl-3-(3-dimethylaminopropyl) carbodiimide (treatment in tissue engineering) |
| **EGDE** | ethylene-glycol-diglycidylether (treatment in tissue engineering) |
| **ELAC** | electrochemically aligned collagen (yarn for zig-zag-pattern-weaving) |

**Biomechanics of Tendons and Ligaments. http://dx.doi.org/10.1016/B978-0-08-100489-0.00007-7**

| | |
|---|---|
| **Fb** | fibroblasts (cells in living beings and element in tissue engineering) |
| **g** | gel (collagen) (element in tissue engineering) |
| **GAG** | glycosaminoglycan (element of tendons and ligaments) |
| **gel** | gelatin (element in tissue engineering) |
| **hCADs** | human cadavers (specimens for lab experiments) |
| **HCl** | hydrogen chloride (treatment in tissue engineering) |
| **hDMG** | human dermal matrix graft (element in tissue engineering) |
| **hum** | human (description for tissue origins) |
| **l** | long (feature of grafts) |
| **LFT** | long flexor tendon (tendons and ligaments) |
| **MCA** | matrices commercially available (collagen matrices in tissue engineering) |
| **mMSC** | mouse mesenchymal stem cells (cell type) |
| **MSC** | mesenchymal stem cells (cell type) |
| **MSt** | mechanical stimulation (mechanical treatment in tissue engineering) |
| **MTM** | mechanical tensile machine (mechanical treatment in tissue engineering) |
| **Ø** | mean, average of values (statistical element) |
| **Ohm** | unity for resistance to electrical current (treatment in tissue engineering) |
| **PBS** | phosphate buffered saline (treatment in tissue engineering) |
| **PD** | porcine dermal patch (element in tissue engineering) |
| **PDS** | polydioxanone (element in tissue engineering) |
| **PEG** | polyethylene glycol (treatment in tissue engineering) |
| **PLA** | platelet (cells in living beings) |
| **PT** | patellar tendon (tendons and ligaments) |
| **RC** | rotator cuff (tendons and ligaments) |
| **rMSC** | rabbit mesenchymal stem cells (cell type) |
| **s** | short (feature of grafts) |
| **sCAD** | sheep cadaver (specimens for lab experiments) |
| **SD** | standard deviation (statistical element) |
| **SEM** | scanning electron microscopy (imaging technique) |
| **SEM** | standard error of mean (statistical element) |
| **SIS** | small intestinal submucosa (tissue) |
| **sp** | sponge (collagen) (element in tissue engineering) |
| **Sut** | suture (surgical technique) |
| **T&L** | tendon and ligaments (tendons and ligaments) |
| **Tb** | tenoblasts (cells in living beings and element in tissue engineering) |
| **TEC** | tissue-engineered construct (element in tissue engineering) |
| **UV** | ultra violet (physical treatment in tissue engineering) |

## 7.1 Introduction

As tendon and ligament (T&L) tissue is mainly composed of collagen (Sharma and Maffulli, 2006), it is self-evident that many approaches aiming at tendon reconstructions are based on collagen sponges, gels or collagen-based scaffolds. Isolation and purification techniques are well established for collagen, particularly for collagen type I. There are 29 natural types of collagen occurring in biological tissues (Kumbar et al., 2014), with collagen type I being predominant in healthy tendons and ligaments, while collagen type III predominates during the first 6 weeks after rupture or lacerations in the remodelling phase (Sharma and Maffulli, 2006). Important amino acid sequences in collagen allow the cells to adhere via integrins and also to proliferate. In tissue engineering, however, pure collagen constructs may limit the cell migration caused by too strong cell attachment to the material. When used as an implant material, collagen scaffolds often exclude water and contract—with the risk of an impaired function, lowered nutrient supply and necrotic cores caused by cell death.

## 7.2 External or internal collagen scaffolds contraction: Impact on biomechanical properties

Cell-seeded collagen scaffolds may be contracted either endogenously by cell-mediated contractions or externally by mechanical stimulation—having different impacts on the constructs' mechanical in vitro performance. For example, in a study by Berry et al. (2009), human neonatal dermal fibroblasts (Fbs) were seeded in type I collagen and the gels were cast in a racetrack-shaped mould containing a removable central island. Two of the four models were mechanically stressed (extended to 20 and 10 mm), as complete contraction was prevented by the presence of the island. In the third model, the central island was removed (released) and the fourth model allowed multiaxial contraction. As such, these four models differed with respect to the ability of the cells to contract. After 6-day cultivation, the mechanical properties of the resultant gels were assessed. The viability of the cells in Fb-seeded collagen gels depended on the cell-generated stress as a function of construct dimension—with higher stresses leading to lower cell viability. As for the mechanical properties of the resultant collagen gels, higher cell-generated stress led to a decreased mechanical performance of the whole scaffold. In the 20 mm model (most stressed of the four models), failure load was only 63% of the 10 mm model (lower internal stress) as well as of the released model. Furthermore, the stiffness was also only 68% of the 10 mm model and only 62% of the released model (Table 7.1a). As in the 10 mm model the cell alignment and matrix orientation was higher compared to the released models, it was suggested that slight stress was the best option for tissue-engineered T&L repair involving an orientated Fb-seeded collagen gel.

In contrast, external mechanical stimulation is reported to be beneficial with respect to the biomechanical properties of the tissue-engineered constructs (TECs). Tendon tissue engineering based on a commercially available collagen type I-gel has been reported to lead to 2.9 times higher ultimate stress and to 3.7 times higher elastic modulus (Table 7.1a) when avian tendon internal Fbs were seeded and uniaxially loaded for 1 h/day and 1% elongation at 1 Hz in a bioreactor (Garvin et al., 2003). Also, Feng et al. (2006) were able to significantly improve rat type I collagen gel properties embedded with Fbs, when the constructs aimed at tendon repair were uniaxially contracted with 0.01 Hz for 9 days; a 5.5 times higher ultimate stress and a 15 times higher (!) elastic modulus were obtained (Table 7.1a).

**Table 7.1 In vitro-properties of scaffolds of collagen: Mechanical and structural properties**

| Target tendon or lig. | Scaffold material | Specifications | Ultimate stress (MPa) | Elastic modulus (MPa) | Ultimate load (N) | Stiffness (N/mm) | Ref. |
|---|---|---|---|---|---|---|---|
| **(a) Collagen fibers and gels** | | | | | | | |
| T&L | Coll I-gel | Coll+Fb | 0.06±0.02 | 0.04±0.02 | | | Feng et al. (2006) |
| | | Coll+Fb+MSt | 0.35±0.04 | 0.54±0.04 | | | |
| | Coll-gel-mats (multilayer) | Coll(4 conc), 1–12 layers | 0.6–1.8 | 2.0–6.3 | | | Kumar et al. (2013) |
| | | Coll(4 conc)+CL, 1–4 L | 4.7–10.5 | 52–114 | | | |
| | 3D-Coll extruded | + Dest water dry | 112±*5* | 154±*11*[a] | 2.6±*0.06* | | Zeugolis et al. (2008a,b) |
| | | + Dest water wet | 3±*0.4* | 4±*1*[a] | 0.2±*0.02* | | |
| | | + PBS dry | 170±*19* | 1251±*184*[a] | 2.7±*0.09* | | |
| | | + PBS wet | 4±*0.6* | 17±2[a] | 0.2±*0.03* | | |
| | | + Isopropanol dry | 356±*25* | 1966±*188*[a] | 3.1±*0.12* | | |
| | | + Isopropanol wet | 3±*0.4* | 6±*1*[a] | 0.2±*0.01* | | |
| | | + Tris–HCl dry | 199±*22* | 790±*101*[a] | 3.8±*0.1* | | |
| | | + Tris–HCl wet | 0.1±*0.01* | 0.1±*0.02*[a] | 0.03±*0.003* | | |
| | | + Vegetable oil dry | 182±*18* | 1077±*92*[a] | 3.2±*0.1* | | |
| | | + Vegetable oil wet | 6±*1* | 5±*0.4*[a] | 0.2±*0.03* | | |
| | | Cited: Coll extr dry | 115–224 | | | | |
| | | Cited: Coll extr wet | 0.7–2.4 | | | | |
| | | Coll+EDC-CL+Tb | 4.6±*0.4* | 19±*2* | | | Enea et al. (2011) |
| | | Coll+EDC/EGDE-CL+Tb | 11±*1* | 46±*5* | | | |
| Flexors | 3D-Coll I-gel | Coll+Tb | 0.11±0.01 | 0.49±0.24 | | | Garvin et al. (2003) |
| | | Coll+Tb+MSt | 0.33±0.18 | 1.80±1.82 | | | |

| | | | | | | | |
|---|---|---|---|---|---|---|---|
| RC | Coll MCA | GraftJacket extr. 2.0 (hum der) | | | 229 ± 72 | | Barber et al. (2006) |
| | | MaxForce 1.4 (hum dermis) | | | 182 ± 50 | | |
| | | GraftJacket 1.0 (hum dermis) | | | 157 ± 38 | | |
| | | Permacol 1.0 (porcine dermis) | | | 128 ± 26 | | |
| | | TissueMend 1.1 (foetal bov. der) | | | 76 ± 22 | | |
| | | TissueMend 1.2 (foetal bov. der) | | | 70 ± 13 | | |
| | | Restore 1.0 (porcine SIS) | | | 38 ± 3 | | |
| | | CuffPatch 1.0 (porcine SIS) | | | 32 ± 4 | | |
| | | Native canine infraspin. | | 405 ± 86 | | | Derwin et al. (2006) |
| | | GraftJacket (hum dermis) | | 23 ± 5 | | 16 ± 6 | |
| | | TissueMend (foetal bov. der) | | 15 ± 4 | | 7 ± 1 | |
| | | Restore (porcine SIS) | | 36 ± 9 | | 7 ± 2 | |
| | | CuffPatch (porcine SIS) | | 40 ± 15 | | 7 ± 2 | |
| | | SIS patches Cook Biotech | | 26 ± 14 | | | Raghavan et al. (2005) |
| | | Permacol (porcine dermis) | 33 ± 7 | 69 ± 8 | 50 ± 12 | | Chaudhury et al. (2012) |
| | | Restore (porcine SIS) | 27 ± 4 | 71 ± 19 | 13 ± 1 | | |
| | | GraftJacket (hum der) | 27 ± 7 | 48 ± 50 | 28 ± 7 | | |
| | | SportMesh (synthetic PU) | 12 ± 5 | 14 ± 2 | 18 ± 8 | | |

*Continued*

**Table 7.1 Continued**

| Target tendon or lig. | Scaffold material | Specifications | Ultimate stress (MPa) | Elastic modulus (MPa) | Ultimate load (N) | Stiffness (N/mm) | Ref. |
|---|---|---|---|---|---|---|---|
| ACL/ T&L | 3D-Coll extruded | 3D-Coll dry | 112 ± 10 | 154 ± 22[a] | 2.5 ± 0.1 | | Zeugolis et al. (2009) |
| | | 3D-Coll wet | 3 ± 1 | 4 ± 1[a] | 0.2 ± 0.04 | | |
| | | 3D-Coll dry + chemical CL | 20–403 | 15–1786[a] | 1.5–3.2 | | |
| | | 3D-Coll dry + biological CL | 173 ± 47 | 352 ± 124[a] | 3.3 ± 0.2 | | |
| | | 3D-Coll dry + physical CL | 47–106 | 36–583[a] | 1.0–2.6 | | |
| ACL | 3D-Coll extruded (div. C = content. mg/ml, + div. P = % PEG-treat.) | Bovine AT coll, C1 | 168 ± 62 | 1490 ± 570[b] | 1.8 ± 0.2 | | Zeugolis et al. (2008a) |
| | | Bovine AT coll, C2 | 108 ± 19 | 560 ± 460[b] | 3.1 ± 0.2 | | |
| | | Rat tail tendon, C3 | 255 ± 89 | 3310 ± 1090[b] | 0.5 ± 0.1 | | |
| | | Rat tail tendon, C4 | 134 ± 43 | 1590 ± 610[b] | 1.6 ± 0.1 | | |
| | | 3D-Coll + P5 | 139 ± 33 | 1200 ± 500[b] | 1.8 ± 0.2 | | |
| | | 3D-Coll + P20 | 136 ± 57 | 900 ± 700[b] | 1.4 ± 0.2 | | |
| | | 3D-Coll + P40 | 57 ± 40 | 400 ± 500[b] | 1.4 ± 0.3 | | |
| | Coll I-gel with centr. island (ci) | Coll g (20 mm, ci) + Fb | | | 0.05–0.07 | ~0.01 | Berry et al. (2009) |
| | | Coll g (10 mm, ci) + Fb | | | 0.10–0.12 | 0.01-0.02 | |
| | | Coll g (released ci) + Fb | | | 0.09–0.11 | 0.01-0.02 | |
| | Coll yarn | Native human ACL (younger) | 38 ± 9 | 111 ± 26 | | | Noyes and Grood (1976), Walters et al. (2012) |
| | | Coll g | 0.8 ± 0.1 | 2 ± 0.2 | | | |
| | | Coll g + EDC-CL + gelatin | 1.1 ± 0.1 | 7 ± 1 | | | |
| | | Coll g + EDC-CL | 19 ± 3 | 148 ± 17 | | | |
| PT | Coll I-gel | Coll g + MSC, static. | | 0.005 ± *0.001* | | 0.020 ± 0.002 | Butler et al. (2009) |
| | | Coll g + MSC + MSt | | 0.020 ± *0.010* | | 0.050 ± 0.010 | |

| | | | | | | | |
|---|---|---|---|---|---|---|---|
| **(b) Collagen sponges** | | | | | | | |
| T&L knee | Coll I-sp | Coll 11 mm+rMSC | 0.007±*0.001* | 0.024±*0.004* | 0.07±*0.01* | 0.018±*0.004* | Nirmalanandhan et al. (2007b) |
| | | Coll 51 mm+rMSC | 0.015±*0.001* | 0.064±*0.009* | 0.14±*0.01* | 0.031±*0.005* | |
| | | Coll+rMSC (static) | | 0.005±*0.001* | | 0.015±*0.004* | Nirmalanandhan et al. (2007a) |
| | | Coll+rMSC+dyn. | | 0.016±*0.004* | | 0.048±*0.009* | |
| | | Coll+mMSC (static) 0 d | | | | 0.049±0.008 | Chokalingam et al. (2009) |
| | | Coll+mMSC (static) 7 d | | | | 0.015±0.003 | |
| | | Coll+mMSC (static) 14 d | | | | 0.017±0.004 | |
| | | Coll+mMSC (dyn) 7 d | | | | 0.035±0.006 | |
| | | Coll+mMSC (dyn) 14 d | | | | 0.029±0.004 | |
| PT | Coll I-sp | Coll+rMSC (static) | 0.003±0.001 | 0.005±0.002 | 0.08±0.02 | 0.02±0.01 | Juncosa-Melvin et al. (2007) |
| | | Coll+rMSC+dyn. | 0.005±0.002 | 0.02±0.01 | 0.11±0.03 | 0.05±0.02 | |
| | | Coll ac | | | | 0.027±*0.004* | Kinneberg et al. (2010) |
| | | Coll+MSC (static) | | | | 0.053±*0.006* | |
| | | Coll+MSC+MSt (100 cy) | | | | 0.050±*0.004* | |
| | | Coll+MSC+MSt (3000 cy) | | | | 0.041±*0.005* | |
| | | Coll/GAG | | | | 0.025±*0.003* | |
| | | Coll/GAG+MSC (static) | | | | 0.057±*0.009* | |
| | | Coll/GAG+MSC +MSt(100) | | | | 0.080±*0.006* | |
| | | Coll/GAG+MSC +MSt(3000) | | | | 0.032±*0.003* | |

*Continued*

**Table 7.1 Continued**

| Target tendon or lig. | Scaffold material | Specifications | Ultimate stress (MPa) | Elastic modulus (MPa) | Ultimate load (N) | Stiffness (N/mm) | Ref. |
|---|---|---|---|---|---|---|---|
| PT +AT | Coll I- sp. + g short/long | Coll ac g (short) | | | | 0.025 ± 0.005 | Nirmalanandhan et al. (2008) |
| | | Coll g (long) | | | | 0.047 ± 0.004 | |
| | | Coll g + MSC | | | | 0.024 ± 0.005 | |
| | | Coll sp + MSC | | | | 0.045 ± 0.004 | |
| | | Coll g, static | | | | 0.024 ± 0.004 | |
| | | Coll g + MSt | | | | 0.045 ± *0.005* | |
| | | Coll g(s) + MSC, static | | | | 0.011 ± *0.003* | |
| | | Coll g (s) + MSC + MSt | | | | 0.021 ± *0.010* | |
| | | Coll g (l) + MSC, static | | | | 0.024 ± *0.007* | |
| | | Coll g (l) + MSC + MSt | | | | 0.041 ± *0.004* | |
| | | Coll sp (s) + MSC, static | | | | 0.016 ± *0.006* | |
| | | Coll sp (s) + MSC + MSt | | | | 0.048 ± *0.011* | |
| | | Coll sp (l) + MSC, static | | | | 0.047 ± *0.005* | |
| | | Coll sp (l) + MSC + MSt | | | | 0.069 ± *0.006* | |
| **(c) Other scaffolds based on collagen** | | | | | | | |
| RC | Coll w (ELAC) | Coll w | | | 52 ± 12 | 22 ± 2 | Islam et al. (2015) |

[a] At 2.2% strain.
[b] At 2% strain.

*3D*, three dimensional; *ACL*, anterior cruciate ligament; *AT*, Achilles tendon; *ci*, central island; *CL*, cross-linking; *Coll MCA*, collagen matrices commercially available; *Coll*, collagen; *cy*, cycles; *d*, days; *EDC*, 1-ethyl-3-(3-dimethylaminopropyl) carbodiimide; *EGDE*, ethylene-glycol-diglycidylether; *ELAC*, electrochemically aligned collagen (yarn for zig-zag-pattern-weaving); *Fb*, fibroblasts; *g*, gel; *GAG*, glycosaminoglycan; *gel*, gelatin; *HCl*, hydrogen chloride; *hum der*, human dermis; *l*, long; *mMSC*, mouse MSC; *MSC*, mesenchymal stem cells; *MSt*, mechanical stimulation; *Ø ± SD*, mean ± standard deviation; *Ø ± SEM*, mean ± *standard error of the mean; PBS*, phosphate buffered saline; *PEG*, polyethylene glycol; *PT*, patellar tendon; *PU*, polyurethane urea; *RC*, rotator cuff; *rMSC*, rabbit MSC; *s*, short; *SIS*, small intestinal submucosa; *sp*, sponge; *T&L*, *tendon and* ligaments; *w*, woven.

## 7.3 Processing collagen plays a pivotal role in terms of biomechanics

There are many different processing conditions and techniques including layer-by-layer techniques as well as cross-linking. Thereby, a wide range of biomechanical properties of collagen scaffolds materials can be generated and the strength can be tuned for T&L tissue engineering (Kumar et al., 2013). For example, robust planar collagen fiber constructs were fabricated by drying collagen gels to form dense collagen mats that were layered (Fig. 7.1a).

The collagen mats were mechanically tuned as a function of concentration, thickness and layering. Ultimate stress of such layered construct ranged from 0.5 to 11 MPa and values of linear modulus were between 1 and 115 MPa under preservation of native fibrillar, D-periodic structure were reported (Kumar et al., 2013) (Fig. 7.2, "Mats" and "Mats + CL"). Not only was the wide range of biomechanical properties convincing and impressive (orders of magnitudes), but also bio-integration was shown to be remarkable in an in vivo ventral hernia repair rat model where the reconstituted collagen with its specific hierarchical structure and organisation was found to mimic the structure of the native tissue better than other grafts such as decellularized matrices.

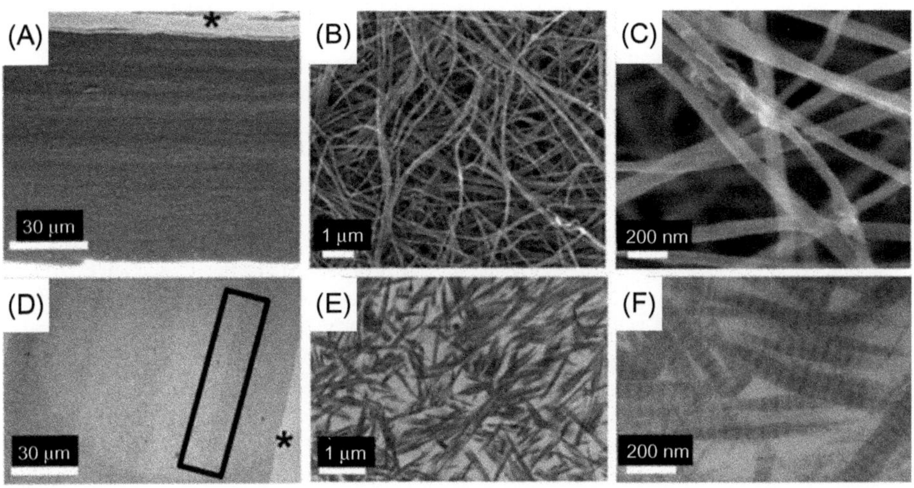


**Fig. 7.1** Ultrastructure of collagen matrices: scanning electron microscopy (SEM) of side views of four layered collagen matrix (A), showing indistinguishable interface between layers, magnified image to show dense fibrillar structure 10,000 × (B), and 50,000 × (C). Transmission electron micrographs (TEM) of four layered collagen matrix (D), magnified image to show dense fibrillar structure at interface of collagen layers (E), and magnified image showing D-periodic banding pattern (F). *Marks the top of layered construct. Box shows faint interface region between two collagen layers.
Reproduced with permission from Kumar, V.A., Caves, J.M., Haller, C.A., Dai, E., Liu, L., Grainger, S., Chaikof, E.L., 2013. Collagen-based substrates with tunable strength for soft tissue engineering. Biomater. Sci. 1, 1193–1202. The Royal Society of Chemistry.

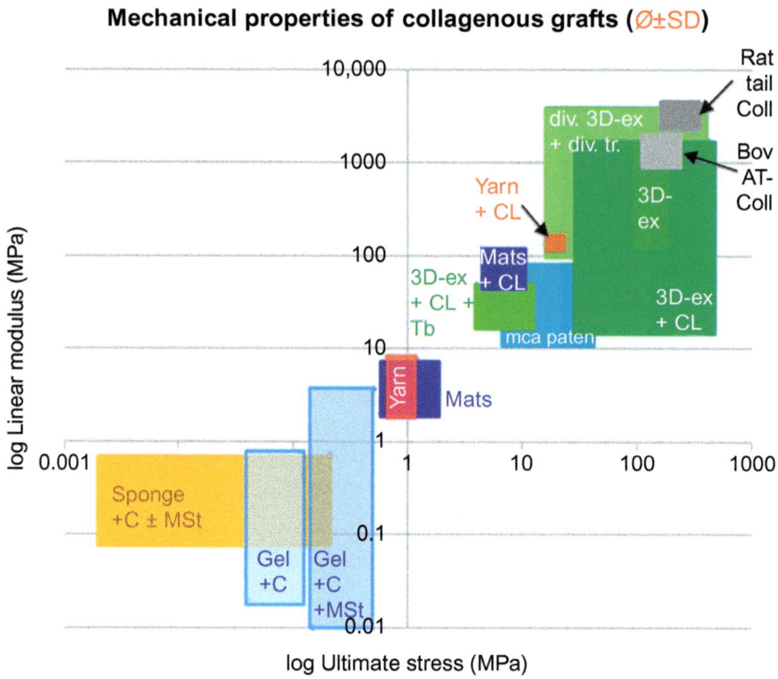


**Fig. 7.2** In vitro biomechanical correlations of various collagen grafts: sponges, gels, mats, patches, yarns, in comparison with rat tail collagen and bovine Achilles tendon collagen. Keys: *3D-ex*, three-dimensional extruded collagen; *C*, cells for cell seeding (tenocytes and fibroblasts); *CL*, cross-linked collagen; *div. tr.*, diverse treatments; *MCA*, matrices commercially available; *MSt*, mechanic stimulation; *Tb*, tenoblasts Data: *Ø* ± *SD*, mean ± standard deviation.

Data from Feng, Z., Ishibashi, M., Nomura, Y., Kitajima, T., Nakamura, T., 2006. Constraint stress, microstructural characteristics, and enhanced mechanical properties of a special fibroblast-embedded collagen construct. Artif. Organs, 30, 870–877; Garvin, J., Qi, J., Maloney, M., Banes, A.J., 2003. Novel system for engineering bioartificial tendons and application of mechanical load. Tissue Eng. 9, 967–979; Kumar, V.A., Caves, J.M., Haller, C.A., Dai, E., Liu, L., Grainger, S., Chaikof, E.L., 2013. Collagen-based substrates with tunable strength for soft tissue engineering. Biomater. Sci. 1, 1193–1202; Nirmalanandhan, V.S., Rao, M., Sacks, M.S., Haridas, B., Butler, D.L., 2007b. Effect of length of the engineered tendon construct on its structure-function relationships in culture. J. Biomech., 40, 2523–2529; Juncosa-Melvin, N., Matlin, K.S., Holdcraft, R.W., Nirmalanandhan, V.S., Butler, D.L., 2007. Mechanical stimulation increases collagen type I and collagen type III gene expression of stem cell—collagen sponge constructs for patellar tendon repair. In: ASME Summer Bioengineering Conference, Jun 24–28. Amer Soc Mechanical Engineers, Keystone, CO, pp. 981–982; Enea, D., Henson, F., Kew, S., Wardale, J., Getgood, A., Brooks, R., Rushton, N., 2011. Extruded collagen fibres for tissue engineering applications: effect of crosslinking method on mechanical and biological properties. J. Mater. Sci. Mater. Med. 22, 1569–1578; Zeugolis, D.I., Paul, G.R., Attenburrow, G., 2009. Cross-linking of extruded collagen fibers—a biomimetic three-dimensional scaffold for tissue engineering applications. J. Biomed. Mater. Res. A 89, 895–908; Zeugolis, D.I., Paul, R.G., Attenburrow, G., 2008a. Extruded collagen-polyethylene glycol fibers for tissue engineering applications. J. Biomed. Mater. Res. B Appl. Biomater. 85B, 343–352; Zeugolis, D.I., Paul, R.G., Attenburrow, G., 2008b. Post-self-assembly experimentation on extruded collagen fibres for tissue engineering applications. Acta Biomater. 4, 1646–1656; Walters, V.I., Kwansa, A.L., Freeman, J.W., 2012. Design and analysis of braid-twist collagen scaffolds. Connect. Tissue Res. 53, 255–266; Chaudhury, S., Holland, C., Thompson, M.S., Vollrath, F., Carr, A.J., 2012. Tensile and shear mechanical properties of rotator cuff repair patches. J. Shoulder Elb. Surg. 21, 1168–1176.

## 7.4 Commercially available collagen matrices

There are extracellular matrices for tendon augmentation which are commercially available (MCA) and that are primarily composed of collagen type I. These include different sources (dermis or small intestinal submucosa (SIS)), species (human, porcine, bovine), age of donor (foetal or adult) and processing methods and were evaluated in a biomechanical study for potential rotator cuff (RC) repairs (Derwin et al., 2006; Barber et al., 2006; Chaudhury et al., 2012; Table 7.1a). However, the elastic modulus of these materials was an order of magnitude lower than that of natural tendon, suggesting a limited mechanical role in augmentation of tendon repair. In the same sense, xenograft and allograft collagen matrices were compared with respect to their ultimate load (Barber et al., 2006; Table 7.1a). Although ultimate loads differed up to a factor of 7 with SIS having the lowest and human dermis the highest failure loads, the authors claimed that their study was not to suggest which graft material intended at tendon augmentation was superior to another for clinical practice, but rather, to demonstrate the wide range of biomechanical properties found for different commercially available collagen matrices—where the surgeons can choose what is appropriate for their specific needs (Barber et al., 2006).

Two application studies were followed by this research team. In a first study, human cadaver (hCAD) RC tendons were augmented with a human dermis collagen matrix, GraftJacket (Barber et al., 2008a). One shoulder from a matched pair was augmented with the human dermal allograft secured to the *humerus* and the *supraspinatus* tendon using the same sutures and suture anchors as the control (no collagen matrix). This led to a statistically significant 1.2 times increased ultimate load (Table 7.2b), which was claimed to be clinically relevant because the augmentation significantly increases the time zero strength in RC repair. In a second study, the same augmentation was used for hCAD Achilles tendons (ATs). Here, the augmentation led even to a twofold increase in time zero failure load (Barber et al., 2008b). Moreover, the stiffness of GraftJacket augmented ATs was three times higher compared to non-augmented controls—offering a possible beneficial treatment option in clinics (Table 7.2b).

However, depending on the tendon augmentation studies, biomechanical outcomes using GraftJacket are controversial. For example, while stiffness increased threefold in a GraftJacket augmented human AT as mentioned above, it did not change significantly in a GraftJacket augmented human RC (Omae et al., 2012); although a 1.6 times higher ultimate load resulted in the RC study, the stiffness did not change significantly (Table 7.2b). Furthermore, in another cadaver study, however, a human dermal allograft (Arthrex and Lifenet Health, Virginia Beach, Virginia) used for human RC augmentation did not enhance ultimate load, but just resulted in less variability of failure load values (Table 7.2b; Van Der Meijden et al., 2013). As such, the value and significance of cadaver studies using dermal collagen matrices to augment tendons have to be considered carefully—and the exact experimental conditions be closely studied in order to conclude about the benefits in terms of time zero strength and stiffness of the corresponding augmented tendons. For potential subsequent clinical

**Table 7.2 Properties of reconstructed tendons with collagen patches**

| Model | Target tendon or ligament | Augmented/ reconstr. by graft | Specifications | Healing time (weeks) | Ultimate stress (MPa) | Elastic modulus (MPa) | Ref. |
|---|---|---|---|---|---|---|---|
| **(a) Mechanical properties** | | | | | | | |
| Rabbit | PT | – | Native rabbit PT | – | 101 ± *16* | 861 ± *106* | Juncosa-Melvin et al. (2006b) |
| | | Coll gel | Sut only | 12 | 14 ± 7 | 155 ± 91 | Awad et al. (2003) |
| | | | Sut+Coll+MSC | | 19 ± 5 | 301 ± 93 | |
| | | | Sut only | 26 | 18 ± 9 | 238 ± 92 | |
| | | | Sut+Coll+MSC | | 26 ± 8 | 359 ± 129 | |
| | | Coll sponge | IV: Coll sp+MSC | – | 0.003 ± 0.001 | 0.005 ± 0.002 | Juncosa-Melvin et al. (2006c) |
| | | | IV: Coll sp+MCS+MSt | | 0.005 ± 0.002 | 0.020 ± 0.010 | |
| | | | Sut+Coll sp+MSC | 12 | 50 ± *9* | 343 ± *21* | |
| | | | Sut+Coll sp+MCS +MSt(100 cy/d) | | 72 ± *11* | 441 ± *26* | |
| | | Coll sponge +cross-linking | IV: Coll sp+MSC+MSt (100 cy/d) | – | 0.005 ± *0.001* | 0.020 ± *0.004* | Nirmalanandhan et al. (2009) |
| | | | IV: Coll sp/CL+MSC MSt (100) | | 0.015 ± *0.001* | 0.038 ± *0.004* | |
| | | | IV: Coll sp/CL+MSC +MSt (3000) | | 0.017 ± *0.003* | 0.046 ± *0.010* | |
| | | | Sut+Coll sp/CL+MSC +MSt(100) | 12 | 10 ± *2* | 47 ± *8* | |
| | | | Sut+Coll sp/CL+MSC +MSt(3000) | | 14 ± *4* | 51 ± *12* | |

| | | | | | | | |
|---|---|---|---|---|---|---|---|
| | | Coll gel/sponge composite | Sut+coll g/sp | 12 | 24 ± 2 | 160 ± 12 | Juncosa-Melvin et al. (2006b) |
| | | | Sut+coll g/sp+MCS | | 50 ± 4 | 269 ± 20 | |
| | | | Sut+Coll g/sp+MSC | 12 | 44 ± 15 | 277 ± 38 | Shearn et al. (2007) |
| | | | Sut+Coll g/sp+MSC +MSt | | 62 ± 14 | 341 ± 47 | |
| | | | IV: Coll g/s+MSC | | | 0.07 ± *0.01* | Butler et al. (2008) |
| | | | IV: Coll g/s+MSC +MSt | | | 0.14 ± *0.04* | |
| | | | Sut+Coll g+MSC | 12 | 30% | 15%[a] | |
| | | | Sut+Coll g/sp | | 25% | 20%[a] | |
| | | | Sut+Coll g/sp+MSC | | 50% | 30%[a] | |
| | | | Sut+Coll sp+MSC | | 50% | 40%[a] | |
| | | | Sut+Coll g/sp+MSC +MSt | | 60% | 40%[a] | |
| | | | Sut+Coll sp+MSC +MSt | | 70% | 50%[a] | |
| | AT | Coll gel (Coll: cell-ratios) | Native rabbit AT | – | 33 ± *4* | 180 ± *13* | Juncosa-Melvin et al. (2006a) |
| | | | Sut+Coll g+MSC (C: c = 0.04) | 12 | 31 ± *6* | 168 ± *32* | |
| | | | Sut+Coll g+MSC (C: c = 0.08) | | 26 ± *4* | 139 ± *20* | |
| | | – | Native rabbit AT | – | 20–34 | 1.5–3.2 | Meimandi-Parizi et al. (2013), Oryan et al. (2013), Moshiri et al. (2013b) |

*Continued*

Table 7.2 **Continued**

| Model | Target tendon or ligament | Augmented/ reconstr. by graft | Specifications | Healing time (weeks) | Ultimate stress (MPa) | Elastic modulus (MPa) | Ref. |
|---|---|---|---|---|---|---|---|
| | | Collagen | IV: Coll-TEC | – | 8 ± 1 | 1.0 ± 0.1 | Meimandi-Parizi |
| | | 3D-electrosp. | Sut only | 8.5 | 4 ± 2 | 0.2 ± 0.1 | et al. (2013) |
| | | | Sut + Coll-TEC | | 11 ± 0.2 | 0.8 ± 0.1 | |
| | | | Sut only | 17 | 11 ± 1 | 0.5 ± 0.1 | Oryan et al. |
| | | | Sut + Coll | | 19 ± 2 | 1.4 ± 0.2 | (2013) |
| | | Collagen | IV: Coll | – | 0.4 ± 0.1 | 0.004 ± 0.001 | Moshiri et al. |
| | | 2D-sheath | IV: Coll + CL | | 2.7 ± 0.5 | 0.044 ± 0.004 | (2013a) |
| | | 3D-electrosp. | IV: Coll + CL + PDS | | 9.1 ± 1.8 | 0.610 ± 0.057 | |
| | | | Sut only | 8.5 | 6 ± 1 | 0.24 ± 0.09 | |
| | | | Sut + 3D-Coll | | 10 ± 2 | 0.62 ± 0.14 | |
| | | | Sut + 3D-Coll/2D-PDS | | 13 ± 2 | 0.87 ± 0.17 | |
| | | | Sut only | 17 | 20 ± 1 | 1.0 ± 0.2 | |
| | | | Sut + 3D-Coll | | 27 ± 2 | 2.0 ± 0.2 | |
| | | | Sut + 3D-Coll/2D-PDS | | 27 ± 3 | 2.1 ± 0.3 | |
| Dog | RC | hDMG | Sut only | 0 | ~2.9 | | Adams et al. |
| | Infraspin. | | Sut + DMG | | ~6.9 | | (2006) |
| | | | Sut only | 6 | ~4 | | |
| | | | Sut + DMG | | ~3.5 | | |
| | | | Sut only | 13 | ~4.8 | | |
| | | | Sut + DMG | | ~5.5 | | |
| | | | Sut only | 26 | ~9 | | |
| | | | Sut + DMG | | ~8.5 | | |

| | | | | | | | |
|---|---|---|---|---|---|---|---|
| **(b) Structural properties** | | | | | | | |
| Human CAD + MTM | RC supraspin. | Coll MCA | Sut only | 0 | 273±116 | | Barber et al. (2008b) |
| | | | Sut+GraftJacket | | 325±74 | | |
| | RC | Coll MCA | Sut only | 0 | 346±61 | 77±16 | Omae et al. (2012) |
| | | | Sut+GraftJacket | | 560±96 | 65±16 | |
| | | | Native human RC (max) | 0 | 935±*5* | 127±*1* | Van Der Meijden et al. (2013) |
| | | | Sut only (single-row) (max) | | 260±*1* | 97±*1* | |
| | | | Sut only (ext. double-row) (max) | | 715±*4* | 114±*1* | |
| | | | Sut+hDMG (ext. d-row) (max) | | 670±*3* | 101±*1* | |
| | AT | Coll MCA | Sut only | 0 | 217±31 | 4±1 | Barber et al. (2008a) |
| | | | Sut+GraftJacket+MSt (cy) | | 455±76 | 13±5 | |
| | | Coll MCA | Sut only | 0 | 392±46 | 38±10 | Magnussen et al. (2011) |
| | | | Sut+TissueMend | | 821±123 | 18±3 | |
| | | Coll MCA ribbon | Sut only | 0 | 98±18 | | Berlet et al. (2014) |
| | | | Sut+Trellis | | 392±75 | | |
| Sheep CAD + MTM | AT | Coll MCA | Sut only | 0 | 86–276 | 37±7 | Song et al. (2010) |
| | | | Sut+TissueMend | | 508±93 | 31±3 | |
| | | | Sut+GraftJacket | | 460±88 | | |
| Rabbit | RC Infraspin. | Coll w (ELAC) | IV: Coll w | – | 52±12 | 22±2 | Islam et al. (2015) |
| | | | Native rabbit infraspin. | – | 130±25 | 34±8 | |
| | | | Sut only | 0 | 55±15 | 8±2 | |
| | | | Sut+Coll w | 0 | 65±10 | 7±1 | |
| | PT | – | Native rabbit PT | – | 471±*67* | 160±*11* | Juncosa-Melvin et al. (2006b) |

*Continued*

**Table 7.2 Continued**

| Model | Target tendon or ligament | Augmented/ reconstr. by graft | Specifications | Healing time (weeks) | Ultimate stress (MPa) | Elastic modulus (MPa) | Ref. |
|---|---|---|---|---|---|---|---|
| | | Coll gel | Sut only | 12 | 54 ± 32 | 38 ± 24 | Awad et al. (2003) |
| | | | Sut + Coll g + MSC | | 86 ± 34 | 85 ± 48 | |
| | | | Sut only | 26 | 38 ± 19 | 32 ± 16 | |
| | | | Sut + Coll g + MSC | | 105 ± 41 | 91 ± 50 | |
| | | Coll sponge | IV: Coll sp + MSC | – | 0.08 ± 0.02 | 0.02 ± 0.01 | Juncosa-Melvin et al. (2006c) |
| | | | IV: Coll sp + MCS + MSt | | 0.11 ± 0.03 | 0.05 ± 0.02 | |
| | | | Sut + Coll sp + MSC | 12 | 271 ± *18* | 87 ± *10* | |
| | | | Sut + Coll sp + MCS + MSt | | 339 ± *11* | 142 ± *11* | |
| | | Coll sponge + cross-linking | IV: Coll sp + MSC + MSt (100 cy/d) | – | 0.11 ± *0.01* | 0.05 ± *0.01* | Nirmalanandhan et al. (2009) |
| | | | IV: Coll sp/CL + MSC + MSt (100) | | 0.34 ± *0.01* | 0.07 ± *0.01* | |
| | | | IV: Coll sp/CL + MSC + MSt (3000) | | 0.35 ± *0.01* | 0.07 ± *0.01* | |
| | | | Sut + Coll sp/CL + MSC + MSt(100) | 12 | 166 ± *23* | 59 ± *11* | |
| | | | Sut + Coll sp/CL + MSC + MSt(3000) | | 218 ± *35* | 60 ± *10* | |
| | | Coll gel/sponge composite | Sut + coll g/sp | 12 | 198 ± *15* | 50 ± *3* | Juncosa-Melvin et al. (2006b) |
| | | | Sut + coll g/sp + MSC | | 278 ± *20* | 120 ± *10* | |
| | | | IV: Coll g/sp + MSC | – | | 0.08 ± 0.2 | Shearn et al. (2007) |
| | | | IV: Coll g/sp + MSC + MSt | | | 0.15 ± 0.1 | |
| | | | Sut + Coll g/sp + MSC | 12 | 270 ± 17 | 83 ± 10 | |
| | | | Sut + Coll g/sp + MSC + MSt | | 308 ± 13 | 125 ± 11 | |

| | | | | | | | |
|---|---|---|---|---|---|---|---|
| | | | IV: Coll g/sp+MSC | | | 0.08±0.01 | Butler et al. (2008) |
| | | | IV: Coll g/sp+MSC +MSt | | | 0.15±0.04 | |
| | | | Sut+Coll g+MSC | 12 | 30%[a] | 25%[a] | |
| | | | Sut+Coll g/sp | | 40%[a] | 30%[a] | |
| | | | Sut+Coll g/sp+MSC | | 60%[a] | 75%[a] | |
| | | | Sut+Coll sp+MSC | | 55%[a] | 55%[a] | |
| | | | Sut+Coll g/sp+MSC +MSt | | 65%[a] | 80%[a] | |
| | | | Sut+Coll sp+MSC +MSt | | 70%[a] | 85%[a] | |
| | AT | Coll gel (Coll : cell-ratios) | Native rabbit AT | – | 390±*50* | 93±*9* | Juncosa-Melvin et al. (2006a) |
| | | | Sut+Coll g+MSC (C: c=0.04) | 12 | 203±*32* | 60±*12* | |
| | | | Sut+Coll g+MSC (C: c=0.08) | | 185±*36* | 50±*10* | |
| | | – | Native rabbit AT | – | 212–351 | 57–212 | Meimandi-Parizi et al. (2013), Oryan et al. (2013), Moshiri et al. (2013b) |
| | | Collagen 3D-electrosp. | IV: Coll-TEC (final) | – | | 5±1 | Meimandi-Parizi et al. (2013) |
| | | | Sut only | 8.5 | 6±1 | 1.3±0.3 | |
| | | | Sut+Coll-TEC | | 26±3 | 5±2 | |
| | | | Sut only | 17 | 41±6 | 12±2 | Oryan et al. (2013) |
| | | | Sur+Coll | | 119±16 | 59±5 | |
| | | Collagen 2D-sheath 3D-electrosp. | IV: Coll | – | 5±2 | | Moshiri et al. (2013a) |
| | | | IV: Coll+CL | | 28±2 | | |
| | | | IV: Coll+CL+PDS | | 95±5 | | |
| | | | Sut only | 8.5 | 11±4 | 3±1 | |
| | | | Sut+3D-Coll | | 64±3 | 24±2 | |
| | | | Sut+3D-Coll/2D-PDS | | 81±8 | 32±4 | |

*Continued*

Table 7.2 **Continued**

| Model | Target tendon or ligament | Augmented/ reconstr. by graft | Specifications | Healing time (weeks) | Ultimate stress (MPa) | Elastic modulus (MPa) | Ref. |
|---|---|---|---|---|---|---|---|
| | | | Sut only | 17 | 31 ± 6 | 12 ± 2 | |
| | | | Sut + 3D-Coll | | 130 ± 16 | 65 ± 5 | |
| | | | Sut + 3D-Coll/2D-PDS | | 153 ± 8 | 78 ± 4 | |
| Rat | RC | Coll MCA | Native rat RC | 0 | 18 ± 3 | | Ide et al. (2009) |
| | | | Sut only | 2 | 3 ± 3 | | |
| | | | Sut + GraftJacket | | 6 ± 2 | | |
| | | | Sut only | 6 | 3 ± 1 | | |
| | | | Sut + GraftJacket | | 7 ± 2 | | |
| | | | Sut only | 12 | 9 ± 2 | | |
| | | | Sut + GraftJacket | | 13 ± 2 | | |
| Chicken | LFT | Collagen-GAG | Sut only | 3 | 14 ± 2 | | Bhavsar et al. (2010) |
| | | | Sut + Coll-GAG | | 12 ± 2 | | |
| Dog | FDP | Coll patches between stumps | Sut only | 0 | 47 ± 5 | 8 ± 2 | Zhao et al. (2014) |
| | | | Sut only | 6 | 70 ± 19 | 26 ± 13 | |
| | | | Sut + TEC (Coll + cd-HA + gelatin + GDF-5 + MSC) | | 45 ± 9 | 16 ± 5 | |
| | RC Infraspin. | Coll hDMG (human) | Sut only | 0 | ~65 | | Adams et al. (2006) |
| | | | Sut + hDMG | | ~63 | | |
| | | | Sut only | 6 | ~605 | | |
| | | | Sut + hDMG | | ~267 | | |
| | | | Sut only | 13 | ~528 | | |
| | | | Sut + hDMG | | ~539 | | |
| | | | Sut only | 26 | ~919 | | |
| | | | Sut + hDMG | | ~552 | | |

| | | | | | | | |
|---|---|---|---|---|---|---|---|
| Goat | ACL | Coll ECM (pig) | Native ACL | – | 1385–1634 | 112–125 | Fisher et al. (2012) |
| | | | Sut only | 12 | 95 ± 90 | 24 ± 21 | |
| | | | Sut+ECM | | 249 ± 129 | 53 ± 19 | |
| Sheep | RC Infraspin. | Porcine Dermal patch (cross-linked) | Sut only | 9 | 201 ± 60[b] | | Nicholson et al. (2007) |
| | | | Sut+SIS | | 137 ± 16 | | |
| | | | Sut+PD-CL | | 182 ± 63 | | |
| | | SIS | Sut only | 24 | 331 ± 55 | | |
| | | | Sut+SIS | | 340 ± 33 | | |
| | | | Sut+PD-CL | | 316 ± 97 | | |
| Pig | ACL | Collagen slurry composite | Native pig ACL | – | 713 ± 118 | 154 ± 20 | Joshi et al. (2009) |
| | | | Sut only | 4 | 55 ± 53 | 18 ± 17 | |
| | | | Sut+Coll+PLA | | 80 ± 40 | 21 ± 10 | |
| | | | Sut only | 6 | 30 ± 19 | 8 ± 5 | |
| | | | Sut+Coll+PLA | | 47 ± 36 | 17 ± 12 | |
| | | | Sut only | 14 | 80 ± 45 | 13 ± 6 | |
| | | | Sut+Coll+PLA | | 126 ± 53 | 40 ± 22 | |

*2D*, two dimensional; *3D*, three dimensional; *ACL*, anterior cruciate ligament; *AT*, Achilles tendon; *CAD*, cadaver; *cd-HA*, carbodiimide-derivatized hyaluronic acid; *CL*, cross-linking; *Coll MCA*, collagen matrices commercially available; *Coll*, collagen; *cy*, cyclic loads; *d*, days; *ECM*, extra cellular matrix; *ELAC*, electrochemically aligned collagen (yarn for zig-zag-pattern-weaving); *g*, gel; *GAG*, glycosaminoglycan; *hDMG*, human dermal matrix graft; *LFT*, long flexor tendon; *MSC*, mesenchymal stem cells; *MSt*, mechanical stimulation; *MTM*, mechanical tensile machine; *Ø ± SD*, mean ± standard deviation; *Ø ± SEM*, mean ± *standard error of the mean*; *PD*, porcine dermal patch; *PDS*, polydioxanone; *PLA*, platelet; *PT*, patellar tendon; *RC*, rotator cuff; *SIS*, small intestine submucosa; *sp*, sponge; *Sut*, suture; *TEC*, tissue-engineered construct; *w*, woven.

[a] % From native rabbit PT.

[b] [lb].

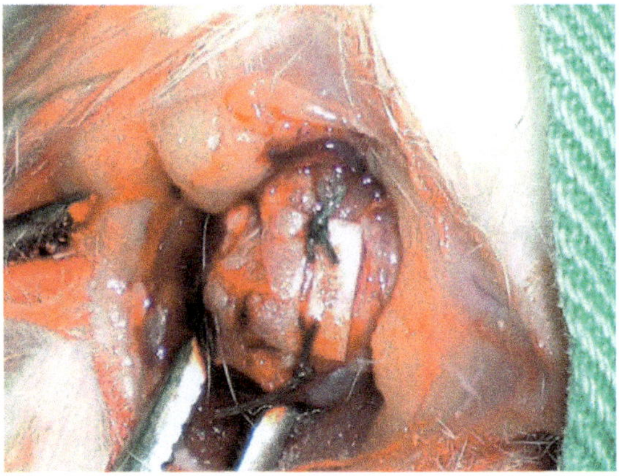

**Fig. 7.3** Large rotator cuff tear in a rat model reconstructed with a 3 × 5 mm acellular dermal matrix patch. The proximal tendon ends were affixed with a 5–0 nonabsorbable mattress suture and the opposite free end of the implant on the greater tuberosity was fixed in transosseous mattress fashion with a 5–0 nonabsorbable suture passing through the bone tunnel. Reproduced with permission from Ide, J., Kikukawa, K., Hirose, J., Iyama, K.-I., Sakamoto, H., Mizuta, H., 2009. Reconstruction of large rotator-cuff tears with acellular dermal matrix grafts in rats. J. Shoulder Elbow Surg. 18, 288–295, reprinted from Journal of Shoulder and Elbow Surgery, Copyright 2009, with permission from Elsevier.

applications, in vivo studies are more significant and meaningful with respect to translation into the clinical setting.

Besides, GraftJacket was not only applied as an augmentation device, but also to bridge a large gap in rat RC defects in vivo (Fig. 7.3). Twelve weeks postoperation, ultimate strength was 1.4 times higher in the GraftJacket group compared to the nonaugmented group (Ide et al., 2009). Moreover, not only were biomechanical properties significantly improved by GraftJacket application, but also the tissue integration as well as the tissue structure including collagen fiber alignment and the repopulation of the primarily cell-free graft were histologically examined and found to be beneficial and comparable to natural tendon tissue 12 weeks postsurgery (Table 7.2b).

The human dermal decellularized skin allograft GraftJacket was also compared with a collagen matrix derived from foetal bovine dermis, TissueMend. In a sheep cadaver study, TissueMend and GraftJacket augmentation of AT defects were found not to be significantly different with respect to ultimate load, however, the stiffness of TissueMend augmented ATs was higher, which implies that clinical application of TissueMend in ruptured tendons might be favourable (Table 7.2b). Clearly, both collagen matrices, GraftJacket as well as TissueMend, respectively, improved biomechanics significantly compared to nonaugmented reconstructions (Song et al., 2010). This beneficial biomechanical effect of augmentation by TissueMend was corroborated by Magnussen et al. (2011); TissueMend augmentation of hCAD AT were reported to also decrease gap formation (Table 7.2b).

Finally, collagen ribbon (Trellis Collagen Ribbon) augmentation of hCAD AT led to significant higher failure loads (four times higher) and lower gap formation (seven times lower) (Berlet et al., 2014; Table 7.2b). The collagen ribbons were woven in a box pattern and applied in addition to a conventional suture (4-strand Krakow). Although the biomechanical time zero improvement and the reduction in gap formation were convincing, it remains to be elucidated how well such a collagen ribbon will tissue-integrate in vivo and adapt functional tendon properties and characteristics.

## 7.5 Extrusion of collagen fibers

In 1989, Kato et al. for the first time reported on extruded collagen fibers for tissue engineering applications having similar physical and mechanical characteristics as natural rat tail tendon tissue (Kato et al., 1989). Since then, extruded collagen fibers have been widely examined as a graft material amongst others for tendon rupture repair. For example, extruded collagen fibers were analysed biomechanically, and dry and wet properties were compared (Zeugolis et al., 2008b). Obviously, various rehydration processes increased collagen fiber diameter significantly, going along with a significant decrease in ultimate stress (decrease by factors up to around 2000), ultimate load (more than 100) and elastic modulus (up to around 8000), respectively, for all treatments (including PBS = phosphate buffered saline, isopropanol, Tris–HCl and vegetable oil) (Table 7.1a). The same research group sought optimum tissue engineering application by combining extruded collagen fibers and PEG (polyethylene glycol) in various ratios. The 20:80 PEG:collagen-ratio was found to be the best for the production of reproducible extruded fibers. An increase of the collagen concentration resulted in fibers with higher diameter; force and strain at break values, whereas the stress at break and the modulus values decreased (Table 7.1a). When the internal diameter of the extruded fiber was increased, this had a significant influence on all the investigated mechanical properties (Zeugolis et al., 2008a).

## 7.6 Cross-linking of extruded collagen fibers

Chemical cross-linking of extruded collagen fibers by chromium sulphate, aldehydes, isocyanates was performed among others by Zeugolis et al. (2009). The authors compared biomechanical characteristics with the corresponding values by biological cross-linking (microbial transglutaminase) or physical cross-linking (ultraviolet (UV), photooxidation). They reported that chemical cross-linking was by far the most potent approach with respect to biomechanical properties (Table 7.1a). For example, immersion of collagen fibers into 0.05% sodium borohydride solution in distilled water at room temperature (reductive treatment) and an overnight washing in 0.01 M PBS led to force at break of $3.2 \pm 0.3$ N and a modulus of $1786 \pm 44$ MPa (at a strain of 2.2%), while a photo-oxidative treatment led to values of the same parameters of only $2.5 \pm 0.14$ N and $45 \pm 19$ MPa, respectively. The collection of such a variety of chemical, biological, as well as physical cross-linking treatments as well as the comparison of dry and wet fiber properties for the same treatment renders the study by Zeugolis et al. (2009) a very valuable compilation because it enables the reader a direct comparison of all these treatments with respect to fiber dimensions as well as biomechanics.

Other chemical cross-linking approaches such as 1-ethyl-3-(3-dimethylaminopropyl) carbodiimide (EDC) and EDC combined with ethylene-glycol-diglycidylether (EDC/EGDE) were investigated (Enea et al., 2011). Each of these approaches had advantages and disadvantages; while EDC was reported to have average biomechanical properties, but high biocompatibility, EDC/EGDE had improved biomechanical characteristics (Table 7.1a), but lower biocompatibility. Finally, EDC and UV cross-linking of braid-twist collagen yarn (± gelatine) were compared (Walters et al., 2012). It was found that EDC-cross-linked collagen scaffolds without gelatine performed best with respect to biomechanics (19 times higher ultimate stress than with gelatine and 22 times higher elastic modulus), approaching natural ACL (anterior cruciate ligament) properties the nearest (Table 7.1a).

## 7.7 Collagen sponges and gels

Not only fibers of collagen, but also sponges have been checked with respect to their suitability for tendon rupture repair and tissue engineering based on their biomechanical properties. From this point of view, the size of a sponge is crucial. For example, the length of a rat-MSC (mesenchymal stem cells) seeded collagen sponge that had been fixed between two posts in order to prevent the constructs from shrinking has been shown to be essential (MSC). When 11 and 51 mm long TECs were compared after 2 weeks in culture, ultimate stress, linear stiffness, ultimate force as well as linear modulus were significantly increased in the long construct compared to the short ones (Nirmalanandhan et al., 2007b; Table 7.1b). In contrast to these findings, when cadaver AT tendon grafts were used for posterior cruciate ligament reconstructions, short grafts were reported to have higher stiffness compared to long grafts (Defrate et al., 2004; Table 6.1c). According to the authors' interpretation (Nirmalanandhan et al., 2007b), the reason why the long tissue-engineered cell-collagen sponges had higher values for the above mentioned parameters was that the long construct had more homogenously aligned fibers than the short ones—resembling natural tendons more adequately.

MSC-collagen sponges are reported to be superior compared to MSC-collagen gels, because biomechanical properties of cell-loaded sponges improve upon mechanical stimulation, while the corresponding gels are not affected by these measures (Nirmalanandhan et al., 2007a). Moreover, expression of collagen type I and collagen type III is increased in rabbit MSC-collagen sponges due to mechanical loading (Juncosa-Melvin et al., 2007) as well as in murine MSC-collagen sponges (Chokalingam et al., 2009; Table 7.1b). In the same research group, interaction effects of varying length, collagen type (sponge vs gel) and mechanical stimulation (loading vs no loading) on the stiffness of the TECs revealed an interaction between length and stimulation; longer and stimulated MSC-collagen sponges showed significantly higher in vitro stiffness compared to all other combinations (Nirmalanandhan et al., 2008; Table 7.1b). Finally, different bioreactor systems to produce such functional tissue-engineered tendon constructs with various levels of costs and sophistication are presented by Butler et al. (2009) (Table 7.1a).

Not only mechanical loading, but also incorporation of typical components of natural tendon tissue into collagen sponges was reported to be beneficial with respect to biomechanical outcome. When chondroitin-6-sulphate, a glycosaminoglycan (GAG), was incorporated into a rabbit MSC-collagen type I sponge and the construct was loaded by 100 or 3000 cycles/day, respectively, it was found that the combination of the GAG and 100 cycles/day increased the linear stiffness of the construct significantly by a factor of 3.2 after 12 days of in vitro cultivation (Kinneberg et al., 2010; Table 7.1b). Besides, collagen-GAG membranes wrapped around chicken flexor tendons are reported to be beneficial with respect to antiadhesion and lead to lower work of flexion (Bhavsar et al., 2010; Table 7.2b).

## 7.8 In vivo experiments using different collagen scaffolds as tendon graft

A series of in vivo experiments were performed with collagen-based tendon grafts (Butler et al., 2008). In a study where the ratio of rabbit bone MSCs to collagen gel was varied, lower cell-to-collagen ratios were found to be beneficial when full AT defects were reconstructed and analysed biomechanically 12 weeks postoperation (Juncosa-Melvin et al., 2006a). For example, while stiffness, strain energy, maximum displacement and modulus were not affected by increasing the cell concentration from 0.04 to 0.08 million $MSC/mg_{Collagen}$, the ultimate load was decreased from $203 \pm 32$ N to $185 \pm 36$ (Table 7.2a and b). Compared to the central third of normal patellar tendon (PT) having $471 \pm 67$ N, both values were significantly lower. It was stressed by the authors, however, that peak in vivo forces were nevertheless achieved, which—to their opinion—was a more important prerequisite for later potential clinical application than having a similar load until failure (Juncosa-Melvin et al., 2006a). Moreover, also cell-free collagen constructs were implanted in full PT defects and compared to their cell-seeded analogues. Ultimate load and stiffness were lower in the cell-free constructs; however, the modulus was significantly increased compared to the cell-seeded collagen gels.

Next, collagen gel/sponge composites were either seeded or not with rabbit bone MSCs and implanted bilaterally into full-thickness, full-length defects in the central third of rabbit PT. The analysis at 12 weeks postsurgery revealed improved biomechanics, with 2.1, 1.6, 1.4, and 2.4 times increased ultimate stress, modulus, ultimate force and stiffness, respectively, compared to cell-free constructs (Juncosa-Melvin et al., 2006b). These biomechanical properties reached in vivo peak values and were considered beneficial in terms of tendon graft properties. In the same study, histological analysis revealed strong staining for fibronectin and decorin, and the alignment of the cells was very similar to native tendon tissue. Moreover, when the rabbit bone MSC-collagen sponges were mechanically stimulated prior to implantation as a PT graft in rabbits, biomechanical properties even further improved and fair in vitro-to-in vivo correlations of stiffness ($R=0.6$) and modulus ($R=0.7$) were gathered, with $R$ being the square root of the linear correlation coefficient ($R^2$) (Juncosa-Melvin et al., 2006c; Shearn et al., 2007; Table 7.2a and b). Such in vitro-to-in vivo correlations are interesting with respect to the 3Rs (reduction, refinement, replacement; Rusche, 2001); if high positive linear correlations between properties of the in vitro construct and the extracted ex vivo constructs are achieved, this may lead to the fact that less animals are needed because in vivo properties can be predicted. With less animal experiments, there is also a welcome reduction of costs. Because in the above mentioned study, the linear correlation coefficients were only moderately positive, it was concluded, however, that in vivo experiments for this specific purpose of tendon grafting are nevertheless needed before making a decision about using it in clinical trials.

Finally, collagen sponges that were additionally dehydrothermally cross-linked before being seeded with rabbit bone MSC and mechanical stimulation were examined in vitro and in vivo (Nirmalanandhan et al., 2009; Table 7.2a and b). Although the additional cross-linking improved in vitro biomechanics, the in vivo biomechanical outcome was significantly degraded (Fig. 7.4). This lack of in vitro-to-in vivo correlation might be explained by the mechanisms of dehydrothermally cross-linking. It causes dehydration of the fibers, drawing the collagen molecules closer together. While this action increases temperature stability, the act of cross-linking remains a delicate balance between removing intra-fibrillar water content and maintaining thermal stability of collagen fibers. Therefore, the thermal damage to collagen fiber subunits could have altered the potential of collagen to effectively bear load in this rabbit PT repair model (Nirmalanandhan et al., 2009). In sum, while the mechanical stimulation of the mentioned collagen scaffolds significantly improved in vitro and in vivo outcomes, the cross-linking procedure was only beneficial on the in vitro level due to dehydration of the fibers which destabilised the tendon graft.

Pig ACL augmentation by a rat tail tendon collagen-platelet composite scaffold was analysed histologically and biomechanically 4, 6, and 14 weeks postsurgery (Joshi et al., 2009). The composite scaffold was fabricated by dissolving rat tail tendon collagen in an acidic enzymatic solution; during surgery, blood was taken from the corresponding rat and after centrifugation the platelet rich plasma solution was added to the dissolved collagen solution. Augmentation of the porcine ACL by collagen-platelet composite increased ultimate load and stiffness at all the mentioned time points significantly (Table 7.2b). Noteworthy, however, was the determination of decreased biomechanics at 6 weeks, which the authors explained to be caused by revascularization ending up in hyperemia—and that this indicates a need for protection of the repair site especially during this regeneration period (Joshi et al., 2009).

Bovine type I collagen was extracted from AT, decellularized and furthermore underwent electrospinning (Oryan et al., 2013). These electrospun collagen grafts were then used successfully for AT reconstruction in rabbits as a full graft. They were shown to be biocompatible and biodegradable. Biomechanical properties compared to

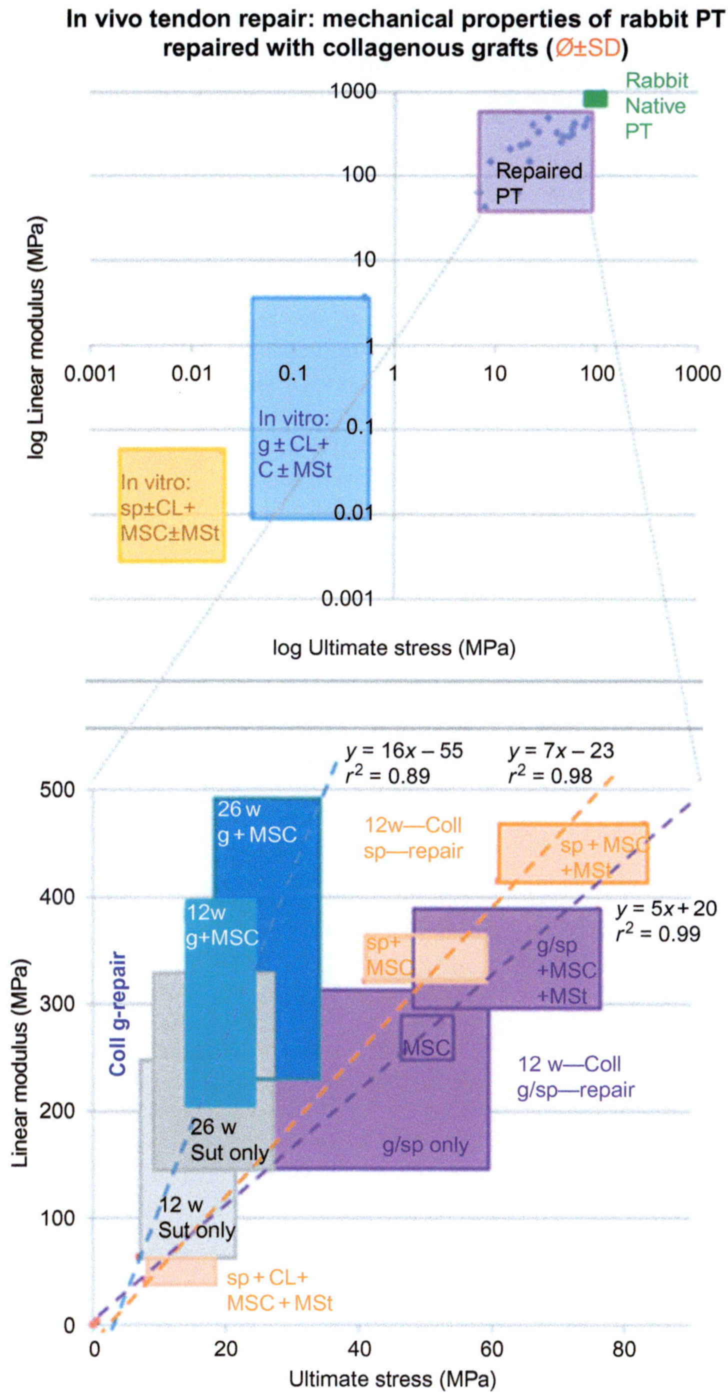


**Fig. 7.4** In vivo tendon repair with tissue engineering in comparison with in vitro collagen grafts und native tendon: correlation of mechanical properties of rabbit patellar tendons repaired

nongraft treated specimen 120 days postinjury improved, with 1.7, 2.8, 2.9, and 5.0 times higher ultimate stress, elastic modulus, ultimate load, and stiffness, respectively (Table 7.2a and b).

Furthermore, the electrospun collagen graft was modified by UV cross-linking and it was checked whether a polydioxanone sheet wrapped around the electrospun collagen graft would add benefit to in vitro as well as in vivo biomechanical outcome in the rabbit AT model (Moshiri et al., 2013a). The samples were analysed at 8.5 and 17 weeks postinjury, respectively. Briefly, while UV treatment had a beneficial impact on biomechanical properties of the electrospun meshes in vitro and to a lesser extent in vivo at both time points postoperatively, polydioxanone wrapping further increased biomechanics in vitro, however, in vivo the effect of such a sheet was only significant at 8.5 weeks postsurgery, while it disappeared after 17 weeks (Table 7.2a and b). In the same comprehensive study, also other aspects were investigated; the adhesion formation to the surrounding tissue was assessed by the flexion degree and in vivo with ultrasound—measuring echogenicity of the regenerated tissue. Moreover, a gross pathological examination focused on hyperemia and muscle atrophy as well as fibrosis, while analysis of histological sections included many parameters

**Fig. 7.4 Continued.** with collagenous grafts. Keys: *Collagen types*: *G*, gel; *G/SP*, gel/sponge combination; *SP*, sponge. *Treatments of collagen*: *CL*, cross-linking. *Cells for tissue engineering*: *MSC*, mesenchymal stem cells. *Mechanical treatment*: *MSt*, mechanical stimulation. *Healing time*: *w*, weeks; *Sut*, suture. *Data*: *Ø* ± *SD*, mean ± standard deviation. Correlation lines through middle points of rectangles.
From Awad, H.A., Boivin, G.P., Dressler, M.R., Smith, F.N.L., Young, R.G., Butler, D.L., 2003. Repair of patellar tendon injuries using a cell-collagen composite. J. Orthop. Res. 21, 420–431; Juncosa-Melvin, N., Boivin, G.P., Galloway, M.T., Gooch, C., West, J.R., Butler, D.L., 2006a. Effects of cell-to-collagen ratio in stem cell-seeded constructs for Achilles tendon repair. Tissue Eng. 12, 681–689; Juncosa-Melvin, N., Boivin, G.P., Gooch, C., Galloway, M.T., West, J.R., Dunn, M.G., Butler, D.L., 2006b. The effect of autologous mesenchymal stem cells on the biomechanics and histology of gel-collagen sponge constructs used for rabbit patellar tendon repair. Tissue Eng. 12, 369–379; Juncosa-Melvin, N., Shearn, J.T., Boivin, G.P., Gooch, C., Galloway, M.T., West, J.R., Nirmalanandhan, V.S., Bradica, G., Butler, D.L., 2006c. Effects of mechanical stimulation on the biomechanics and histology of stem cell-collagen sponge constructs for rabbit patellar tendon repair. Tissue Eng. 12, 2291–2300; Nirmalanandhan, V.S., Juncosa-Melvin, N., Shearn, J.T., Boivin, G.P., Galloway, M.T., Gooch, C., Bradica, G., Butler, D.L., 2009. Combined effects of scaffold stiffening and mechanical preconditioning cycles on construct biomechanics, gene expression, and tendon repair biomechanics. Tissue Eng. A, 15, 2103–2111; Shearn, J.T., Juncosa-Melvin, N., Boivin, G.P., Galloway, M.T., Goodwin, W., Gooch, C., Dunn, M.G., Butler, D.L., 2007. Mechanical stimulation of tendon tissue engineered constructs: effects on construct stiffness, repair biomechanics, and their correlation. J. Biomech. Eng. 129, 848–854; Feng, Z., Ishibashi, M., Nomura, Y., Kitajima, T., Nakamura, T., 2006. Constraint stress, microstructural characteristics, and enhanced mechanical properties of a special fibroblast-embedded collagen construct. Artif. Organs, 30, 870–877; Garvin, J., Qi, J., Maloney, M., Banes, A.J., 2003. Novel system for engineering bioartificial tendons and application of mechanical load. Tissue Eng. 9, 967–979.

(total cells, total Fbs, immature Fbs, mature Fbs, diameter of Fbs (mature and immature), neutrophils, lymphocytes, plasma cells, macrophages, vessels (small, medium, large), diameter of vessels (small, medium, large), collagen density and cell density). In addition, SEM imaging (scanning electron microscopy) was performed and water uptake assessed. A special measurement was also the direct transitional electrical current (DTEC) and the tissue resistance to electrical current (Ohm). If treated tendon lesions are in an advanced stage of healing, hydration is usually reduced going along with beneficial electrical properties, in other words reduced transitional current. In this study, the DTEC was significantly reduced as a function of time, showing a beneficial reduction of water content in the so far regenerated tendon tissue. All those parameters are listed here in order to show that Moshiri et al. offer a very concise and thorough data set that might be interpreted in relation to the biomechanical outcome. In other words, such a huge data set comprising the tissue mechanics before and after implantation, histological cell and tissue analysis and electrical studies for the same type of experiment allow correlations between the parameters assessed. This study showed clearly—at all levels—that it is possible to apply a collagen scaffold as an implant material beneficially with or without the polydioxanone sheet wrapped around in the reconstruction of a large AT defect successfully.

A parallel study also investigated the biomechanical properties of the dried collagen scaffolds in vitro and confirmed the 8.5 weeks postinjury in vivo results. There, the collagen scaffolds were implanted in a large rabbit AT defect model but without application of the polydioxanone wrapping (Meimandi-Parizi et al., 2013). As control, the defect was left untreated. While the gross appearance of a normal healthy tendon was brilliant white, having dense connective tissue, the regenerated tendon where a collagen scaffold was implanted looked similar in size and shape, however, hyperemia was denoted turning the tissue slightly reddish (Fig. 7.5). In contrast to the healthy and the collagen-treated tendons, the control tendons showed that the gap had been filled with quite loose fibrous tissue. Moreover, the diameter of the scarry fibrous tissue in the control sample was significantly smaller compared to the healthy and collagen treated and also it was longer and more translucent.

As for the biomechanics, the collagen-treated tendons had a 4.6 times higher load until failure compared to the control specimen (Table 7.2a and b). Also other biomechanical properties were significantly better in the collagen-treated samples; a 3.8 times higher stiffness, a 2.3 times higher maximum stress and a 4.4-fold elastic modulus. The providing of collagen at the defect site was also depicted by the fact that the dry matter content of the collagen-treated specimen was 2.1 times higher than that of the nontreated control specimen, which is nicely visible in Fig. 7.5. Although the improvement of filling a gap with collagen is obvious when comparing the biomechanical characteristics to a gap that was not filled, the properties were far below the ones observed for healthy tendons; with 8.1 times lower maximum load, 11 times lower stiffness, 1.9 times lower maximum stress, 1.8 times lower elastic modulus—and 2.1 times lower dry matter content. Accidentally, the dry matter content increased from nontreated to collagen treated by a factor of 2.1 and from collagen treated to healthy tendons also by 2.1. This enables a direct comparison of the changed biomechanical properties. Although the dry matter content of healthy tendons is 2.1 times

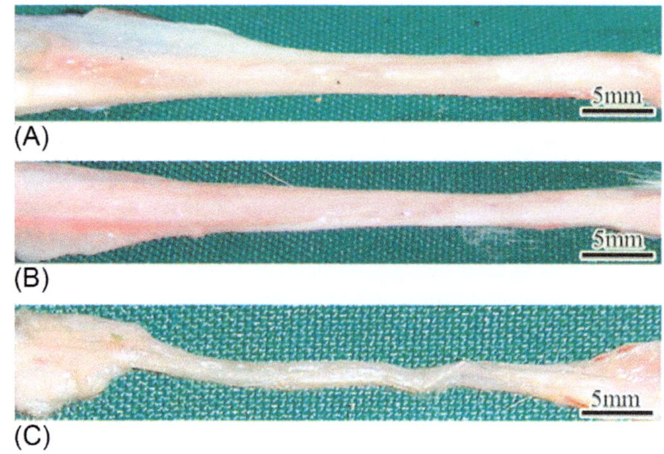


**Fig. 7.5** Extracted healthy rabbit Achilles tendon AT (A), collagen-treated rabbit AT 8.5 weeks postinjury (B) and control specimen which had not undergone gap filling, 8.5 weeks postinjury (C).
Reproduced with permission from Meimandi-Parizi, A., Oryan, A., Moshiri, A., Silver, I.A., 2013. Novel application of a tissue-engineered collagen-based three-dimensional bio-implant in a large tendon defect model: a broad-based study with high value in translational medicine. Tissue Cell 45, 282–294, reprinted from Tissue and Cell, Copyright 2013, with permission from Elsevier.

higher compared to collagen treated, the load is 8.1 times increased. In the comparison of control to collagen treated an increase of also 2.1 times dry matter content "only" leads to a 4.6 times higher failure load (1.8 times less increased). Also the comparison of the stiffness results had the same trend; while it was 11 times higher for healthy tendons compared to collagen treated, it was only 3.8 times higher for collagen treated compared to control tendons. This corroborates the findings of macroscopic observations, namely that the dry matter of natural, collagen treated and control tendon tissue is not the same and therefore crucially influences the biomechanics of the regenerated rabbit ATs (Meimandi-Parizi et al., 2013).

## 7.9 Summary

Because collagen is the main component of T&L tissue making up around 86% of the dry weight, it is an obvious approach to take collagen as a scaffold material in T&L tissue engineering aimed at reconstruction of large defects or as a means of augmentation. In this chapter, we addressed not only commercially available collagen matrices, but also the different ways of manufacturing and processing collagen, among them extrusion and cross-linking of collagen fibers. Furthermore, collagen sponges and gels are discussed and their suitability is shown in case they are cell-seeded. Like this, wide ranges of different collagen grafts are presented that do vary in their mechanical properties by orders of magnitudes.

# References

Adams, J.E., Zobitz, M.E., Reach Jr., J.S., An, K.-N., Steinmann, S.P., 2006. Rotator cuff repair using an acellular dermal matrix graft: an in vivo study in a canine model. Arthroscopy 22, 700–709.

Awad, H.A., Boivin, G.P., Dressler, M.R., Smith, F.N.L., Young, R.G., Butler, D.L., 2003. Repair of patellar tendon injuries using a cell-collagen composite. J. Orthop. Res. 21, 420–431.

Barber, F.A., Herbert, M.A., Coons, D.A., 2006. Tendon augmentation grafts: biomechanical failure loads and failure patterns. Arthroscopy 22, 534–538.

Barber, F.A., Herbert, M.A., Boothby, M.H., 2008a. Ultimate tensile failure loads of a human dermal allograft rotator cuff augmentation. Arthroscopy 24, 20–24.

Barber, F.A., Mcgarry, J.E., Herbert, M.A., Anderson, R.B., 2008b. A biomechanical study of Achilles tendon repair augmentation using GraftJacket matrix. Foot Ankle Int. 29, 329–333.

Berlet, G.C., Hyer, C.F., Lee, T.H., Blum, B.E., 2014. Collagen ribbon augmentation of Achilles tendon tears: a biomechanical evaluation. J. Foot Ankle Surg. 53, 298–302.

Berry, C.C., Shelton, J.C., Lee, D.A., 2009. Cell-generated forces influence the viability, metabolism and mechanical properties of fibroblast-seeded collagen gel constructs. J. Tissue Eng. Regen. Med. 3, 43–53.

Bhavsar, D., Shettko, D., Tenenhaus, M., 2010. Encircling the tendon repair site with collagen-GAG reduces the formation of postoperative tendon adhesions in a chicken flexor tendon model. J. Surg. Res. 159, 765–771.

Butler, D.L., Juncosa-Melvin, N., Boivin, G.P., Galloway, M.T., Shearn, J.T., Gooch, C., Awad, H., 2008. Functional tissue engineering for tendon repair: a multidisciplinary strategy using mesenchymal stem cells, bioscaffolds, and mechanical stimulation. J. Orthop. Res. 26, 1–9.

Butler, D.L., Hunter, S.A., Chokalingam, K., Cordray, M.J., Shearn, J., Juncosa-Melvin, N., Nirmalanandhan, S., Jain, A., 2009. Using functional tissue engineering and bioreactors to mechanically stimulate tissue-engineered constructs. Tissue Eng. A 15, 741–749.

Chaudhury, S., Holland, C., Thompson, M.S., Vollrath, F., Carr, A.J., 2012. Tensile and shear mechanical properties of rotator cuff repair patches. J. Shoulder Elb. Surg. 21, 1168–1176.

Chokalingam, K., Juncosa-Melvin, N., Hunter, S.A., Gooch, C., Frede, C., Florert, J., Bradica, G., Wenstrup, R., Butler, D.L., 2009. Tensile stimulation of murine stem cell-collagen sponge constructs increases collagen type I gene expression and linear stiffness. Tissue Eng. A 15, 2561–2570.

Defrate, L.E., Van Der Ven, A., Gill, T.J., Li, G., 2004. The effect of length on the structural properties of an Achilles tendon graft as used in posterior cruciate ligament reconstruction. Am. J. Sports Med. 32, 993–997.

Derwin, K.A., Baker, A.R., Spragg, R.K., Leigh, D.R., Iannotti, J.P., 2006. Commercial extracellular matrix scaffolds for rotator cuff tendon repair. Biomechanical, biochemical, and cellular properties. J. Bone Joint Surg. Am. 88, 2665–2672.

Enea, D., Henson, F., Kew, S., Wardale, J., Getgood, A., Brooks, R., Rushton, N., 2011. Extruded collagen fibres for tissue engineering applications: effect of crosslinking method on mechanical and biological properties. J. Mater. Sci. Mater. Med. 22, 1569–1578.

Feng, Z., Ishibashi, M., Nomura, Y., Kitajima, T., Nakamura, T., 2006. Constraint stress, microstructural characteristics, and enhanced mechanical properties of a special fibroblast-embedded collagen construct. Artif. Organs 30, 870–877.

Fisher, M.B., Liang, R., Jung, H.-J., Kim, K.E., Zamarra, G., Almarza, A.J., Mcmahon, P.J., Woo, S.L.Y., 2012. Potential of healing a transected anterior cruciate ligament with genetically modified extracellular matrix bioscaffolds in a goat model. Knee Surg. Sports Traumatol. Arthrosc. 20, 1357–1365.

Garvin, J., Qi, J., Maloney, M., Banes, A.J., 2003. Novel system for engineering bioartificial tendons and application of mechanical load. Tissue Eng. 9, 967–979.

Ide, J., Kikukawa, K., Hirose, J., Iyama, K.-I., Sakamoto, H., Mizuta, H., 2009. Reconstruction of large rotator-cuff tears with acellular dermal matrix grafts in rats. J. Shoulder Elbow Surg. 18, 288–295.

Islam, A., Bohl, M.S., Tsai, A.G., Younesi, M., Gillespie, R., Akkus, O., 2015. Biomechanical evaluation of a novel suturing scheme for grafting load-bearing collagen scaffolds for rotator cuff repair. Clin. Biomech. 30, 669–675.

Joshi, S.M., Mastrangelo, A.N., Magarian, E.M., Fleming, B.C., Murray, M.M., 2009. Collagen-platelet composite enhances biomechanical and histologic healing of the porcine anterior cruciate ligament. Am. J. Sports Med. 37, 2401–2410.

Juncosa-Melvin, N., Boivin, G.P., Galloway, M.T., Gooch, C., West, J.R., Butler, D.L., 2006a. Effects of cell-to-collagen ratio in stem cell-seeded constructs for Achilles tendon repair. Tissue Eng. 12, 681–689.

Juncosa-Melvin, N., Boivin, G.P., Gooch, C., Galloway, M.T., West, J.R., Dunn, M.G., Butler, D.L., 2006b. The effect of autologous mesenchymal stem cells on the biomechanics and histology of gel-collagen sponge constructs used for rabbit patellar tendon repair. Tissue Eng. 12, 369–379.

Juncosa-Melvin, N., Shearn, J.T., Boivin, G.P., Gooch, C., Galloway, M.T., West, J.R., Nirmalanandhan, V.S., Bradica, G., Butler, D.L., 2006c. Effects of mechanical stimulation on the biomechanics and histology of stem cell-collagen sponge constructs for rabbit patellar tendon repair. Tissue Eng. 12, 2291–2300.

Juncosa-Melvin, N., Matlin, K.S., Holdcraft, R.W., Nirmalanandhan, V.S., Butler, D.L., 2007. Mechanical stimulation increases collagen type I and collagen type III gene expression of stem cell—collagen sponge constructs for patellar tendon repair. In: ASME Summer Bioengineering Conference, Jun 24–28. Amer Soc Mechanical Engineers, Keystone, CO, pp. 981–982.

Kato, Y.P., Christiansen, D.L., Hahn, R.A., Shieh, S.J., Goldstein, J.D., Silver, F.H., 1989. Mechanical-properties of collagen-fibers—a comparison of reconstituted and rat tail tendon fibers. Biomaterials 10, 38–41.

Kinneberg, K.R.C., Nirmalanandhan, V.S., Juncosa-Melvin, N., Powell, H.M., Boyce, S.T., Shearn, J.T., Butler, D.L., 2010. Chondroitin-6-sulfate incorporation and mechanical stimulation increase MSC-collagen sponge construct stiffness. J. Orthop. Res. 28, 1092–1099.

Kumar, V.A., Caves, J.M., Haller, C.A., Dai, E., Liu, L., Grainger, S., Chaikof, E.L., 2013. Collagen-based substrates with tunable strength for soft tissue engineering. Biomater. Sci. 1, 1193–1202.

Kumbar, S.G., Laurencin, C.T., Deng, M. (Eds.), 2014. Natural and Synthetic Biomedical Polymers. Elsevier Science, Amsterdam.

Magnussen, R.A., Glisson, R.R., Moorman 3rd., C.T., 2011. Augmentation of Achilles tendon repair with extracellular matrix xenograft: a biomechanical analysis. Am. J. Sports Med. 39, 1522–1527.

Meimandi-Parizi, A., Oryan, A., Moshiri, A., Silver, I.A., 2013. Novel application of a tissue-engineered collagen-based three-dimensional bio-implant in a large tendon defect model: a broad-based study with high value in translational medicine. Tissue Cell 45, 282–294.

Moshiri, A., Oryan, A., Meimandiparizi, A., 2013a. A novel application of biosynthetic tissue engineered tridimensional implant on large tendon defects: a comprehensive detailed in vivo investigation with significant clinical value. Connect. Tissue Res. 54 (3), 227–243.
Moshiri, A., Oryan, A., Meimandi-Parizi, A., 2013b. Role of tissue-engineered artificial tendon in healing of a large Achilles tendon defect model in rabbits. J. Am. Coll. Surg. 217, 421–441.
Nicholson, G.P., Breur, G.J., Van Sickle, D., Yao, J.Q., Kim, J., Blanchard, C.R., 2007. Evaluation of a cross-linked acellular porcine dermal patch for rotator cuff repair augmentation in an ovine model. J. Shoulder Elb. Surg. 16, 184S–190S.
Nirmalanandhan, V.S., Dressler, M.R., Shearn, J.T., Juncosa-Melvin, N., Rao, M., Gooch, C., Bradica, G., Butler, D.L., 2007a. Mechanical stimulation of tissue engineered tendon constructs: effect of scaffold materials. J. Biomech. Eng. 129, 919–923.
Nirmalanandhan, V.S., Rao, M., Sacks, M.S., Haridas, B., Butler, D.L., 2007b. Effect of length of the engineered tendon construct on its structure-function relationships in culture. J. Biomech. 40, 2523–2529.
Nirmalanandhan, V.S., Rao, M., Shearn, J.T., Juncosa-Melvin, N., Gooch, C., Butler, D.L., 2008. Effect of scaffold material, construct length and mechanical stimulation on the in vitro stiffness of the engineered tendon construct. J. Biomech. 41, 822–828.
Nirmalanandhan, V.S., Juncosa-Melvin, N., Shearn, J.T., Boivin, G.P., Galloway, M.T., Gooch, C., Bradica, G., Butler, D.L., 2009. Combined effects of scaffold stiffening and mechanical preconditioning cycles on construct biomechanics, gene expression, and tendon repair biomechanics. Tissue Eng. A 15, 2103–2111.
Noyes, F.R., Grood, E.S., 1976. The strength of the anterior cruciate ligament in humans and rhesus monkeys. J. Bone Joint Surg. Am. 58, 1074–1082.
Omae, H., Steinmann, S.P., Zhao, C., Zobitz, M.E., Wongtriratanachai, P., Sperling, J.W., An, K.-N., 2012. Biomechanical effect of rotator cuff augmentation with an acellular dermal matrix graft: a cadaver study. Clin. Biomech. (Bristol, Avon) 27, 789–792.
Oryan, A., Moshiri, A., Parizi Meimandi, A., Silver, I.A., 2013. A long-term in vivo investigation on the effects of xenogenous based, electrospun, collagen implants on the healing of experimentally-induced large tendon defects. J. Musculoskelet. Neuronal Interact. 13, 353–367.
Raghavan, D., Kropp, B.P., Lin, H.K., Zhang, Y., Cowan, R., Madihally, S.V., 2005. Physical characteristics of small intestinal submucosa scaffolds are location-dependent. J. Biomed. Mater. Res. A 73, 90–96.
Rusche, B., 2001. The 3Rs and animal welfare—conflict or the way forward. ALTEX 18, 48.
Sharma, P., Maffulli, N., 2006. Biology of tendon injury: healing, modeling and remodeling. J. Musculoskelet. Neuronal Interact. 6, 181–190.
Shearn, J.T., Juncosa-Melvin, N., Boivin, G.P., Galloway, M.T., Goodwin, W., Gooch, C., Dunn, M.G., Butler, D.L., 2007. Mechanical stimulation of tendon tissue engineered constructs: effects on construct stiffness, repair biomechanics, and their correlation. J. Biomech. Eng. 129, 848–854.
Song, L., Olsen, R.E., Spalazzi, J.P., Davisson, T., 2010. Biomechanical evaluation of acellular collagen matrix augmented Achilles tendon repair in sheep. J. Foot Ankle Surg. 49, 438–441.
Van Der Meijden, O.A., Wijdicks, C.A., Gaskill, T.R., Jansson, K.S., Millett, P.J., 2013. Biomechanical analysis of two-tendon posterosuperior rotator cuff tear repairs: extended linked repairs and augmented repairs. Arthroscopy 29, 37–45.
Walters, V.I., Kwansa, A.L., Freeman, J.W., 2012. Design and analysis of braid-twist collagen scaffolds. Connect. Tissue Res. 53, 255–266.

Zeugolis, D.I., Paul, R.G., Attenburrow, G., 2008a. Extruded collagen-polyethylene glycol fibers for tissue engineering applications. J. Biomed. Mater. Res. B Appl. Biomater. 85B, 343–352.

Zeugolis, D.I., Paul, R.G., Attenburrow, G., 2008b. Post-self-assembly experimentation on extruded collagen fibres for tissue engineering applications. Acta Biomater. 4, 1646–1656.

Zeugolis, D.I., Paul, G.R., Attenburrow, G., 2009. Cross-linking of extruded collagen fibers—a biomimetic three-dimensional scaffold for tissue engineering applications. J. Biomed. Mater. Res. A 89, 895–908.

Zhao, C., Ozasa, Y., Reisdorf, R.L., Thoreson, A.R., Jay, G.D., An, K.-N., Amadio, P.C., 2014. CORR® ORS Richard A Brand Award for Outstanding Orthopaedic Research: Engineering flexor tendon repair with lubricant, cells, and cytokines in a canine model. Clin. Orthop. Relat. Res. 472, 2569–2578.

# Synthetic polymer scaffolds for tendon and ligament repair: Materials and biomechanics

8

## Abstract

Polymeric synthetic scaffold materials for tendon and ligament rupture repair and augmentation include inert and biodegradable materials. As for the inert materials such as polyethylene terephthalate, approaches of covering the surface with more biocompatible materials are investigated. On the other hand, for the biodegradable polymeric scaffolds like PLGA (poly(lactic-*co*-glycolic acid)), different processing methods are in use in order to tailor the biomechanical properties of the tendon and ligament grafts according to their future destination. Biomechanics of polymeric grafts depend on cell seeding, including the type of cell and its cultivation method (static or dynamic). Moreover, also appropriate (growth) factors or other biological components serving as tendon healing supportive entities may be incorporated in synthetic polymeric scaffold materials, and their release kinetics are investigated to be tuned according to their aim, mostly mimicking natural healing patterns.

**Keywords:** Synthetic scaffold materials (PGA, PLGA, PCL, PU, PET, DegraPol®, etc.), Biodegradability, Manufacture techniques, Cell seeding, Mechanical stimulation.

## Abbreviations

| | |
|---|---|
| **3D** | three-dimensional |
| **ACL** | anterior cruciate ligament (tendons and ligaments) |
| **Ant.** | anterior (position) |
| **ASCs** | adipose-derived stem cells (cells for seeding) |
| **AT** | Achilles tendon (tendons and ligaments) |
| **ATR** | Artelon tissue reinforcement (synthetic scaffold material) |
| **bFGF** | basic fibroblast growth factor (growth factor) |
| **br** | braided (manufacture technique) |
| **CAD** | cadaver |
| **Coll sp** | collagen sponge (natural scaffold material) |
| **d** | days |
| **degr** | degraded (process of biodegradable products) |
| **DGN** | dextran nanoparticles (chemical agent in tissue engineering) |
| **DP** | DegraPol® (synthetic scaffold material) |
| **ECM** | extracellular matrix (natural graft material) |
| **es** | electrospun (manufacture technique) |
| **ex** | extruded (manufacture technique) |
| **Fb** | fibroblasts (cells for seeding) |

**Biomechanics of Tendons and Ligaments. http://dx.doi.org/10.1016/B978-0-08-100489-0.00008-9**

| | |
|---|---|
| **FDP** | flexor digitorum profundus (tendons and ligaments) |
| **fi** | fiber (manufacture technique) |
| **HA** | hyaluronic acid (natural agent in tissue engineering) |
| **Hum** | human |
| **Ibp** | ibuprofen (drug) |
| **IGF** | insulin-like growth factor (growth factor) |
| **infrasp.** | infraspinatus tendon (rotator cuff) (tendons and ligaments) |
| **IV** | in vitro (laboratory method) |
| **kn** | knitted (manufacture technique) |
| **MCL** | medial collateral ligament (tendons and ligaments) |
| **MSC** | mesenchymal stem cells (cells for seeding) |
| **MSt** | mechanical stimulation (mechanical treatment in tissue engineering) |
| **Ø** | mean (element of statistics) |
| **PCL** | polycaprolactone (synthetic scaffold material) |
| **PDGF-β** | platelet-derived growth factor (growth factor) |
| **PDLA** | poly(D-lactide acid) (synthetic scaffold material) |
| **PDLLA** | poly(D,L-lactide acid) (synthetic scaffold material) |
| **PEG** | polyethylene glycol (synthetic scaffold material) |
| **PELA** | poly(L-lactic acid)-polyethylene glycol (synthetic scaffold material) |
| **PEO** | polyethylene oxide (synthetic scaffold material) |
| **PET** | polyethylene terephthalate (synthetic scaffold material) |
| **PGA** | poly(glycolide) acid (synthetic scaffold material) |
| **PLA** | polylactic acid (synthetic scaffold material) |
| **PLAGA** | polylactic-*co*-glycolic acid 82:18 (synthetic scaffold material) |
| **PLCL** | polylactic caprolactone (synthetic scaffold material) |
| **PLDLA** | poly(L-lactide-*co*-D,L-lactide acid) (synthetic scaffold material) |
| **PLGA** | poly(lactic-*co*-glycolic acid) (synthetic scaffold material) |
| **PLLA** | poly(L-lactide acid) (synthetic scaffold material) |
| **Post.** | posterior (position) |
| **PU** | polyurethane (synthetic scaffold material) |
| **RC** | rotator cuff (tendons and ligaments) |
| **RTFs** | rat tendon fibroblasts (cells for seeding) |
| **SD** | standard deviation (element of statistics) |
| **SEM** | standard error of the mean (element of statistics) |
| **Si-np** | silver nanoparticles (chemical agent in tissue engineering) |
| **SIS** | small intestine submucosa (natural graft material) |
| **Sup.** | superior (position) |
| **supraspin.** | supraspinatus tendon (rotator cuff) (tendons and ligaments) |
| **Sut** | suture (surgical technique) |
| **T&L** | tendons and ligaments (tendons and ligaments) |
| **Tb** | tenoblasts (cells for seeding) |
| **Tc** | tenocytes (cells for seeding) |
| **tw** | twisted (manufacture technique) |
| **un** | unwoven (manufacture technique) |
| **wo** | woven (manufacture technique) |

## 8.1 Introduction

The development of highly stable synthetic polymeric materials intended at tendon and ligament rupture repair has been a major research challenge for the last 20 years. On the one hand side, there are polymers available for clinical purposes that are nondegradable such as polyethylene terephthalate (PET); on the other hand side, degradable synthetic polymeric materials are getting more and more attractive in tendon tissue engineering because of their biodegradation going along with the regeneration of the natural tissue (e.g., poly(glycolide) acid (PGA), poly(lactic-*co*-glycolic acid) (PLGA), and polylactic acid (PLA)). Degradable synthetic polymers must fulfill more stringent requirements in terms of biocompatibility compared with nondegradable ones because of the potential toxicity of degradation products leaching from the implanted polymer such as monomers, stabilizers, polymerization initiators, and their subsequent metabolites. As a consequence, only a limited number of nontoxic monomers have been successfully applied in the synthesis of degradable polymeric scaffolds for tendon tissue engineering and tendon and ligament augmentation. Finally, tensile properties of synthetic scaffolds depend also on the manufacture techniques.

## 8.2 Polyglycolic acid

PGA is the simplest linear aliphatic polyester ($[O—CH_2—CO—]_n$) (Kumbar et al., 2014). It is highly crystalline and has a high melting point of 150–160°C (Dumitriu and Popa, 2013). Compared with other polymers such as poly(L-lactide acid) (PLLA), PLGA, or polycaprolactone (PCL) (see succeeding text), PGA has a significantly higher elastic modulus, which makes it an attractive choice when synthetic polymers are needed for tendon or ligament reconstruction (Petrigliano et al., 2006; Table 8.1). Moreover, its ultimate stress is 28–50 MPa (Dumitriu and Popa, 2013). Therefore, many investigations have been performed using PGA as a scaffold material for tendon tissue engineering. For example, in 2006, unwoven PGA was either seeded or not with avian tenocytes, and the tensile stress of cell-seeded scaffolds was measured after 4, 6, and 10 weeks in vitro either with or without constant strain (Cao et al., 2006). The maximum tension force was applied empirically to a point that would not be high enough to break the construct because the strain on the constructs was difficult to be controlled exactly. As a result, a threefold higher ultimate stress was determined for the cell-seeded PGA under constant strain compared to the nonstretched scaffold (Table 8.1). Despite of the significantly better stress values obtained in the stretched group, the authors concluded that a constant strain may not be a physiological way of mechanical loading, because it makes the tissue engineered tendon thinner, possibly caused by cell apoptosis due to the long-term constant strain; they rather suggested dynamic loading. Moreover, the cell-free PGA scaffolds had degraded at 4 weeks postincubation (Cao et al., 2006).

Table 8.1 **In vitro properties of synthetic scaffolds, mechanical, and structural properties**

| Scaffold material | Specifications | Target tissue | Ultimate stress (MPa) | Elastic modulus (MPa) | Ultimate load (N) | Stiffness (N/mm) | Ref. |
|---|---|---|---|---|---|---|---|
| PGA | PGA fi | ACL | | $7.0 \times 10^3$ | | | Middleton and Tipton (2000) |
| | PGA film | – | 28–50 | | | | Dumitriu and Popa (2013) |
| | PGA 3D br | ACL | 378 ± 18 | | 502 ± 24 | | Lu et al. (2005) |
| | PGA + Tc | T&L | 0.3–0.4 | | | | Cao et al. (2006) |
| | PGA + Tc + MSt | | 1.3 ± 0.4 | | | | |
| | PGA + Fb (static) | T&L | 0.2–1.1 | | | | Deng et al. (2009) |
| | PGA + Fb + MSt (tension) | | 1.3–4.7 | | | | |
| | PGA + Tc + MSt (tension) | | 3.1 ± 0.3 | | | | |
| PGA/SIS | PGA un/SIS | FDP | | | 9 ± 1 | | Cao et al. (2002) |
| PLGA | PLGA fi | ACL | | $2.0 \times 10^3$ | | | Middleton and Tipton (2000) |
| | PLGA es (degr 7 d) | T&L | | | 18 ± 4 | 4 ± 1 | Sahoo et al. (2006) |
| | PLGA es (degr 14 d) | | | | 2 ± 1 | 1 ± 0.2<br>6 ± 1 | |
| | PLGA kn + FG (initial) | | | | 56 ± 7 | 3 ± 1<br>1 ± 0.2 | |
| | PLGA kn + FG (degr 7 d) | | | | 15 ± 3 | | |
| | PLGA kn + FG (degr 14 d) | | | | 2 ± 1 | | |
| | PLGAsut thread | | 591 ± 63 | $(4.9 \pm 0.4) \times 10^3$ | | | Vaquette et al. (2010) |
| | PLGA kn (Ø) | | 219 ± 24 | 762 ± 85 | | | |
| | PLGA kn (initial) | | 164 ± 23 | 765 ± 108 | | | |

| | | | | | | | |
|---|---|---|---|---|---|---|---|
| | PLGA kn (degr 14 d) | | 105±28 | 604±101 | | | |
| | PLGA kn (degr 35 d) | | 10±2 | 128±9 | | | |
| | PLGA es | ACL | | 18±*3* | 1.5±*0.2* | | Surrao et al. (2012) |
| PLA | PLA wo | RC | | | 1214±93 | 70±12 (N %) | MacGillivray et al. (2006) |
| PLLA | PLLA fi | ACL | | $2.7\times10^3$ | | | Middleton and Tipton (2000) |
| | PLLA 3D br | ACL | 165±33 | | 298±59 | | Lu et al. (2005) |
| | PLLA br | ACL | 44–82 | 425–810 | | | Freeman et al. (2007) |
| | PLLA tw | | 81±7 | 772–888 | | | |
| | PLLA es | ACL | | 3±*0.3* | 0.2±*0.02* | | Surrao et al. (2012) |
| | PLLA es | T&L | 4.2±0.3 | 61±6 | | | Liu et al. (2013b) |
| | PLLA es+Si-np | | 3.8–4.0 | 63–69 | | | |
| | PLLA es | AT | 4.4±0.3 | | | | Liu et al. (2013c) |
| | PLLA es+bFGF | | 4.1±0.2 | | | | |
| | PLLA es+ bFGF/DGNs | | 3.5±0.3 | | | | |
| | PLLA wo device | RC | | | 796±34 | 195±2 | Derwin et al. (2009) |
| | Sut thread | | | | 397±18 | | |
| PLLA/ECM | PLLA/ECM (hum FL) | RC | | | 153±27 | | Baker et al. (2012) |
| PCL | PCL fi | ACL | | $0.4\times10^3$ | | | Holland and Tighe (1992) |
| | PCL es (wrapped) | FDP | 2.1±0.3 | 20±3 | | | Liu et al. (2012) |
| | PCL es+HA (wrapped) | | 1.6–1.9 | 24–30 | | | |
| | PCL es | FDP | 1.4±0.1 | 7.1±0.2 | | | Chen et al. (2014) |
| | PCL es+HA | | 1.6±0.2 | 8.0±1.1 | | | |
| PLAGA | PLAGA 3D br | ACL | 177±12 | | 215±23 | | Lu et al. (2005) |

*Continued*

**Table 8.1 Continued**

| Scaffold material | Specifications | Target tissue | Ultimate stress (MPa) | Elastic modulus (MPa) | Ultimate load (N) | Stiffness (N/mm) | Ref. |
|---|---|---|---|---|---|---|---|
| PLDLA | PLDLA 98 kDa, es | ACL | | 3±*0.3* | 0.3±*0.02* | | Surrao et al. (2012) |
| PDLLA | PLDLA 250 kDa, es | | | 19±*4* | 1.3±*0.2* | | |
| | PLDLA 250 kDa | | | 11±*1* | 1.3±*0.1* | | |
| | +Fb[a] (1 w) | | | 36±*2* | 3.9±*0.2* | | |
| | PLDLA 250 kDa | | | 14±*2* | 1.0±*0.2* | | |
| | +Fb[a] (8 w) | | | | | | |
| | PDLLA es | | | | | | |
| PELA | PELA es | FDP | 3.7±0.3 | 52±5 | | | Liu et al. (2013a) |
| | PELA es+Ibp | | 2.9–3.4 | 60–69 | | | |
| | (var. %) | | | | | | |
| PLCL | 1 layer br | ACL | | | 16±*1* | 14±*1* | Laurent et al. (2012) |
| (85:15) | 3 layers br | | | | 42±*2* | 26±*2* | |
| | 5 layers br | | | | 62±*4* | 31±*1* | |
| | Combinations | | | | 38–209 | 58–324 | |

[a]Fb, Tb (fibroblast from bovine central ligament in lower metacarpal joint).
3D, three-dimensional; *ACL*, anterior cruciate ligament; *AT*, Achilles tendon; *bFGF*, basic fibroblast growth factor; *br*, braided; *d*, days; *degr*, degraded; *DGNs*, dextran nanoparticles; *ECM*, extracellular matrix; *es*, electrospun; *Fb*, fibroblasts; *FDP*, flexor digitorum profundus; *FG*, fibrin gel; *fi*, fiber; *HA*, hyaluronic acid; hum FL, human fascia lata; *Ibp*, ibuprofen; *kn*, knitted; *MSC*, mesenchymal stem cells; *MSt*, mechanical stimulation; *Ø±SD*, mean±standard deviation; *Ø±SEM*, *mean±standard error of the mean*; *PCL*, polycaprolactone; *PDLLA*, poly(D,L-lactide acid); *PELA*, poly(L-lactic acid)-polyethylene glycol; *PGA*, poly(glycolide) acid; *PLA*, polylactic acid; *PLAGA*, polylactic-*co*-glycolic acid 82:18; *PLCL*, polylactic caprolactone; *PLDLA*, poly(L-lactide-*co*-D,L-lactide acid); *PLGA*, poly(lactic-*co*-glycolic acid); *PLLA*, poly(L-lactide acid); *RC*, rotator cuff; *Si-np*, silver nanoparticles; *SIS*, small intestine submucosa; *Sut*, suture; T&L, tendons and ligaments; *Tb*, tenoblast; *Tc*, tenocytes; *tw*, twisted; *un*, unwoven; *w*, week; *wo*, woven.

Which cell type is the best to seed PGA for tendon tissue engineering? Is it tenocytes? In a study where cell-seeded PGA scaffolds were cultivated in vitro under static strain up to 14 weeks, human dermal fibroblasts were compared with tenocytes (Deng et al., 2009). Because the ultimate stress of the construct did not change as a function of cell type, the authors concluded that human dermal fibroblast-seeded PGA scaffolds were indistinguishable from the counterpart engineered with human tenocytes. This supports the viewpoint that human dermal fibroblasts were likely to replace tenocytes in tendon graft development (Table 8.1). Such a result may be regarded as a great benefit because usually human dermal fibroblasts are much more easily accessed in daily clinical life by taking skin biopsies than human tenocytes. Moreover, also stem cells were compared with tenocytes. In a rat Achilles tendon (AT) model, bone marrow stromal mesenchymal stem cells (MSCs) were compared with tenocytes seeded on either electrospun PGA scaffolds or collagen scaffolds with respect to ultimate load 16 weeks after implantation (Pietschmann et al., 2013). Here, the fully differentiated tenocytes led to significantly higher ultimate loads compared with the MSCs; tenocyte-seeded PGA scaffolds had around 50% higher ultimate loads. Not surprisingly, the natural polymer collagen that is the primary component of native tendons performed better than the synthetic PGA, with 17% higher failure loads (Table 8.2b). In addition to the better biomechanical properties in the case of tenocyte-seeded scaffolds compared to their MSC counterparts, histological analysis revealed a lower ossification of the regenerated tendon tissue and more aligned collagen fibers. Ossification may be an unwanted result of the pluripotent bone marrow MSCs undergoing osteogenesis when uniaxial stretching is applied, which is the case in several physiological bone environments (Koike et al., 2005; Chen et al., 2008). Therefore, fully differentiated cell types such as dermal fibroblasts or tenocytes may be advantageous over stem cells due to the risk of differentiation to cell types that are not wanted in the regenerated tendon tissue.

Another approach is transduction of cells in order to incorporate growth factors stimulating the tendon regeneration. In this regard, PGA was reported to be a useful polymer carrier for rat tendon fibroblasts (RTFs) that were transduced with insulin-like growth factor-1 (IGF-1) (Dines et al., 2007). In a transwell system, the authors were able to show that transduced cells stimulated native RTFs by the secretion of the corresponding growth factor, which resulted in a 2.3-fold increase of ultimate loads in vitro (Table 8.2b).

## 8.3 Poly(lactic-*co*-glycolic acid) (PLGA)

Copolymers of glycolic acid with the more hydrophobic lactic acid are intensively investigated in terms of synthetic polymer scaffold materials intended at tendon rupture repair. The chemical structure of the polymer is described as $HO[CO{-}CH(CH_3){-}O{-}]_x[CO{-}CH_2{-}O]_y{-}H$. The ratio of glycolic acid to lactic acid ($y$:$x$) influences the mechanical properties of the corresponding PLGA; however, there is no linear relationship between the ratio and the physical properties of the corresponding copolymers (Kumbar et al., 2014). Whereas PGA is highly crystalline, the crystallinity is

**Table 8.2 Properties of reconstructed tendons with scaffolds of synthetic origin**

| Model | Target tendon or ligam. | Augmented/ reconstr. by graft | Specifications | Healing time (weeks) | Ultimate stress (MPa) | Elastic modulus (MPa) | Ref. |
|---|---|---|---|---|---|---|---|
| (a) Mechanical properties | | | | | | | |
| Rat | RC Suprasp. | PCL, PCL/PEO | Sut only | 4 | 1.6±0.7 | 16±9 | Beason et al. (2012) |
| | | | Sut+PCL | 8 | 1.0±0.4 | 10±8 | |
| | | | Sut+PCL/PEO | | 1.3±0.6 | 11±7 | |
| | | | Sut only | | 2.3±1.1 | 22±12 | |
| | | | Sut+PCL | | 1.8±0.9 | 15±8 | |
| | | | Sut+PCL/PEO | | 1.7±0.5 | 16±6 | |
| | AT | PGA es | Native rat AT | | 28±8 | | Pietschmann et al. (2013) |
| | | | Sut+PGA | 16 | 6±2 | | |
| | | | Sut+PGA+MSC | 16 | 5±2 | | |
| | | | Sut+PGA+Tc | | 14±2 | | |
| | | Collagen sponge (biphasic) | Sut+Coll sp | | 5±2 | | |
| | | | Sut+Coll sp+MSC | | 5±1 | | |
| | | | Sut+Coll sp+Tc | | 18±5 | | |
| Rabbit | AT | PLGA kn | Native rabbit AT | | | 100% | Ouyang et al. (2003) |
| | | | Sut+PLGA | 12 | | 53% | |
| | | | Sut+PLGA+MSC | | | 63% | |
| | | | Native rabbit AT | 13 | | 100% | Vaquette et al. (2010) |
| | | | Sut only | | | 42±3% | |
| | | | Sut+PLGA/Alginate | | | 42±7% | |
| | | | Sut+PLGA/Alginate +MSC | | | 62±5% | |
| Chicken | FDP | PELA es | IV: PELA | – | 3.7±0.3 | 52±5 | Liu et al. (2013a) |
| | | | IV: PELA+Ibp (var. %) | 3 | 2.9–3.4 | 60–69 | |
| | | | Sut. only | | 2.5±0.5 | | |
| | | | Sut+PELA | | 2.3±0.4 | | |
| | | | Sut+PELA+Ibp (6%) | | 2.2±0.6 | | |

| | | | | | | | |
|---|---|---|---|---|---|---|---|
| Pig | FDS | PGA un | Native pig FDS | – | 36–42 | | Liu et al. (2006) |
| | | | Sut+PGA | 6 | 15 ± 2 | | |
| | | | Sut+PGA+Tc | 14 | 19 ± 3 | | |
| | | | Sut+PGA+Fb | | 19 ± 2 | | |
| | | | Sut+PGA | | 21 ± 2 | | |
| | | | Sut+PGA+Tc | | 25 ± 2 | | |
| | | | Sut+PGA+Fb | | 25 ± 3 | | |
| | | | Sut+PGA | 26 | 27 ± 3 | | |
| | | | Sut+PGA+Tc | | 32 ± 4 | | |
| | | | Sut+PGA+Fb | | 31 ± 2 | | |

| **Animal model** | **Target tendon or ligam.** | **Augmented/ reconstr. by graft** | **Specifications** | **Healing time (weeks)** | **Ultimate load (N)** | **Stiffness (N/mm)** | **Ref.** |
|---|---|---|---|---|---|---|---|
| (b) Structural properties | | | | | | | |
| Hum CAD | RC | PLLA wo | IV: PLLA devoce | – | 796 ± 34 | 195 ± 2 | McCarron et al. (2010) |
| | | | IV: Sut thread | 0 | 397 ± 18 | 100 ± 14 | |
| | | | Ant. suprasp.: sut only | | 358 ± 68 | 115 ± 17 | |
| | | | Ant. suprasp.: sut+PLLA | | 673 ± 209 | 81 ± 13 | |
| | | | Post. suprasp.: sut only | | 279 ± 32 | 96 ± 26 | |
| | | | Post. suprasp.: sut+PLLA | | 437 ± 78 | 130 ± 31 | |
| | | | Sup. infrasp.: sut only | | 406 ± 100 | 113 ± 14 | |
| | | | Sup. infrasp.: sut+PLLA | | 694 ± 191 | | |
| | AT | Artelon ATR-polymer | Native human AT | | 843–5579 | | Giza et al. (2011) |
| | | | Sut only | 0 | 248 ± 20 | 10 ± 3 | |
| | | | Sut+ATR | | 370 ± 25 | 12 ± 3 | |

*Continued*

Table 8.2 **Continued**

| Animal model | Target tendon or ligam. | Augmented/ reconstr. by graft | Specifications | Healing time (weeks) | Ultimate load (N) | Stiffness (N/mm) | Ref. |
|---|---|---|---|---|---|---|---|
| Pig | AT | PEG/PCL+Coll | Sut only | 0 | 105±25 | 18±4 | Brodie et al. |
| CAD | | +muscle adhesive | Sut+construct/bov. peric. | | 151±37 | 24±4 | (2011) |
| | | prot. | Sut+construct/Biotape | | 130±46 | 22±5 | |
| Rat | AT | PGA es | Native rat AT | | 53±8 | | Pietschmann |
| | | Collagen sponge | Sut+PGA | 16 | 43±8 | | et al. (2013) |
| | | (biphasic) | Sut+PGA+MSC | 16 | 39±9 | | |
| | | | Sut+PGA+Tc | | 58±11 | | |
| | | | Sut+Coll sp | | 51±14 | | |
| | | | Sut+Coll sp+MSC | | 44±15 | | |
| | | | Sut+Coll sp+Tc | | 67±11 | | |
| | | PLLA es | Sut only | 3 | 31±6 | 21±5 | Liu et al. (2013c) |
| | | | Sut+PLLA | | 26±6 | 15±6 | |
| | | | Sut+PLLA+bFGF | | 27±8 | 17±8 | |
| | | | Sut+PLLA+bFGF/DGN | | 29±6 | 19±7 | |
| Rat | RC | PGA un | Sut only | 6 | 1.2±*0.6* | | Dines et al. |
| | | | Sut+PGA | | 1.2±*0.6* | | (2007) |
| | | | Sut+PGA+Tb (RTF) | | 1.4±*0.6* | | |
| | | | +PDGF-β | | 2.8±*0.6* | | |
| | | | Sut+PGA+Tb (RTF) | | | | |
| | | | +IGF-1 | | | | |
| | RC, | PCL es, PCL/PEO | Sut only | 4 | 11±5 | 21±13 | Beason et al. |
| | supraspin. | es | Sut+PCL | 8 | 9±3 | 17±8 | (2012) |
| | | | Sut+PCL/PEO | | 11±4 | 20±9 | |
| | | | Sut only | | 15±7 | 32±14 | |

| | | | | | | | |
|---|---|---|---|---|---|---|---|
| | | | Sut + PCL | | 18 ± 9 | 31 ± 12 | |
| | | | Sut + PCL/PEO | | 17 ± 5 | 34 ± 14 | |
| Rabbit | AT | PLGA kn | Native rabbit AT | 12 | | 32 ± 9 | Ouyang et al. (2003) |
| | | | Sut + PLGA | | | 18 ± 5 | |
| | | | Sut + PLGA + MSC | | | 28 ± 6 | |
| Chicken | FDP | PGA/SIS un | IV: PGA un/SIS | – | 9 ± 1 | | Cao et al. (2002) |
| | | | Native FDP | 14 | 110 ± 9 | | |
| | | | Sut + PGA/SIS + Tc | | 92 ± 12 | | |
| | | PCL/HA (wrapped) | Sut only | – | 6 ± 2 | | Liu et al. (2012) |
| | | | Sut + PCL | 3 | 5 ± 2 | | |
| | | | Sut + PCL/HA | | 5 ± 1 | | |
| Dog CAD | RC | – | Native dog infraspinatus | | 1595 ± 285 | 515 ± 141 | Derwin et al. (2009) |
| | Infrasp. | PLLA woven | IV: PLLA device | – | 796 ± 34 | 195 ± 2 | Derwin et al. (2009) |
| | | | IV: Sut thread | 0 | 397 ± 18 | 214 ± 72 | |
| | | | Sut only | 12 | 668 ± 146 | 207 ± 39 | |
| | | | Sut + PLLA | | 801 ± 145 | 199 ± 53 | |
| | | | Sut only | | 990 ± 386 | 243 ± 45 | |
| | | | Sut + PLLA | | 1223 ± 316 | | |
| | | PLLA/ECM (fascia lata) | IV: PLLA/ECM + sut fi | – | 153 ± 27 | 211 ± 22 | Baker et al. (2012) |
| | | | Sut only | 0 | 670 ± 112 | 232 ± 43 | |
| | | | Sut + PLLA/ECM/Sut | 12 | 966 ± 160 | 265 ± 49 | |
| | | | Sut only | | 1228 ± 115 | 248 ± 43 | |
| | | | Sut + PLLA/ECM/Sut | | 1037 ± 189 | | |
| Goat | RC Infrasp. | PLA woven | IV: PLA woven | – | 1214 ± 93 | 70 ± 12 (N %) | MacGillivray et al. (2006) |
| | | | Sut only | 3 | 201 ± 152 | | |
| | | | Sut + PLA | 6 | 272 ± 56 | | |
| | | | Sut only | 12 | 609 ± 193 | | |
| | | | Sut + PLA | 24 | 543 ± 183 | | |
| | | | Sut only | | 849 ± 269 | | |

*Continued*

**Table 8.2 Continued**

| Animal model | Target tendon or ligam. | Augmented/ reconstr. by graft | Specifications | Healing time (weeks) | Ultimate load (N) | Stiffness (N/mm) | Ref. |
|---|---|---|---|---|---|---|---|
| | | | Sut+PLA | | 658±362 | | |
| | | | Sut only | | 1064±301 | | |
| | | | Sut+PLA | | 1046±248 | | |
| Sheep | RC Infrasp. | PU mech | Native sheep infraspinatus | 12 | 3516±*279* | 244±*23* | Santoni et al. (2010) |
| | | | Sut only | | 762±*167* | 64±*10* | |
| | | | Sut+PU | | 1327±*142* | 99±*18* | |
| | ACL | PET | Native sheep ACL | 16 | 1400±100 | 230±30 | Seitz et al. (2013) |
| | | | Sut only | | 400±80 | 55±25 | |
| | | | Sut+PET | | 600±50 | 90±15 | |

*Ant.*, anterior; *AT*, Achilles tendon; *ATR*, Artelon tissue reinforcement; *bFGF*, basic fibroblast growth factor; *CAD*, human cadavers; *Coll* sp, collagen sponge; *DGN*, dextran nanoparticles; *ECM*, extracellular matrix; *es*, electrospun; *Fb*, fibroblasts; *FDP*, flexor digitorum profundus; *FDS*, flexor digitorum superficialis; *fi*, fiber; *HA*, hyaluronic acid; *Hum*, human; *Ibp*, ibuprofen; *IGF*, insulin-like growth factor; *infrasp.*, infraspinatus tendon; *IV*, in vitro; *kn*, knitted; *MCL*, medial collateral ligament; *MSC*, mesenchymal stem cells; *Ø±SD*, mean±standard deviation; *Ø±SEM*, mean±*standard error of the mean*; *PCL*, polycaprolactone; *PDGF-β*, platelet-derived growth factor beta; *PEG*, polyethylene glycol; *PELA*, poly(L-lactic acid)-polyethylene glycol; *PEO*, polyethylene oxide; *PET*, polyethylene terephthalate; *PGA*, poly(glycolide) acid; *PLA*, polylactic acid; *PLGA*, poly (lactic-*co*-glycolic acid); *PLLA*, poly(L-lactide acid); *post.*, posterior; *PU*, polyurethane; *RC*, rotator cuff; *RTF*, rat tendon fibroblasts; *SIS*, small intestine submucosa; *Sup.*, superior; *supraspin.*, supraspinatus tendon; *Sut*, suture; *Tb*, tenoblasts; *Tc*, tenocytes; *un*, unwoven; *wo*, woven.

lost in PLGA. Such morphological changes lead to increased rates of hydrolysis—with a 50:50 PLGA copolymer degrading faster compared with either PGA or PLA.

In a rabbit AT defect model, either bone marrow stromal cell-seeded or nonseeded knitted PLGA scaffolds were implanted and evaluated 12 weeks postoperation (Ouyang et al., 2003). Compared with other processing methods like electrospinning, knitting of PLGA leads to higher mechanical strength of the material. At the time of the study (2003), PLGA had not been widely used in tendon repair application; therefore, the study showed primarily that PLGA is a biocompatible and biodegradable scaffold material when applied in AT regeneration. The formation of neotendon went along with PLGA degradation, and only little inflammatory response was observed. In terms of cell seeding, the authors showed that the seeding of MSCs clearly enhanced biomechanical properties of extracted tendons compared to cell-free constructs. While there was an increase of 18% in the elastic modulus, the stiffness was increased by 55% in the presence of the stem cells, nearly reaching natural values (Table 8.2a and b). In contrast to the study by Pietschmann et al. (2013), no ossification was reported although MSCs were used.

Because knitted PLGA has quite high mechanical strength, a composite material based on knitted PLGA and alginate gel was developed with alginate acting as a cell carrier (Vaquette et al., 2010). Alginate is a natural polysaccharide, and it is the structural element of cell walls in algae. The PLGA/alginate composite scaffolds were implanted either with or without autologous bone MSCs into a rabbit AT defect. Compared to the cell-free construct, the elastic modulus increased by a factor of 1.5 in the cell-seeded constructs 13 weeks postoperation (Vaquette et al., 2010; Tables 8.1, 8.2a and Fig. 8.2).

Although the tensile behavior of the knitted PLGA/alginate composite was determined at different stretch rates (10, 30, 50, and 100 mm $min^{-1}$, respectively), very similar stress-strain curves were obtained. The curves were also similar to native rabbit AT stress-strain curves, with a typical toe region, subsequent linear region ending up in the failure region. Moreover, also the relaxation behavior of the knitted scaffolds was assessed. Here, the stretch rate had an influence; increasing the stretch rate resulted in a faster relaxation (Vaquette et al., 2010).

In the same study, the pH was measured as a function of time. It stayed constant during the first 14 days at pH 7.3; however, it decreased significantly to pH 7.1 by day 21 and furthermore to pH 5 by day 42. Not only PLGA but also PGA and PLA among other synthetic polymeric materials do degrade with a pH drop (Vieira et al., 2009), which causes inflammatory reactions toward the implant material. Mechanistically, the degradation of polyesters is initiated by extracellular hydrolases. Biodegradation of a polymer in the form of hydrolysis is controlled by its chemical structure like esters, ethers, and amides, which are susceptible to hydrolysis. Moreover, for aliphatic polyesters such as PLGA, PGA, or PLA, the temperature and the ester group density are important. Hydrolysis is also accelerated by local pH drop caused by the degradation itself. In the study (Vaquette et al., 2010), the high cell density found in the regenerated tendons was explained by an inflammatory response probably caused by acidic degradation products of PLGA. Nevertheless, the tendon seemed to have started the remodeling phase as the extracellular matrix appeared aligned and to have a crimp pattern like the natural equivalent (Vaquette et al., 2010).

## 8.4 Polylactic acid

As lactic acid is a chiral molecule, two enantiomeric molecules exist (D- and L-lactic acid). Therefore, PLA can be either PLLA or poly(D-lactide acid) (PDLA); also the racemic form poly(L-lactide-*co*-D,L-lactide acid) (PLDLA) can be fabricated; however, it is rarely used in practice. The chemical composition of PLA is $[O—CH(CH_3)—CO]_n$. PLLA is more often used in medical applications than PDLA because upon degradation of the polymer the L-lactic acid is obtained, which is the naturally occurring enantiomer of the lactic acid. While the PLDLA is an amorphous polymer, PLLA is semicrystalline. Amorphous polymers are used preferably in drug delivery applications where it is important to have a homogenous dispersion, while semicrystalline and crystalline polymers do exhibit higher mechanical strength and toughness, so their field of application lies predominantly in orthopedic applications. Not surprisingly, for tendon tissue engineering and tendon repair augmentation, PLLA has been chosen often up to date.

With the intention to mimic the mechanical properties of natural collagen, Surrao et al. fabricated a series of PL(G)A-based electrospun meshes (Surrao et al., 2012). They used PLGA, PLDLA (98 or 250 kDa), and PLLA. In order to mimic natural collagen crimp pattern, the authors first did the electrospinning on a rotating mandrel to get aligned fibers and then immersed the mesh in heated water in order to obtain a crimp-like pattern similar to the pattern found in native anterior cruciate ligament (ACL) collagen. In terms of biomechanical properties of the such prepared scaffolds, PLGA, PLDLA (250 kDa), and PLLA had similar elastic moduli; PLDLA (98 kDa) and PLLA had six times smaller moduli and up to six times smaller ultimate loads in vitro (Table 8.1). Therefore, for the cell seeding with bovine fibroblasts, the authors chose PLDLA (250 kDa) and found a threefold increase in elastic modulus and ultimate load during 8 weeks in static culture compared with the first week (Surrao et al., 2012). The authors concluded—as confirmed by a more recent study dealing with crimps in PLA scaffolds (Liu et al., 2015)—that the geometric microenvironment of the scaffold should mimic natural native microenvironment so that upon cell seeding, beneficial in vitro characteristics (morphology, alignment, and biomechanics) are achieved for further in vivo experiments.

In 2006, a woven PLA patch for rotator cuff (RC) augmentation in a goat model was compared to nonaugmented tendons; at 3, 6, 12, and 24 weeks postoperation, no significant differences were found between the two treatment groups with respect to ultimate load of the specimen (MacGillivray et al., 2006). The authors reported only trends of biomechanical properties being enhanced at early time points for the patch-augmented RCs compared with nonaugmented RCs; however, also observed as trends, at later time points (24 weeks), the patch-augmented group even performed with respect to their biomechanical outcome (Table 8.2b). As such, RC augmentation by the woven PLA patch remained questionable. On the one hand side, early increased strength of the repair was welcome; on the other hand side, it should be at least equally strong compared to nonaugmented specimen at later time points.

Incorporation of nanoparticles into electrospun PLLA meshes has been performed for different purposes including antibacterial agents and also growth factors intended at acceleration of the tendon healing. For example, antibacterial silver nanoparticles were loaded into electrospun PLLA meshes (Liu et al., 2013b). in vitro measurements showed antibacterial effects on *Staphylococcus epidermidis* and *Staphylococcus aureus*, which is a welcome effect. In terms of the mechanical properties such as ultimate stress and elastic modulus, they were unaffected by the different amounts of silver nanoparticles incorporated in the PLLA and also by the presence of such nanoparticles at all—also offering a good starting point for later in vivo experiments (Table 8.1 and Fig. 8.2). Moreover, Liu et al. fabricated a PLLA electrospun mesh with basic fibroblast growth factor (bFGF) or with dextran nanoparticles that were loaded with bFGF (Fig. 8.1). Although cell proliferation in a rat AT model was enhanced, ultimate loads and stiffness 3 weeks postsurgery were not affected; however, the ultimate load and stiffness values were not significantly different from the control healthy tendons (Liu et al., 2013c; Table 8.1).

Furthermore and in contrast to the findings in the PLLA-augmented goat RCs (MacGillivray et al., 2006), besides showing a good biocompatibility and tissue integration a commercially available woven PLLA (X-Repair; Synthasome, San Diego, California, United States) augmentation of RC repair in a canine model showed a 35% higher failure load and a 26% higher stiffness 12 weeks postsurgery compared to nonaugmented *infraspinati* (Derwin et al., 2009). According to the authors, the only limitation that might impede translation into clinical practice is that this PLLA device increased the cross-sectional area by 137%, possibly resulting in an impingement on the acromial arch (Table 8.2b).

Moreover, this PLLA X-Repair device was also applied in a human cadaver RC model. For the anterior and posterior *supraspinatus* and for the superior *infraspinatus*, augmentation had a significant impact on ultimate loads (increase by 60–90%), and the stiffness was also slightly increased (by at maximum 20%) (McCarron et al., 2010). As such, time-zero strength was significantly higher when the PLLA device was applied, which is very important considering early active motion during rehabilitation protocols (Table 8.2b).

Also the processing of PLLA has an impact on its biomechanical properties as shown in a study about PLLA yarns for ACL reconstruction by Freeman et al. (2007) where braided scaffolds, braided-twisted ones, and twisted ones were compared and the influence of the braiding and twisting angles evaluated (Table 8.1 and Fig. 8.2). In the stress-strain curves, the toe region increased with decreasing angle (braiding and twisting) because more force was needed for the alignment of the fibers of the yarn along the loading axis instead of just stretching the fibers. For ACL replacement, the authors concluded that a braid-twist 60–72 PLLA scaffold would be the best option (60–72 denoting the angles involved: fibers twisted to 60 degrees and fiber bundles twisted to 72 degrees).

Also combinations of tissues from natural origin with synthetic polymeric scaffolds are under investigation in the field of tendon repair. The biomechanical impact of

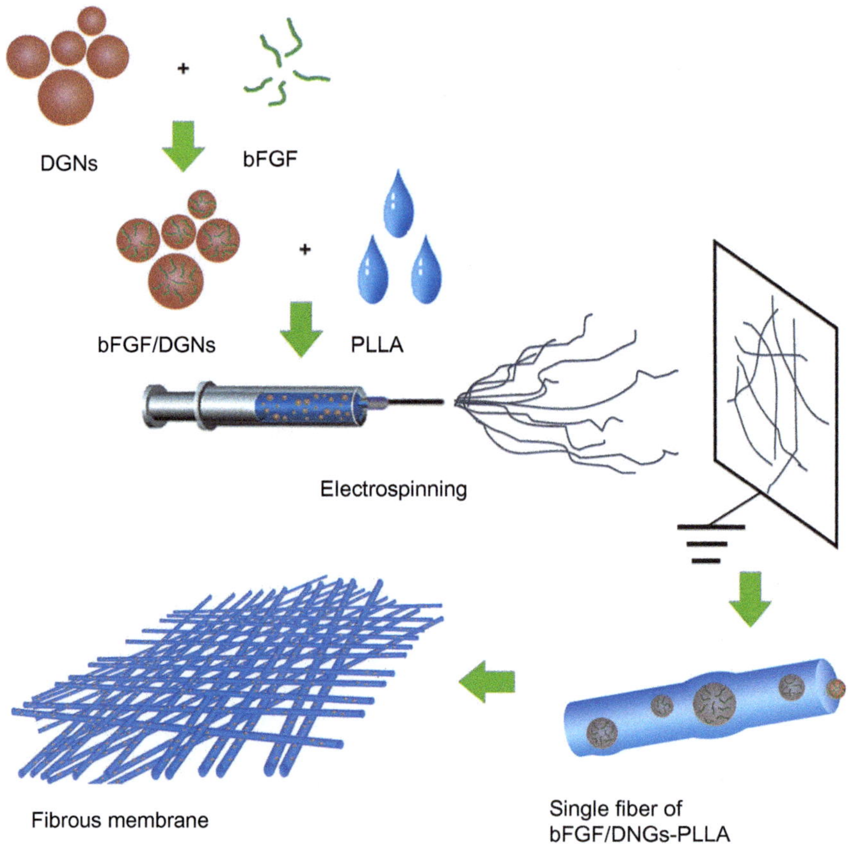


**Fig. 8.1** Fabrication of a PLLA graft for tissue engineering. After incorporation of bFGF into dextran nanoparticles (DGNs) by a freeze-induced particle-forming process, a bFGF/DGNs-loaded PLLA fibrous membrane was fabricated by electrospinning, producing a membrane that can release bFGF sustainably and has the ability to secure the biological activity of bFGF. Key: *bFGF*, basic fibroblast growth factor; *PLLA*, poly(L-lactide acid).
From Liu, S., Qin, M., Hu, C., Wu, F., Cui, W., Jin, T., Fan, C., 2013c. Tendon healing and anti-adhesion properties of electrospun fibrous membranes containing bFGF loaded nanoparticles. Biomaterials 34, 4690–4701, reprinted from Biomaterials, Copyright 2013, with permission from Elsevier.

*fascia lata* that were reinforced with PLLA was assessed in a human cadaver shoulder model; significantly less gap formation and an increase in the number of cycles necessary for failure were observed (McCarron et al., 2012). Such a combination of biological tissue and synthetic tissue was suggested to possibly reduce the incidence of RC repair failure in the clinical setting (McCarron et al., 2012). However, when the PLLA-reinforced fascia patch was tested in vivo in a canine infraspinatus model, failure loads did not reach the ones obtained for nonaugmented tendons, the stiffness being not significantly different from nonaugmented species either (Baker et al., 2012; Table 8.2b).

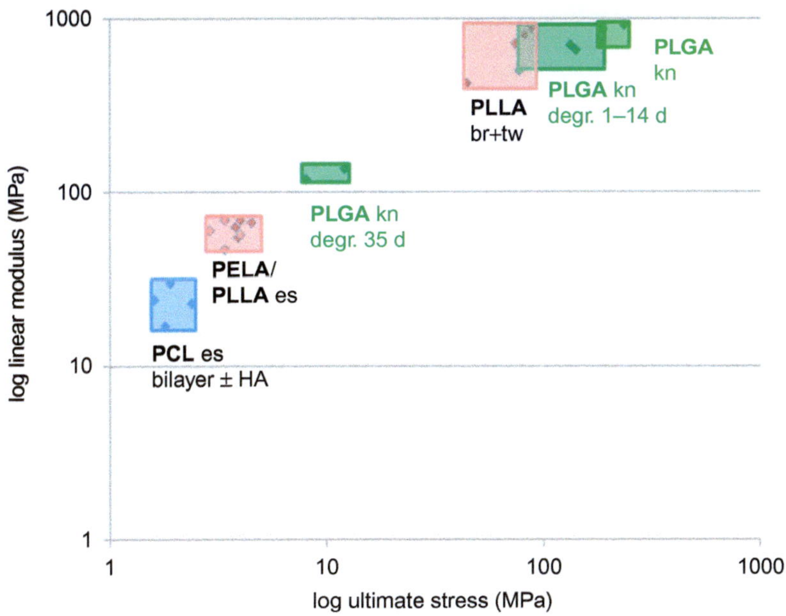


**Fig. 8.2** In vitro mechanical properties of synthetic scaffolds. Key: scaffold materials—*HA*, hyaluronic acid; *PCL*, polycaprolactone; *PELA*, poly(L-lactic acid)-polyethylene glycol; *PLGA*, poly(lactic-*co*-glycolic acid); *PLLA*, poly(L-lactide acid). Manufacture techniques—*br*, braided; *d*, days; *degr*, degraded; *es*, electrospun; *kn*, knitted; *tw*, twisted. Data: $\varnothing \pm SD$, mean ± standard deviation or $\varnothing \pm SEM$, mean ± standard error of the mean.
Data from Freeman, J.W., Woods, M.D., Laurencin, C.T., 2007. Tissue engineering of the anterior cruciate ligament using a braid–twist scaffold design. J. Biomech. 40, 2029–2036; Liu, S., Hu, C.M., Li, F.F., Li, X.J., Cui, W.G., Fan, C.Y., 2013a. Prevention of peritendinous adhesions with electrospun ibuprofen-loaded poly(L-lactic acid)-polyethylene glycol fibrous membranes. Tissue Eng. A. 19, 529–537; Liu, S., Liu, S.H., Liu, X.D., Zhao, J.W., Cui, W.G., Fan, C.Y., 2013b. Antibacterial antiadhesion membranes from silver-nanoparticle-doped electrospun poly(L-lactide) nanofibers. J. Appl. Polym. Sci. 129, 3459–3465; Liu, S., Zhao, J., Ruan, H., Tang, T., Liu, G., Yu, D., Cui, W., Fan, C., 2012. Biomimetic sheath membrane via electrospinning for antiadhesion of repaired tendon. Biomacromolecules 13, 3611–3619; Vaquette, C., Slimani, S., Kahn, C.J., Tran, N., Rahouadj, R., Wang, X., 2010. A poly(lactic-*co*-glycolic acid) knitted scaffold for tendon tissue engineering: an in vitro and in vivo study. J. Biomater. Sci. Polym. Ed. 21, 1737–1760.

## 8.5 Polycaprolactone (PCL)

PCL is a semicrystalline polymer with a low melting point (59–64°C) and an exceptional ability to form blends. The chemical formula of PCL is $[CO{-}(CH_2)_5{-}O{-}]_n$. It degrades at slower rates than PLA and has a high permeability for many drugs, which makes it ideal as a drug delivery device. Upon biodegradation, it is nontoxic.

It is well known that hyaluronic acid (HA) diminishes adhesion formation to the surrounding tissue in the regeneration of intrasynovial tendons (Akasaka et al., 2005). Therefore, electrospun PCL was combined with HA in the way that different percentages of HA were incorporated into the PCL mesh (4, 8, and 12%, respectively). The in vitro ultimate stress decreased with increasing amounts of HA present in the mesh (Liu et al., 2012). Moreover, in a chicken flexor digitorum profundus (FDP) model, these meshes were wrapped around the tendon; 3 weeks postsurgery, ultimate loads were indistinguishable from untreated tendons, for both, HA-loaded and the pure PCL meshes; however, a significant and beneficial decrease in the work of flexion by 40% was found for the 12%-HA-loaded mesh (Liu et al., 2012; Tables 8.1 and 8.2b). In this study, in vitro biomechanical assessment was not predictive for in vivo biomechanical outcome because the PCL/HA mesh was wrapped around the lacerated and sutured tendon and was therefore not primarily used in order to strengthen the repair site.

In a rat *supraspinatus* repair model, tendon augmentation by pure electrospun PCL scaffold was compared with PCL/polyethylene oxide (PEO) scaffold. PEO is a porogen and is used to increase porosity of electrospun meshes after immersion into ethanol/water (Beason et al., 2012). Both scaffold materials were very similar in terms of in vivo biomechanical outcome: ultimate stress, elastic modulus, ultimate load, and stiffness were similar and, however, tendentially lower compared to the control where there was no augmentation (Table 8.2a and b and Fig. 8.3). Nevertheless, the authors intended to use the PCL/PEO scaffold as a carrier for drug delivery (Beason et al., 2012).

## 8.6 Polyurethane

Polyurethanes (PUs) are polymers joined by urethane links whereby a typical composition is [O—$(CH_2)_4$—O—CO—NH—$(CH_2)_6$—NH—CO—]. Most commercially available PUs today are block copolymers indicating that there are two alternating segments in the polymer. One segment is called the *hard* segment that is glassy or crystalline, while the other is called the *soft* segment that is rubbery. As such, the hard segment might provide reinforcing domains that give rise to strength and toughness, while the soft segment leads to elasticity (Kumbar et al., 2014).

For augmentation of a chronic RC tear in a sheep model, Santoni et al. used a commercially available PU-based patch, a material that is not biodegradable (Biomerix RCR Mesh, Biomerix Corp, New York; Santoni et al., 2010). They found that this patch increased the ultimate load 12 weeks postsurgery by a factor of 1.7 and the stiffness by 1.6 (Table 8.2b). The authors claimed that compared with acute transection RC models, the chronic model used in their study more closely simulated the type of RC injury observed in human beings; according to a review including 48 clinical studies, most patients deferred surgical intervention at least 6–12 months after the onset of the symptoms associated with an RC injury (Marx et al., 2009). The beneficial in vivo results demonstrate that the PU patch not only is suitable for a better long-term

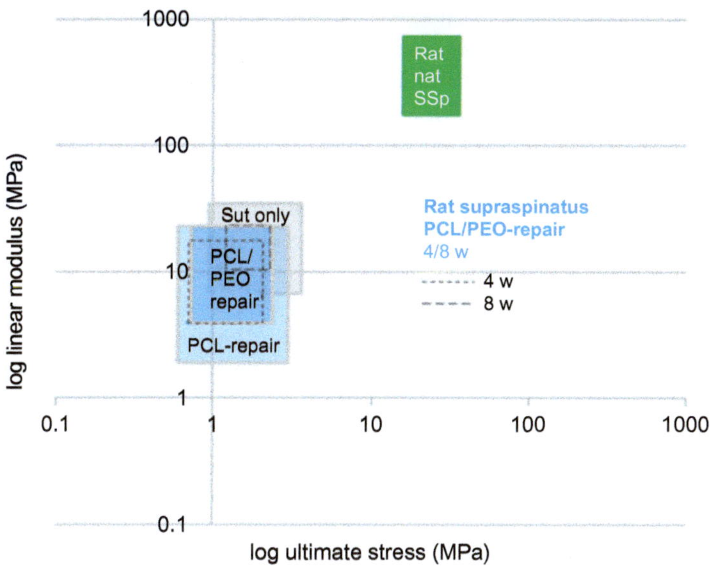


**Fig. 8.3** In vivo tendon repairs with synthetic scaffolds: mechanical properties in an animal model. Key: *PCL*, polycaprolactone; *PEO*, polyethylene oxide; *Suprasp*, supraspinatus tendon (rotator cuff); *w*, weeks. Data: $\varnothing \pm SD$, mean ± standard deviation.
Data from Carpenter, J.E., Thomopoulos, S., Flanagan, C.L., Debano, C.M., Soslowsky, L.J., 1998. Rotator cuff defect healing: a biomechanical and histologic analysis in an animal model. J. Shoulder Elb. Surg. 7, 599–605; Beason, D.P., Connizzo, B.K., Dourte, L.M., Mauck, R.L., Soslowsky, L.J., Steinberg, D.R., Bernstein, J., 2012. Fiber-aligned polymer scaffolds for rotator cuff repair in a rat model. J. Shoulder Elb. Surg. 21, 245–250.

clinical outcome but also suggests to be a strategy to reduce the frequency of rupture in the early postoperative period (Santoni et al., 2010).

The Artelon tissue reinforcement (ATR; Arimplant AB, Västra Frölunda, Sweden) is a synthetic degradable PU urea used for reinforcements of soft tissues established in orthopedics. It acts as a scaffold for cellular ingrowth in order to facilitate healing (allowing angiogenesis and neocollagenesis). Moreover, it is intended at tendon stabilization (Huss et al., 2008). In a human cadaver AT study, it was shown that AT reconstruction with an Artelon augmentation increased ultimate load significantly by 49%, while stiffness was enhanced by 20% (not significantly) (Giza et al., 2011; Table 8.2b). Moreover, the mean creep was significantly reduced from $3.1 \pm 1.1$ mm for the nonaugmented group to $2.0 \pm 0.5$ mm in the Artelon augmented group.

## 8.7 Polylactic caprolactone

Tissue-engineered scaffolds for ACL replacement are critical in terms of deformability. In an in vitro study, Laurent et al. chose a polylactic caprolactone (PLCL) in various ratios (97:3, 85:15, and 70:30). The mechanical behavior is adjustable from elastomeric to rigid by increasing the amount of lactic acid in the copolymer. This copolymer was chosen because (a) of its excellent biocompatibility, (b) of its slow degradation, (c) the blend compensates the slightly brittle behavior of pure PLLA and the rather low stiffness of pure PCL, and (d) it can be processed into fibers by custom plastic extruders that were used in the study. The braided PLCL (85:15) was determined to mimic the best ligament tissue properties, and the architecture of the scaffold was varied by different amounts of the braided layers. As expected, more layers led to increased ultimate loads accompanied by increased stiffness (Laurent et al., 2012). Noteworthy, computer-aided tissue engineering predicted accurately which ratio of PLA to PCL would be best with respect to biomechanics (Table 8.1). Moreover, also the number of layers, fiber diameter, and braiding angle used as parameters in the code aimed at finite element analysis (FEA) were accurately predicting the experimental biomechanical outcomes such as the ultimate load, the stiffness, and the toe region (in %) in the load-strain curve. As such, not only a beneficial scaffold for ACL reconstruction but also a predictive numerical method allowing computer-aided tissue engineering for tendon and ligament applications was presented.

## 8.8 DegraPol®

DegraPol® (DP) is a biodegradable polyester urethane polymer with polyhydroxybutyrate as a crystalline segment (hard segment) and ε-caprolactone as a soft segment (Buschmann et al., 2015). By varying the ratio of hard and soft segment, the elasticity can be tailored and adjusted to the intended purposes. Upon degradation, no pH drop is observed as is the case for other polymers such as PLGA (Vaquette et al., 2010; Vieira et al., 2009).

Because DP had been shown to be biocompatible for fibroblasts, osteoblasts, and tenocytes in vitro and to be biodegradable and cyto- and hemocompatible (Saad et al., 1998; Sukthankar, 1999; Milleret et al., 2009), a DP tube was developed in our research group intended to support tendon regeneration (Fig. 8.4).

In a rabbit AT model, cell response toward the block copolymer DP was shown to have no adversary effects (Buschmann et al., 2013). Furthermore, a beneficial effect with respect to antiadhesion had been determined (Meier Buergisser et al., 2014). In terms of biomechanical properties, neither ultimate load, nor failure stress, nor elastic modulus was affected significantly by the implantation of the tube compared to the conventionally four-strand sutured specimen at any time point considered (Fig. 8.5A–C), which was expected because the material strength of electrospun DP meshes were far below natural tendon biomechanical characteristics. Interestingly, load until failure was higher at 3 weeks postoperation compared with 6 weeks (for both, the four-strand and the DP group), which is confirmed by other studies (Joshi et al., 2009) explaining the relationship of mechanical properties, cellularity, and vascularity over a 3-month healing period. While cellularity peaks at 4 weeks postoperation, vascularity peaks at 6 weeks postsurgery, going along with a significant increase of the ultimate load only after 12 weeks.

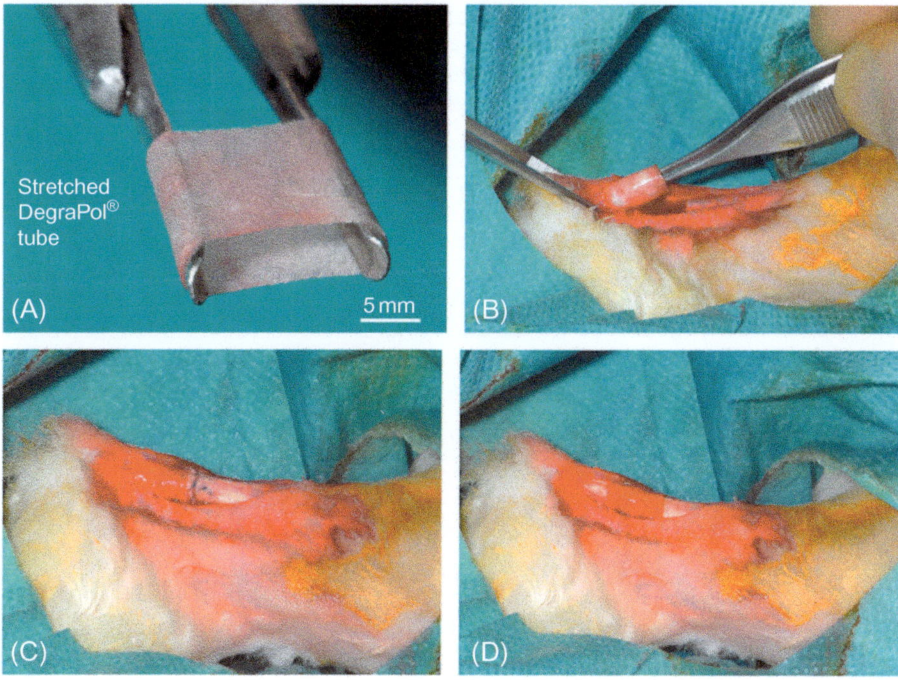


**Fig. 8.4** Application of a highly stretchable DegraPol® (DP) tube (A) in a rabbit Achilles tendon (AT) full defect repair (B) with a four-strand suture (C) and the DP tube pulled over the sutured site (D).
(A, B, and D) From Meier Buergisser, G., Calcagni, M., Muller, A., Bonavoglia, E., Fessel, G., Snedeker, J.G., Giovanoli, P., Buschmann, J., 2014. Prevention of peritendinous adhesions using an electrospun DegraPol polymer tube: a histological, ultrasonographic, and biomechanical study in rabbits. Biomed Res. Int. 2014, 656240, Hindawi open access article. (C) From Buschmann, J., Meier Buergisser, G., Bonavoglia, E., Neuenschwander, P., Milleret, V., Giovanoli, P., Calcagni, M., 2013. Cellular response of healing tissue to DegraPol tube implantation in rabbit Achilles tendon rupture repair: an in vivo histomorphometric study. J. Tissue Eng. Regener. Med. 7, 413–420. Copyright by Journal of Tissue Engineering and Regenerative Medicine, with permission from Wiley.

It was concluded that in order to improve tendon biomechanics, the DP tube should act as a carrier for growth factor(s) stimulating the tendon regeneration process (Evrova et al., 2016). Nevertheless, besides its beneficial effect in terms of anti-adhesion, also a significant reduction of the diameter at the repair site was found during operation (time point 0; Fig. 8.5D). Such a reduction of bulging—caused by suturing the two tendon stumps—is an important feature because it facilitates the gliding of the freshly sutured tendon in the sheath.

## 8.9 Polyethylene terephthalate

PET is a polyester with the chemical formula [O—CO—$C_6H_4$—CO—O—$CH_2$—$CH_2$—]. It is an inert material (not biodegradable) with a melting point of $>250°C$.

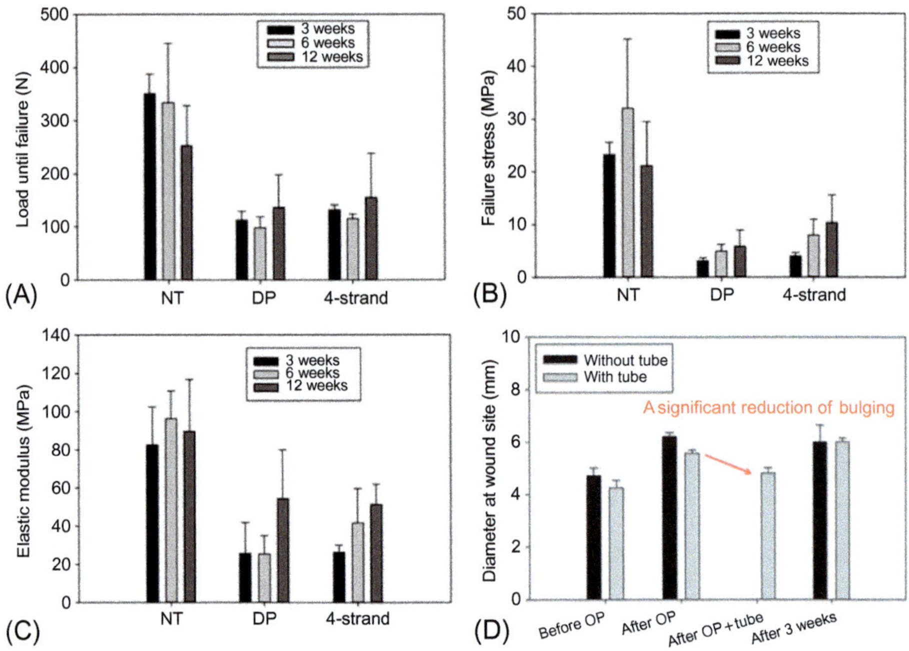


**Fig. 8.5** Biomechanical properties of native, healthy nontreated (NT) rabbit Achilles tendons (AT) compared with fully dissected and four-strand Becker sutured ATs (four-strand) and compared with four-strand Becker sutured tendons that received a DP tube (DP) at three different time points postoperatively (3, 6, and 12 weeks, respectively); load until failure (N) (A), failure stress (MPa) (B), and elastic modulus (MPa) (C). The effect of reducing the bulging at the repair site was determined by the diameter (mm) before (before OP = before operation) and after DP application to a four-strand Becker sutured rabbit AT (after OP + tube = after pulling a DP tube over the sutured tendon). (D) Data for the 6-week experiments are taken from Meier Buergisser et al., 2014; data for 3- and 12-week experiments were performed accordingly.

In tendon repair, PET was used to augment ACL reconstructions in an ovine model (Seitz et al., 2013). Biomechanical evaluation showed that a PET augmentation significantly increased ultimate loads 16 weeks postoperatively by 50% (being 2.3 times smaller than native ACL; Table 8.2b). Moreover, PET-augmented specimen had a 1.6-fold higher stiffness, which was, however, still 2.5 times smaller compared to intact ACL's stiffness. Thus, although very promising with respect to mechanical properties, a PET augmentation does not reach natural values; it allows, however, a more rigorous early active motion during the rehabilitation phase (Seitz et al., 2013).

PET is an inert material, and therefore, cellular attachment is difficult. In order to make it more attractive with respect to cell attachment, cell proliferation, collagen regeneration, and biointegration in tendon repair, Li et al. (2013) covered it with layers of chitosan and HA by a self-assembly technique. While in vitro experiments with human foreskin fibroblasts clearly indicated that cell attachment and proliferation were supported by the layers covering the PET sheet, in vivo implantation for a medical collateral ligament reconstruction (without cell seeding before) was also favorable; not only was neocollagenesis between PET fibers significantly enhanced compared with the noncoated PET, but also was the inflammatory response (assessed by macrophages, lymphocytes, and foreign body giant cells) reduced. It was concluded that such a chitosan-/HA-coated PET graft had a considerable potential as a ligament substitute. Unfortunately, no biomechanical data were reported so far.

## 8.10 Poly(L-lactic acid)-*co*-ethylene glycol

Poly(L-lactic acid)-*co*-ethylene glycol (PELA) is a diblock copolymer of poly (L-lactic)acid and polyethylene glycol. Ibuprofen-loaded electrospun PELA meshes intended at antiadhesion in a FDP chicken model were tested with respect to their biomechanics (Liu et al., 2013a). The ultimate stress of the extracted tendons 3 weeks postoperatively was not affected by PELA application; however, the work of flexion was significantly reduced by 60% (Tables 8.1, 8.2a and Fig. 8.2).

## 8.11 Summary

Synthetic polymeric scaffold materials have been used in tendon tissue engineering as a full replacement of lacerated tendons and in tendon augmentations intended at reinforcing the primary strength at time zero. From a variety of aliphatic polyesters, PGA with its high elastic characteristics has been successfully explored in tendon replacement studies. Moreover, polymers such as PLGA and PLA were also used in tendon tissue engineering, however, having always the problem of biodegradation (hydrolysis) accompanied by a pH drop. In contrast, PCL, PU, or DP degrades at constant pH. Nevertheless, the polymeric scaffold materials PGA, PLGA, and PLA exhibit excellent cell attachment and proliferation characteristics in vitro including several cell types such as tenocytes or fibroblasts and stem cells (MSCs, ASCs, etc.). Approaches dealing with stem cells show that besides the welcome differentiation toward the tenocyte phenotype, also unwanted differentiation pathways may occur such as chondrogenesis under compression. Therefore, cell seeding of the corresponding scaffolds should be planned carefully with respect of cell type and dynamic conditioning before using it in in vivo animal models or even in daily clinical practice.

## References

Akasaka, T., Nishida, J., Araki, S., Shimamura, T., Amadio, P.C., An, K.N., 2005. Hyaluronic acid diminishes the resistance to excursion after flexor tendon repair: an in vitro biomechanical study. J. Biomech. 38, 503–507.

Baker, A.R., Mccarron, J.A., Tan, C.D., Iannotti, J.P., Derwin, K.A., 2012. Does augmentation with a reinforced fascia patch improve rotator cuff repair outcomes? Clin. Orthop. Relat. Res. 470, 2513–2521.

Beason, D.P., Connizzo, B.K., Dourte, L.M., Mauck, R.L., Soslowsky, L.J., Steinberg, D.R., Bernstein, J., 2012. Fiber-aligned polymer scaffolds for rotator cuff repair in a rat model. J. Shoulder Elb. Surg. 21, 245–250.

Brodie, M., Vollenweider, L., Murphy, J.L., Xu, F., Lyman, A., Lew, W.D., Lee, B.P., 2011. Biomechanical properties of Achilles tendon repair augmented with a bioadhesive-coated scaffold. Biomed. Mater. (Bristol, UK) 6, 015014.

Buschmann, J., Calcagni, M., Meier Buergisser, G., Bonavoglia, E., Neuenschwander, P., Milleret, V., Giovanoli, P., 2015. Synthesis, characterization and histomorphometric analysis of cellular response to a new elastic DegraPol (R) polymer for rabbit Achilles tendon rupture repair. J. Tissue Eng. Regener. Med. 9, 584–594.

Buschmann, J., Meier Buergisser, G., Bonavoglia, E., Neuenschwander, P., Milleret, V., Giovanoli, P., Calcagni, M., 2013. Cellular response of healing tissue to DegraPol tube implantation in rabbit Achilles tendon rupture repair: an in vivo histomorphometric study. J. Tissue Eng. Regener. Med. 7, 413–420.

Cao, D., Liu, W., Wei, X., Xu, F., Cui, L., Cao, Y., 2006. In vitro tendon engineering with avian tenocytes and polyglycolic acids: a preliminary report. Tissue Eng. 12, 1369–1377.

Cao, Y., Liu, Y., Liu, W., Shan, Q., Buoncuore, S., Cui, L., 2002. Bridging tendon defects using autologous tenocyte engineered tendon in a hen model. Plast. Reconstr. Surg. 110, 1280–1289.

Chen, S.H., Chen, C.H., Shalumon, K., Chen, J.P., 2014. Preparation and characterization of antiadhesion barrier film from hyaluronic acid-grafted electrospun poly(caprolactone) nanofibrous membranes for prevention of flexor tendon postoperative peritendinous adhesion. Int. J. Nanomedicine 9, 4079–4092.

Chen, Y.-J., Huang, C.-H., Lee, I.C., Lee, Y.-T., Chen, M.-H., Young, T.-H., 2008. Effects of cyclic mechanical stretching on the mRNA expression of tendon/ligament-related and osteoblast-specific genes in human mesenchymal stem cells. Connect. Tissue Res. 49, 7–14.

Deng, D., Liu, W., Xu, F., Yang, Y., Zhou, G., Zhang, W.J., Cui, L., Cao, Y., 2009. Engineering human neo-tendon tissue in vitro with human dermal fibroblasts under static mechanical strain. Biomaterials 30, 6724–6730.

Derwin, K.A., Codsi, M.J., Milks, R.A., Baker, A.R., Mccarron, J.A., Iannotti, J.P., 2009. Rotator cuff repair augmentation in a canine model with use of a woven poly-L-lactide device. J. Bone Joint Surg. Am. 91, 1159–1171.

Dines, J.S., Grande, D.A., Dines, D.M., 2007. Tissue engineering and rotator cuff tendon healing. J. Shoulder Elb. Surg. 16, S204–S207.

Dumitriu, S., Popa, V. (Eds.), 2013. Polymeric Biomaterials. CRC Press, Boca Raton, FL.

Evrova, O., Houska, J., Welti, M., Bonavoglia, E., Calcagni, M., Giovanoli, P., et al., 2016. Bioactive, elastic, and biodegradable emulsion electrospun DegraPol tube delivering PDGF-BB for tendon rupture repair. Macromol. Biosci. 16, 1048–1063.

Freeman, J.W., Woods, M.D., Laurencin, C.T., 2007. Tissue engineering of the anterior cruciate ligament using a braid–twist scaffold design. J. Biomech. 40, 2029–2036.

Giza, E., Frizzell, L., Farac, R., Williams, J., Kim, S., 2011. Augmented tendon Achilles repair using a tissue reinforcement scaffold: a biomechanical study. Foot Ankle Int. 32, S545–S549.

Holland, S.J., Tighe, B.J., 1992. Biodegradable Polymers. Advances in pharmaceutical science, vol. 6. Academic, London, pp. 101–164.

Huss, F.R., Nyman, E., Gustafson, C.-J., Gisselfalt, K., Liljensten, E., Kratz, G., 2008. Characterization of a new degradable polymer scaffold for regeneration of the dermis: in vitro and in vivo human studies. Organogenesis 4, 195–200.

Joshi, S.M., Mastrangelo, A.N., Magarian, E.M., Fleming, B.C., Murray, M.M., 2009. Collagen-platelet composite enhances biomechanical and histologic healing of the porcine anterior cruciate ligament. Am. J. Sports Med. 37, 2401–2410.

Koike, M., Shimokawa, H., Kanno, Z., Ohya, K., Soma, K., 2005. Effects of mechanical strain on proliferation and differentiation of bone marrow stromal cell line ST2. J. Bone Miner. Metab. 23, 219–225.

Kumbar, S.G., Laurencin, C.T., Deng, M. (Eds.), 2014. Natural and Synthetic Biomedical Polymers. Elsevier, Boston, MA.

Laurent, C.P., Durville, D., Mainard, D., Ganghoffer, J.-F., Rahouadj, R., 2012. A multilayer braided scaffold for anterior cruciate ligament: mechanical modeling at the fiber scale. J. Mech. Behav. Biomed. Mater. 12, 184–196.

Li, H., Jiang, J., Ge, Y., Xu, J., Zhang, P., Zhong, W., Chen, S., 2013. Layer-by-layer hyaluronic acid-chitosan coating promoted new collagen ingrowth into a poly(ethylene terephthalate) artificial ligament in a rabbit medical collateral ligament (MCL) reconstruction model. J. Biomater. Sci. Polym. Ed. 24, 431–446.

Liu, S., Hu, C.M., Li, F.F., Li, X.J., Cui, W.G., Fan, C.Y., 2013a. Prevention of peritendinous adhesions with electrospun ibuprofen-loaded poly(L-lactic acid)-polyethylene glycol fibrous membranes. Tissue Eng. A 19, 529–537.

Liu, S., Liu, S.H., Liu, X.D., Zhao, J.W., Cui, W.G., Fan, C.Y., 2013b. Antibacterial anti-adhesion membranes from silver-nanoparticle-doped electrospun poly(L-lactide) nanofibers. J. Appl. Polym. Sci. 129, 3459–3465.

Liu, S., Qin, M., Hu, C., Wu, F., Cui, W., Jin, T., Fan, C., 2013c. Tendon healing and anti-adhesion properties of electrospun fibrous membranes containing bFGF loaded nanoparticles. Biomaterials 34, 4690–4701.

Liu, S., Zhao, J., Ruan, H., Tang, T., Liu, G., Yu, D., Cui, W., Fan, C., 2012. Biomimetic sheath membrane via electrospinning for antiadhesion of repaired tendon. Biomacromolecules 13, 3611–3619.

Liu, W., Chen, B., Deng, D., Xu, F., Cui, L., Cao, Y., 2006. Repair of tendon defect with dermal fibroblast engineered tendon in a porcine model. Tissue Eng. 12, 775–788.

Liu, W., Lipner, J., Moran, C.H., Feng, L., Li, X., Thomopoulos, S., Xia, W., 2015. Generation of electrospun nanofibers with controllable degrees of crimping through a simple, plasticizer-based treatment. Adv. Mater. 27, 2583–2588.

Lu, H.H., Cooper Jr., J.A., Manuel, S., Freeman, J.W., Attawia, M.A., Ko, F.K., Laurencin, C.T., 2005. Anterior cruciate ligament regeneration using braided biodegradable scaffolds: in vitro optimization studies. Biomaterials 26, 4805–4816.

Macgillivray, J.D., Fealy, S., Terry, M.A., Koh, J.L., Nixon, A.J., Warren, R.F., 2006. Biomechanical evaluation of a rotator cuff defect model augmented with a bioresorbable scaffold in goats. J. Shoulder Elb. Surg. 15, 639–644.

Marx, R.G., Koulouvaris, P., Chu, S.K., Levy, B.A., 2009. Indications for surgery in clinical outcome studies of rotator cuff repair. Clin. Orthop. Relat. Res. 467, 450–456.

Mccarron, J.A., Milks, R.A., Chen, X., Iannotti, J.P., Derwin, K.A., 2010. Improved time-zero biomechanical properties using poly-L-lactic acid graft augmentation in a cadaveric rotator cuff repair model. J. Shoulder Elb. Surg. 19, 688–696.

Mccarron, J.A., Milks, R.A., Mesiha, M., Aurora, A., Walker, E., Iannotti, J.P., Derwin, K.A., 2012. Reinforced fascia patch limits cyclic gapping of rotator cuff repairs in a human cadaveric model. J. Shoulder Elb. Surg. 21, 1680–1686.

Meier Buergisser, G., Calcagni, M., Muller, A., Bonavoglia, E., Fessel, G., Snedeker, J.G., Giovanoli, P., Buschmann, J., 2014. Prevention of peritendinous adhesions using an electrospun DegraPol polymer tube: a histological, ultrasonographic, and biomechanical study in rabbits. Biomed Res. Int. 2014, 656240.

Middleton, J.C., Tipton, A.J., 2000. Synthetic biodegradable polymers as orthopedic devices. Biomaterials 21, 2335–2346.

Milleret, V., Simonet, M., Bittermann, A.G., Neuenschwander, P., Hall, H., 2009. Cyto- and hemocompatibility of a biodegradable 3D-scaffold material designed for medical applications. J. Biomed. Mater. Res. B Appl. Biomater. 91B, 109–121.

Ouyang, H.W., Goh, J.C.H., Thambyah, A., Teoh, S.H., Lee, E.H., 2003. Knitted poly-lactide-*co*-glycolide scaffold loaded with bone marrow stromal cells in repair and regeneration of rabbit Achilles tendon. Tissue Eng. 9, 431–439.

Petrigliano, F.A., McAllister, D.R., Wu, B.M., 2006. Tissue engineering for anterior cruciate ligament reconstruction: a review of current strategies. Arthroscopy 22, 441–451.

Pietschmann, M.F., Frankewycz, B., Schmitz, P., Docheva, D., Sievers, B., Jansson, V., Schieker, M., Mueller, P.E., 2013. Comparison of tenocytes and mesenchymal stem cells seeded on biodegradable scaffolds in a full-size tendon defect model. J. Mater. Sci. Mater. Med. 24, 211–220.

Saad, B., Ciardelli, G., Matter, S., Welti, M., Uhlschmid, G.K., Neuenschwander, P., Suter, U.W., 1998. Degradable and highly porous polyesterurethane foam as biomaterial: effects and phagocytosis of degradation products in osteoblasts. J. Biomed. Mater. Res. 39, 594–602.

Sahoo, S., Ouyang, H., Goh, J.C., Tay, T.E., Toh, S.L., 2006. Characterization of a novel polymeric scaffold for potential application in tendon/ligament tissue engineering. Tissue Eng. 12, 91–99.

Santoni, B.G., Mcgilvray, K.C., Lyons, A.S., Bansal, M., Turner, A.S., Macgillivray, J.D., Coleman, S.H., Puttlitz, C.M., 2010. Biomechanical analysis of an ovine rotator cuff repair via porous patch augmentation in a chronic rupture model. Am. J. Sports Med. 38, 679–686.

Seitz, H., Pichl, W., Matzi, V., Nau, T., 2013. Biomechanical evaluation of augmented and non-augmented primary repair of the anterior cruciate ligament: an in vivo animal study. Int. Orthop. 37, 2305–2311.

Sukthankar, A., 1999. DegraPol-btc/btgc: in vitro Untersuchung der Biokompatibilität für Tenozyten. University Hospital Zurich, Zurich.

Surrao, D.C., Waldman, S.D., Amsden, B.G., 2012. Biomimetic poly(lactide) based fibrous scaffolds for ligament tissue engineering. Acta Biomater. 8, 3997–4006.

Vaquette, C., Slimani, S., Kahn, C.J., Tran, N., Rahouadj, R., Wang, X., 2010. A poly(lactic-*co*-glycolic acid) knitted scaffold for tendon tissue engineering: an in vitro and in vivo study. J. Biomater. Sci. Polym. Ed. 21, 1737–1760.

Vieira, A.C., Guedes, R.M., Marques, A.T., 2009. Development of ligament tissue biodegradable devices: a review. J. Biomech. 42, 2421–2430.

# Cell therapies for tendons and ligament repair

# 9

## Abstract

The application of cells to support tendon and ligament healing is one promising option to improve the biomechanical properties of the regenerated tissues. Obviously, tenoblasts and tenocytes might be a useful cell source as they are the predominant cell types in healthy tendons and ligaments. However, also stem cells with their intrinsic capacity for self-renewal and lineage-specific differentiation might be supporting full tissue integration and an adequate generation of new tissue. As such, stem cells of different origin, such as bone marrow or adipose tissue, have been explored in tendon healing. Application methods may also be an important factor regulating the outcome; therefore, cells seeded on different constructs such as decellularized tendon tissue or synthetic polymers like poly(lactic-*co*-glycolic acid) have been explored and investigated in vitro and in vivo in terms of their biomechanical functionality and their capacity to form tendon- and ligament-like connective tissue. Also direct injection at the wound site or intravenous injection of cells with the intention to accelerate and support healing has been reported in preclinical studies. Here, we discuss different approaches and their biological and biomechanical outcome. Furthermore, an overview of current clinical practice is given.

**Keywords:** Tendon and synovial cells, Stem cells, Tendon progenitors, Bone marrow, Adipose tissue.

## Abbreviations

| | |
|---|---|
| **ACL** | anterior cruciate ligament (tendons and ligaments) |
| **AEC** | amniotic epithelial cells (cell type) |
| **ASC** | adipose-derived stem cell (cells for seeding in tissue engineering) |
| **BMP** | bone morphogenetic protein (element in tissues) |
| **bmMSC** | bone marrow-derived mesenchymal stem cells (cells for seeding in tissue engineering) |
| **CD** | cluster of differentiation (cell development) |
| **CTGF** | connective tissue growth factor (protein in living cells) |
| **DAPI** | 4′,6-diamidino-2-phenylindole (fluorescent stain for imaging techniques) |
| **ECM** | extracellular matrix (matrix in living cells) |
| **EGF** | epidermal growth factor (protein in living cells) |
| **ESC** | embryonic stem cells (cells for seeding in tissue engineering) |
| **EGFP** | enhanced green fluorescent protein (labeling method) |
| **FGF** | basic fibroblast growth factor ($n=1, 2, 3$) (protein in living cells) |
| **FTE** | functional tissue engineering (tissue engineering concept) |
| **GFP** | green fluorescent protein (labeling in imaging technique) |
| **HUVEC** | human umbilical vein endothelial cells (cells for cell therapy) |

**Biomechanics of Tendons and Ligaments. http://dx.doi.org/10.1016/B978-0-08-100489-0.00009-0**

| | |
|---|---|
| **IGF** | insulin-like growth factor ($n = 1, 2, 3$) (protein in living cells) |
| **IL** | interleukin-*n* ($n = 1$–$17$) (protein in living cells) |
| **LOX** | lysyl oxidase (enzyme in living cells) |
| **Mkx** | *Mohawk* (transcription factor in tenogenesis) |
| **MMP (−n)** | matrix metalloproteinase 1 (interstitial collagenase) (matrix remodeling marker (primer for coded protein)) |
| **MMP (−n)** | matrix metalloproteinase 3 (stromelysin 1, progelatinase) (matrix remodeling marker (primer for coded protein)) |
| **MMP (−n)** | matrix metalloproteinase-*n* ($n = 1, 2, 3$) (enzymes in living cells) |
| **MP** | metacarpophalangeal joint (body part) |
| **MRI** | magnetic resonance imaging (imaging technique) |
| **MSC** | mesenchymal stem cells (cells for seeding in tissue engineering) |
| **PDGF** | platelet-derived growth factor (protein in living cells) |
| **PDMS** | polydimethylsiloxane (inert substrate) |
| **PIP** | proximal interphalangeal joint (body part) |
| **PLGA** | poly(lactic-*co*-glycolic acid) (synthetic scaffold material) |
| **PRP** | platelet-rich plasma (enriched blood plasma) |
| **PT** | patellar tendon (tendons and ligaments) |
| **RC** | rotator cuff (tendons and ligaments) |
| **SCX** | scleraxis (transcription factor protein, tenogenic marker (primer for coded protein)) |
| **SPIO** | superparamagnetic iron oxide (technical feature of MRI) |
| **TEC** | tissue-engineered constructs (tissue engineering concept) |
| **TGF (−β*n*)** | transforming growth factor beta ($n = 1, 2, 3$) (protein in living cells) |
| **TNMD** | tenomodulin (tenogenic marker (primer for coded protein)) |
| **US** | ultrasound (imaging technique) |
| **VAS** | visual analog scale (clinical questionnaire for pain) |
| **VEGF** | vascular endothelial growth factor (protein in living cells) |
| **VISA-P** | Victorian Institute of Sports Assessment-Patellar questionnaire (Clinical questionnaire) |

## 9.1 Introduction

The main problems occurring after tendon and ligament repair by conventional suture techniques are caused by fibrous tissue leading to scar with poor functional properties and adhesion to the surrounding tissue (Elliot and Giesen, 2013). Fibrous tendons are generally less strong, and adhesions may lead to a reduced range of motion. In order to overcome the inferior structure and function of scar tissue, regenerative medicine proposes different approaches, among them the application of an exogenous cell source that might either help to synthesize an extracellular matrix (ECM) at the repair site resembling more the tendon and ligament tissue than scar and/or having paracrine effects resulting in homing cells for repair that are more adequate for this purpose. In addition, gene-transfected cells in combination either with or without further chemical supplements are under investigation.

The basic questions regarding cell therapies used for tendon and ligament repair therefore addressed in this chapter can be summarized by the following:

**(i)** Which *cell types* have been used for tendon/ligament repair and regeneration?
**(ii)** Why was a certain cell type having a *better* outcome than another?
**(iii)** What was the outcome *biologically* (cellular response of the surrounding tissue toward the applied cell source) and *mechanically* (biomechanical characterization after repair)?
**(iv)** How were the cells *obtained* for that purpose, were they used as extracted? Were they purified or was a subpopulation used after cell-sorting procedures?
**(v)** How were they *applied* (direct injection or via scaffolds or other delivery systems)?
**(vi)** Which *animal* models were used and which *clinical* studies in humans were performed?

In this chapter, also the daily clinical practice in the treatment of tendon and ligament injuries will be discussed in terms of cytotherapeutical aspects, and the current use of them is elucidated.

## 9.2 Cell types used for the repair of tendons and ligaments

### *9.2.1 Tendon- and ligament-derived cells*

Considering the appropriate choice of a cell source for tendon or ligament regeneration, it is self-evident to go for cells that are naturally found in those tissues. In that sense, flexor tendon-derived cells, the tenocytes, were examined and characterized in terms of their local origin. It was found that peritenon-derived tenocytes (Peri Tc) differ from tenocytes that were derived from the core tissue of the tendon (Cadby et al., 2014). As there are still debates about the contributions of intrinsic and extrinsic healing mechanisms to tendon regeneration, information about proliferation, migration, colony-forming units, and differentiation capacity of Peri Tc compared to the core is more valuable for further application in cellular therapies. In the study by Cadby et al., it was found that Peri Tc exhibited a significantly higher migration rate and a significantly higher survival in cell culture after 1 week and that they formed larger colonies (though less in number) than tenocytes derived from the core of the equine superficial digital flexor tendon. Moreover, Figs. 9.1 and 9.2 show that Peri Tc and core-derived Tc differ not only in their tenogenic marker expression but also remarkably in their progenitor characteristic markers (Cadby et al., 2014).

Obviously, peritenon-derived cells (Peri Tc) exhibited a higher progenitor marker expression than tenocytes obtained from the core (Tend Tc), while tenocytes from the core exhibited a significantly higher expression of the gene coding for scleraxis (SCX). These results indicated that in the peritenon-derived cell population, mesenchymal stem cell (MSC) subpopulation has a higher fraction than in the core cell population—and that extrinsic healing might be predominant over intrinsic tendon healing. Such results also implicated that for cellular therapies, the Peri Tc may be more suitable than core tenocytes because of their faster reactive potential (migration and cell survival).

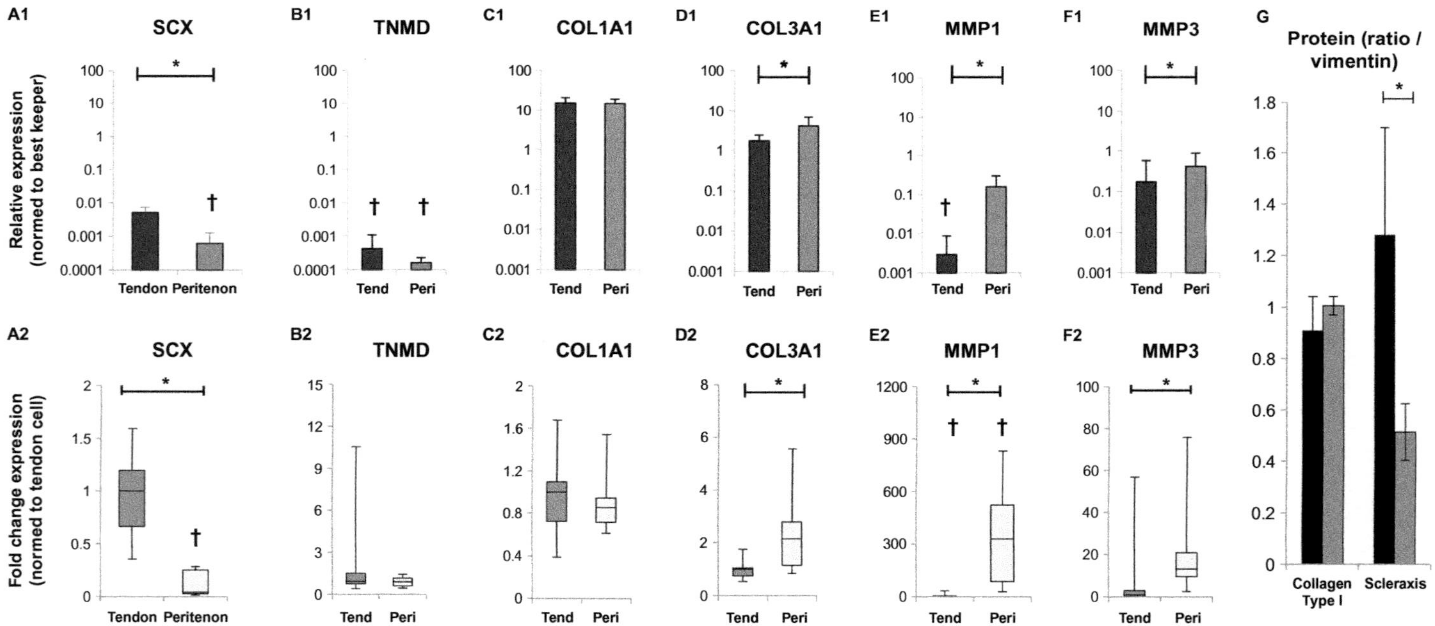


**Fig. 9.1** Gene expression levels of tenogenic markers, matrix remodeling markers, and Western blots on tenogenic markers to confirm gene expression results. Gene expression of (A) SCX = scleraxis, (B) TNMD (tenomodulin), (C) COL1A1 = collagen 1A1, (D) COL3A1 = collagen 3A1, matrix metalloproteinases (E) MMP-1 (interstitial collagenase), and (F) MMP-3 (stromelysin 1, progelatinase) in tendon core cell population (Tend) and peritenon cell population (Peri) cultured in expansion medium. The cells were extracted from eight different horses. Using the $2^{-\Delta\Delta CT}$ method, the data are presented both in terms of "relative expression" normalized to three housekeeping genes and the fold change in gene expression normalized relative to the tendon core expression. (G) Western blots on tenogenic markers to confirm the gene expression results. The graphs display the relative band signal strength in Western blots analyzed by optical densitometry and related to loading control vimentin. Cells from the tendon core population demonstrated a higher level of SCX protein expression compared with cells from the peritenon when cultured in expansion medium for a week. The difference between the two cell populations for collagen protein expression was not significant (key: *, significant difference; †, $p < 0.05$).
Based on Cadby, J.A., Buehler, E., Godbout, C., Van Weeren, P.R., Snedeker, J.G., 2014. Differences between the cell populations from the peritenon and the tendon core with regard to their potential implication in tendon repair. PLoS ONE 9, e92474.

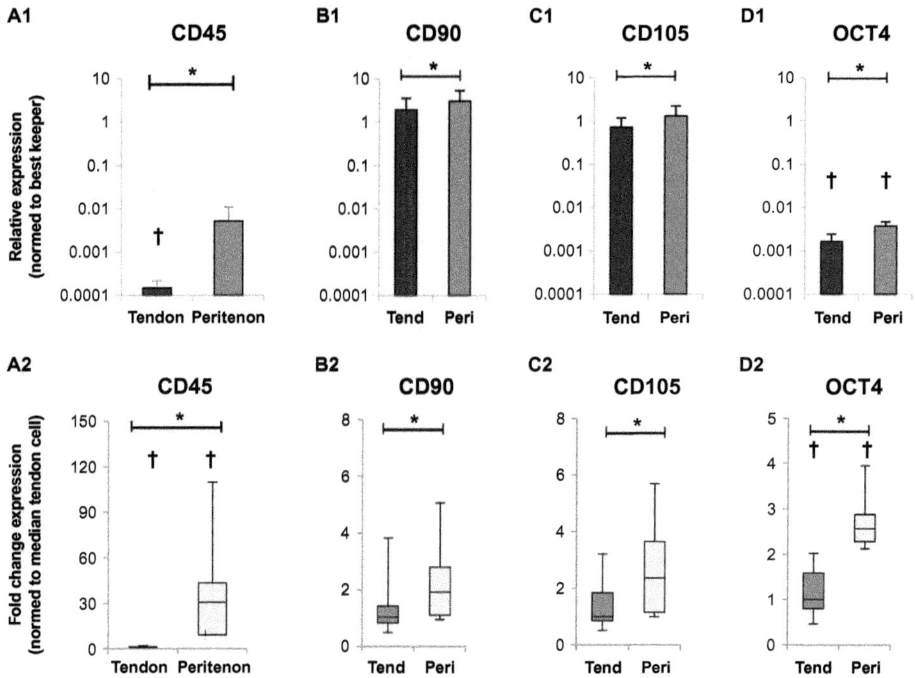


**Fig. 9.2** Gene expression levels of progenitor cell markers indicating higher levels of progenitor cell markers in peritenon cell populations (key: *CD*, cluster of differentiation; see Fig. 9.1).
From Cadby, J.A., Buehler, E., Godbout, C., Van Weeren, P.R., Snedeker, J.G., 2014. Differences between the cell populations from the peritenon and the tendon core with regard to their potential implication in tendon repair. Plos ONE 9, e92474.

## 9.2.2 Synovial cells

Synovial cells are of special interest in intrasynovial tendon regeneration because they secret hyaluronic acid, one of the most important and well-studied antiadhesion substances (Meier Buergisser and Buschmann, 2015). The harvest of synovial cells can be undertaken by using the joint synovium or the surface of tendons as cell source. Morphological assessment of synovial cells and synovial lining cells has been reported (Steinberg and Hodde, 1990). Thus, synovial cells have been used for tissue engineering of a synovial biomembrane (Ozturk et al., 2008). In the latter report, tendon surface synovial cells and knee joint synovium-derived cells formed a multilayer after 2 weeks in vitro cultivation (Fig. 9.3) when seeded on a commercially available tendon collagen matrix.

An in vivo study has been reported where a synovial cell suspension was injected at the repair site of a lacerated and sutured rabbit flexor tendon (Baymurat et al., 2015). Compared to the conventional flexor tendon repair, the synovial cell suspension significantly reduced adhesion formation and increased tendon gliding and the ranges of motion at the joints (PIP = proximal interphalangeal and the MP = metacarpophalangeal joint)

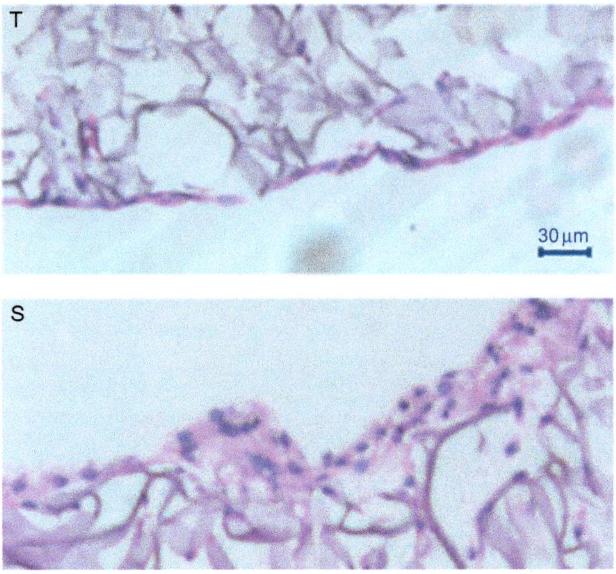


**Fig. 9.3** Surface layer formation by cells at week 2 on hematoxylin-/eosin-stained sections. *T*, tendon surface synovial cells; *S*, knee joint synovium-derived cells.
Based on Ozturk, A.M., Yam, A., Chin, S.I., Heong, T.S., Helvacioglu, F., Tan, A., 2008. Synovial cell culture and tissue engineering of a tendon synovial cell biomembrane. J. Biomed. Mater. Res. A 84A, 1120–1126, © by Journal of Biomedical Materials Research, with permission from Wiley.

(Fig. 9.4). However, when those cells were seeded onto a collagen matrix (biomembrane), the beneficial effects of the synovial cells were even enhanced (Baymurat et al., 2015).

### 9.2.3 Fibroblasts

Fibroblasts are also an option to be used for cellular therapy in tendon regeneration; however, they are more often used as cell type to be seeded on potential substrates for tendon or ligament tissue engineering or to study effects of mechanical stimulation. For example, human foreskin fibroblasts have been successfully used to examine tendon scaffolds intended at anterior cruciate ligament (ACL reconstruction, based on PLGA (poly(lactic-*co*-glycolic acid)), collagen I, and polyurethane, where (i) two different PLGAs were investigated and (ii) the difference of randomly electrospun versus aligned electrospun meshes (Full et al., 2015). For that purpose, the fibroblasts were a useful cell type resembling tenocytes morphologically and in terms of culture conditions and having the advantage of being more easily harvested compared with tenocytes where tendon or ligament biopsies are a prerequisite. Moreover, primary human RC (rotator cuff) fibroblasts have also been used to examine the benefit of mechanical stimulation on RC regeneration. The cells were seeded onto a Flexcell FX5K Tension system, and the relative mRNA levels were compared to nonstimulated cells (Lohberger et al., 2015).

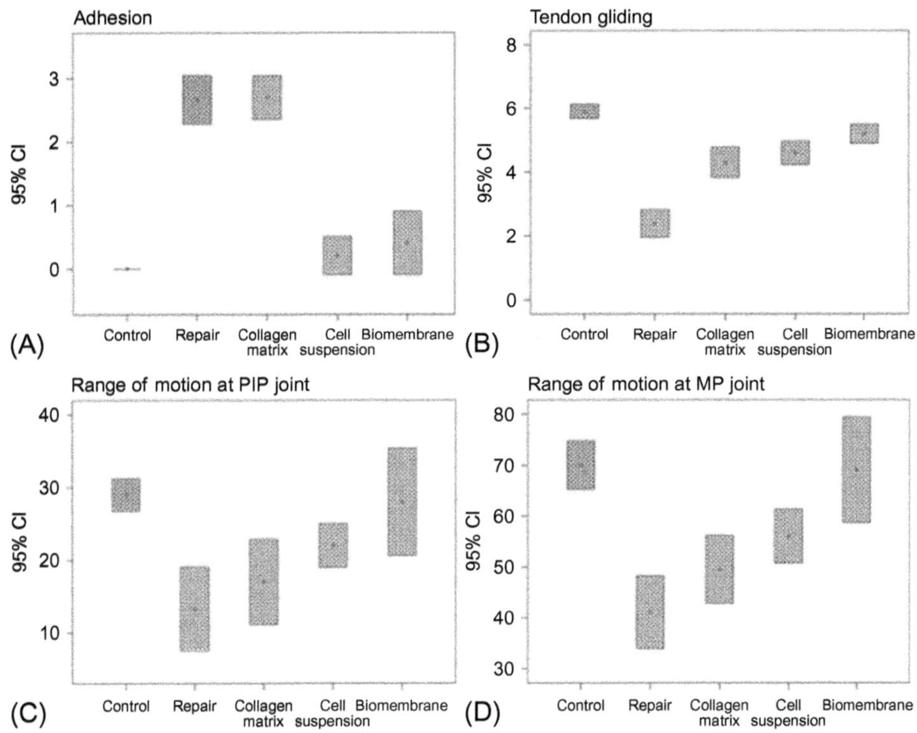


**Fig. 9.4** Bioengineered synovial membrane to prevent tendon adhesions in rabbit flexor tendon model: degree of tendon adhesion (A), tendon gliding (B), range of motion at PIP joint (C), range of motion at MP joint (D) by groups at the 95% confidence interval for mean (key: *PIP*, proximal interphalangeal; *MP*, metacarpophalangeal joint).
From Baymurat, A.C., Ozturk, A.M., Yetkin, H., Ergun, M.A., Helvacioglu, F., Ozkızılcık, A., Tuzlakoğlu, K., Şener, E.E., Erdogan, D., 2015. Bio-engineered synovial membrane to prevent tendon adhesions in rabbit flexor tendon model. J. Biomed. Mater. Res. A 103, 84–90, © by Journal of Biomedical Materials Research, with permission from Wiley.

## 9.2.4 Tendon stem cells

Tendon stem/progenitor cells were identified by their capacity to form colonies in vitro and their potential to differentiate into various lineages such as osteoblasts, chondrocytes, tenocytes, and adipocytes (Bi et al., 2007; Sun et al., 2015), with a higher chondrogenic potential compared to other lineage differentiation capacities caused by a CD105-negative subpopulation (Asai et al., 2014). Depending on the tendon or ligament they are isolated from, differentiation toward selected lineages may differ, as was found for tendon stem cells isolated from human *semitendinosus* and *gracilis* tendons, where the cells were not able to respond to adipogenic stimuli (Stanco et al., 2014). This was suggested to be one of the main differences between tendon stem cells and adipose-derived stem cells (ASCs) (Stanco et al., 2015).

Moreover, application of tendon stem cells in ACL reconstruction has been reviewed and reported to be beneficial (Hirzinger et al., 2014).

Less than 1% of all cells residing in the tendon tissue are tendon stem cells; this is quite a low amount because all cells in tendons make up only 5% of the total tendon volume (Lee et al., 2015). Besides typical surface markers found in MSCs, tendon stem cells exhibit an increased expression of SCX and tenomodulin (Sun et al., 2015). Morphologically, they have a cobblestone-like appearance and are rather small compared with mature tenocytes that are longish like fibroblasts. Tendon stem cells reside in perivascular regions and between the tendon fibrils; however, upon tendon injury, stem cells from other local sources are homed to the tendon injury site—including ASCs from adjacent fat (Buckley et al., 2010), (hematopoietic) stem cells from blood (Kajikawa et al., 2007) among others. The tendon stem cell niche has been found to be based on fibromodulin and biglycan. These two proteins have been reported to regulate the tendon stem cell function by Wnt3a (secreted signaling protein) and bone morphogenetic protein (BMP) signaling pathway modulation (Bi et al., 2007).

The use of tendon progenitors for cellular therapy of injured tendons has been reported for two subpopulations (Lee et al., 2015). In a rat patellar tendon (PT) model, either $CD146^-$ or $CD146^+$ subpopulations of tendon progenitors were applied into the transected site with fibrinogen in addition to conventional sutures. Moreover, the tendon-resident $CD146^+$ progenitor cells were selectively enriched by connective tissue growth factor delivery (CTGF delivery) at the beginning of tendon healing, followed by tenogenic differentiation in the later phase. Tendon regeneration was successful when $CD146^+$ subpopulations were applied in terms of densely aligned collagen fibers, tendon function, and cellularity, while the results for the $CD146^-$ subpopulations were less convincing (Lee et al., 2015).

Other in vivo studies use tendon stem cells to enhance and promote healing deal, for example, with SCX-transduced tendon progenitors applied in rat PT window injuries (Tan et al., 2014). There, it was reported that the SCX transduction leads to a significantly better healing outcome in terms of collagen I expression. Moreover, also allogeneic tendon stem cells were reported to promote tendon repair in the medium to long term and to show only weak immunoreactions and antiinflammatory effects in the host after transplantation (rat PT model) (Lui et al., 2014a). Furthermore, also the regeneration of reconstructed ACL was beneficially promoted by the application of tendon-derived stem cell sheets that were wrapped around ACL and were prepared by applying CTGF and ascorbic acid to rat tendon stem cells (Lui et al., 2014b); as a result, biomechanical properties and histological readouts were significantly better 2 and 6 weeks postsurgery compared to reconstructed ACLs without sheet application.

### 9.2.5 Other stem cells

Not only tendon stem/progenitor cells but also stem cells originating from other tissues have been investigated for tendon regeneration(Ramdass and Koka, 2015; Dhinsa et al., 2015). For example, bone marrow-derived mesenchymal stem cells (bmMSCs) were used for rabbit Achilles tendon (AT) repair; when they were seeded on a PLGA/

silk fibroin (10:90) scaffold filled with collagen as a matrix, the rabbit bmMSCs improved the biomechanical strength of the repaired tendon significantly compared to the cell-free scaffold (Zhang et al., 2015). If stem cells are used for tendon regeneration and if they are seeded onto a scaffold material, the biomechanical properties such as stiffness or elasticity of the material are of high importance in terms of differentiation (Engler et al., 2006; Discher et al., 2005). As, for example, in the regeneration of the bone-tendon enthesis, not only tenogenesis but also osteogenesis is sought—depending on the local position of the cell-seeded scaffold to be implanted; it is of remarkably high relevance that the mechanical properties of the microenvironment the cells encounter provoke the aspired differentiation. Therefore, in vitro studies dealing with different types of substrates regarding the biomechanical properties may predict later in vivo outcome. In that sense, Sharma and Snedeker found that a polyacrylamide scaffold with an elastic modulus ranging from 10 to 90 kPa (resembling the range of natural tendon granulation tissue) that was ligated with either fibronectin, collagen I, or both of them resulted in osteogenesis of human bmMSCs for the rigid substrates (70–90 kPa), with both collagen or fibronectin ligands, while tenogenesis only occurred on 30–50 kP a substrates and only if collagen was present (Sharma and Snedeker, 2012).

Basically, stem cell-mediated tendon tissue modeling and remodeling may be promoted via two different effects. The stem cells may proliferate and differentiate to provide cells needed for the regeneration of the damaged tissue. On the other hand, paracrine signaling that coordinates the local cellular response may also be an important cue during tissue regeneration (Wu et al., 2007). Therefore, to know if the stem cells applied to the site of injury are still there in the postoperative phase has been one question of interest; in other words, stem cell fate is regarded to be one key issue in cellular therapeutic applications. Stem cell tracking has been performed by dually labeling MSCs in an in vivo ovine tendonitis model by using intracellular superparamagnetic iron oxide (SPIO) nanoparticles and simultaneously green fluorescent protein (GFP)-labeled cells. The intralesional stem cell distribution was assessed after 1 and 2 weeks of application, respectively. Interestingly, while after 1 week, the SPIO- and GFP-labeled cells were still at the site of repair, a completely different situation emerged after the second week: the SPIO particles were still found at the repair site; however, the GFP-labeled cells had migrated and were not at the injury site anymore (Scharf et al., 2015). It remained to be elucidated whether the stem cells died and left the nanoparticles embedded at the repair site. Nevertheless, SPIO labeling of cells used as therapeutics may be used to track the cells noninvasively for at least 1 week by magnetic resonance imaging (MRI).

Embryonic stem cells (ESC) have also been used for tendon regeneration. In a study by Chen and coworkers, a knitted silk-collagen scaffold sponge was seeded with human ESCs (hESC) and subjected to mechanical stimulation (Chen et al., 2010). They found that such treatment resulted in the expression of tendon-related markers and in cell morphological changes toward a tenocyte-like appearance. In a rat AT model, it was found that hESC-seeded scaffolds enhanced the tendon regeneration at the wound site, with higher histological scores and better biomechanical performance compared to mere scaffolds (Fig. 9.5). Ultimate load and ultimate stress of

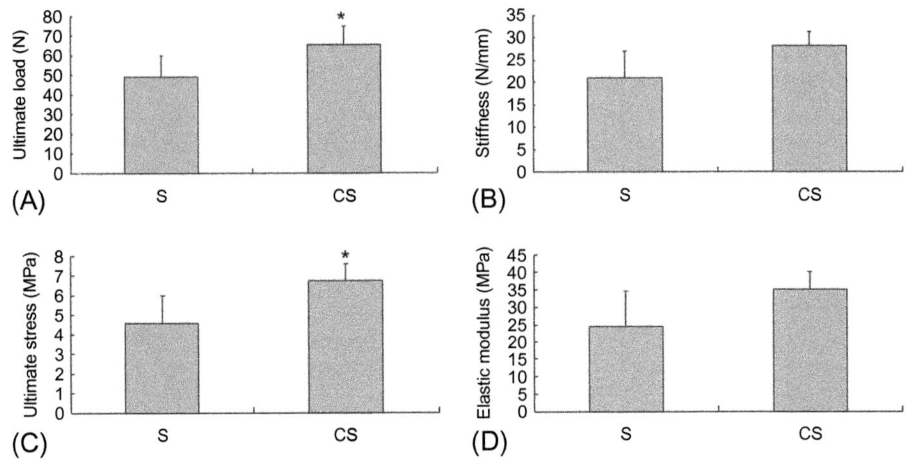


**Fig. 9.5** Mechanical properties of repaired tendon after 4 weeks postsurgery ($n=5$) in a rat Achilles tendon injury model. Samples were carefully fixed for mechanical testing on the Instron machine. The ultimate load (A), stiffness (B), ultimate stress (C), and elastic modulus (D) were compared between the cells+scaffold (CS) group and scaffold (S) group. *Significant difference between two groups at $p<0.05$.
Based on Chen, J.L., Yin, Z., Shen, W.L.A., Chen, X.A., Heng, B.C., Zou, X.A.H., Ouyang, H.W., 2010. Efficacy of hESC-MSCs in knitted silk-collagen scaffold for tendon tissue engineering and their roles. Biomaterials 31, 9438–9451, reprinted from Biomaterials, Copyright 2010, with permission from Elsevier.

the regenerated tendon were significantly higher in the cell-seeded scaffold compared to the cell-free scaffold; stiffness and elastic modulus were tendentially higher, however, not in a statistically significant manner.

As such, ESCs are an attractive source of stem cells for tendon regeneration although their harvesting might rise ethical concerns. Therefore, the comparison of the efficacy of stem cells from further sources is investigated. In addition, studies comparing ESCs and MSCs with regard to their capability for tenogenesis revealed that tenogenic cues such as cyclic tensile loading or the application of transforming growth factor beta 2 (TGF-β2) or basic fibroblast growth factor 4 (FGF-4) can evoke slightly different reactions in embryonic compared to adult stem cells (Brown et al., 2015). While before the application of tenogenic cues, the SCX and elastin levels in adult MSCs were lower than in ESCs, after TGF-β2 alone or in combination with loading, SCX levels were increased in both to the same level. Moreover, while FGF-4 downregulated tendon marker levels in adult stem cells, it did not in the ESCs. Such data may help in choosing the right stem cell source and elaborate novel tendon regeneration strategies in terms of cellular therapies.

Another option for cellular support in tendon wound healing is the application of ASCs. The advantage of choosing ASCs over stem cells from other sources is the fact that they are easily harvested and available in high amounts (Buschmann et al., 2013). In a rat rotator cuff model, not only Lipner and coworkers investigated the influence of

ASCs seeded on an aligned nanofibrous PLGA scaffold that had a mineral gradient to mimic the tendon-to-bone interface, but also they investigated a group of transduced ASCs seeded in the same way. These cells were transduced with BMP-2, having the intention to overcome one of the two main problems in rotator cuffs: the loss of bone (Lipner et al., 2015). The second problem, namely, the formation of poorly aligned scar tissue, was addressed by the aligned fibers in the scaffold material. As such, they investigated and compared four groups: suture only, the acellular scaffold, the ASC-seeded scaffold, and finally the BMP-2-transduced ASC-seeded scaffold. However, against their hypothesis, histological, bone morphological, and biomechanical properties of the specimen analyzed after 2, 4, and 8 weeks, respectively, were poor in the transduced ASC-group compared with their not transduced ASC-seeded counterpart. In addition, regenerative strategies concerning the aligned scaffold application were confronted with the fact that a simple sutured rotator cuff tear had a better outcome than the one where the (acellular) scaffold was used. Therefore, the authors concluded that regenerative medicine may be overwhelmed by the natural scar-mediated response, that BMP-2 is no suitable growth factor even when tendon-to-bone interfaces are under view, and finally that the application of a scaffold (even an aligned one that should theoretically improve the tendon regeneration) may negatively affect tendon healing (Lipner et al., 2015).

Also, bone marrow stromal stem cells were studied intensively in the field of tendon regeneration (Butler et al., 2010). A very detailed protocol was given where the functional tissue engineering approach was used to engineer PT grafts to be applied at a full-length, central-third PT defect with bony troughs at both the patellar and tibial insertions in the New Zealand White rabbit. This approach included as a first step the evaluation of the in vivo mechanical environment acting on the PT in healthy rabbits. Then, tissue-engineered constructs of collagen scaffolds seeded with rabbit bmMSCs harvested from the iliac crest were stimulated mechanically in vitro in a bioreactor. Finally, such tissue-engineered constructs were implanted into rabbits with limb extraction after 3 months. Obviously, this approach led to remarkable outcomes—with biomechanical strength 50% higher than the maximum in vivo forces (IVF) (Fig. 9.6).

Furthermore, also MSCs originating from the human umbilical cord blood were investigated in an ACL reconstruction rabbit model (Jang et al., 2015). Immune rejection of the xenogenous stem cells in the rabbit were assessed by a gross examination of the knee joint, a histological and immunohistochemical evaluation of the regenerated tissue, and an evaluation of bone tunnel widening using micro-computed tomography at 4, 8, and 12 weeks post application. Surprisingly, no evidence of immune rejection was detected. The tendon-bone healing was enhanced through broad fibrocartilage formation with higher histological scores and decreased femoral and tibial tunnel widening compared with the control group where no human umbilical cord blood-derived MSCs were applied (79.2% and 80%, respectively, of the control group tunnel area at 12 weeks). This study demonstrated that the use of nonautologous, not even allogeneic, but xenogenous stem cells may be used in animal models without negative effects in terms of immunorejection. In addition, stem cells originating from the cord blood turned out to be a useful source of cells when ACL reconstruction was investigated with or without such cells (Jang et al., 2015).

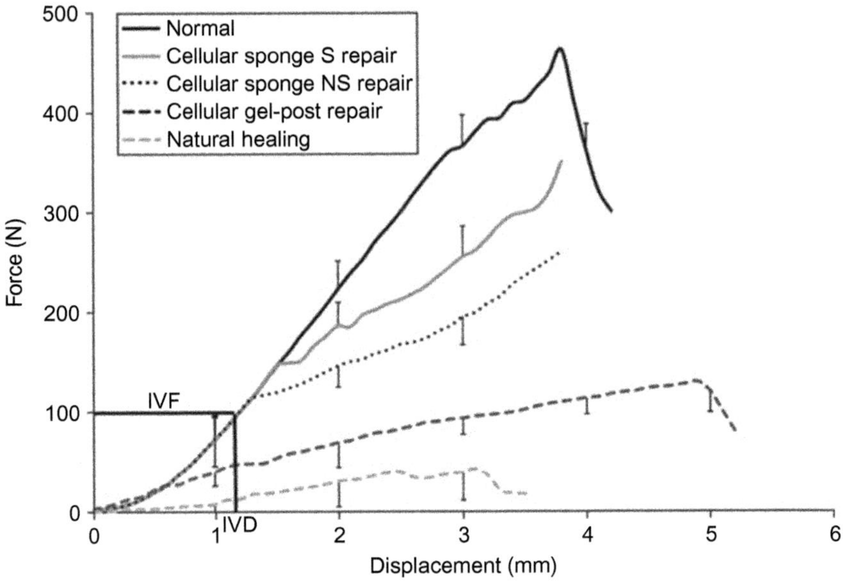


**Fig. 9.6** Average load–displacement curves for functional tissue engineering (FTE) repairs. Twelve-week repairs using MSC-collagen sponge tissue-engineered constructs (TECs) stimulated (S) at 2.4% strain for 100 cycles per day matched normal patellar tendon (PT) tangent stiffness up to 150 N, which is 50% greater than the maximum in vivo forces (IVF) recorded for activities of daily living (IVD = in vivo displacement). Nonstimulated (NS) MSC-collagen sponge TECs and MSC-collagen gel-post TECs also produced significantly greater structural properties compared with natural healing.
Based on Butler, D.L., Gooch, C., Kinneberg, K.R.C., Boivin, G.P., Galloway, M.T., Nirmalanandhan, V.S., Shearn, J.T., Dyment, N.A., Juncosa-Melvin, N., 2010. The use of mesenchymal stem cells in collagen-based scaffolds for tissue-engineered repair of tendons. Nat. Protoc. 5, 849–863, reprinted by permission from Macmillan Publishers Ltd: Nature Protocols, copyright 2010.

Coculture systems have been shown to be superior to monocultures in various cellular therapeutic applications (Buschmann et al., 2011), for bone tissue engineering considering sufficient vascularization (Jin et al., 2015), for the optimization of urothelium grafts (Zhang et al., 2013), or to enhance the regeneration of periodontitis (Wu et al., 2013; Xing et al., 2013). Therefore, also cellular therapeutic approaches regarding tendon and ligament regeneration have tested the benefit of coculture systems, especially with MSCs to be one of the (two) cell types involved because enhanced regenerative response has been reported (Luo et al., 2009; Fan et al., 2008; Bian et al., 2011). For example, an in vitro study with rat tenocytes showed that their proliferation and collagen expression improved when they were cultivated indirectly with ASCs (Veronesi et al., 2015). Moreover, a direct coculture of pig ACL cells and bmMSCs had a beneficial effect on ligament regeneration that was not observed when utilizing either cell source independently (Canseco et al., 2012).

Another example is the study reported by Zhao and coworkers where they showed that rat bmMSCs combined in an indirect coculture system with ligament fibroblasts promoted BMSC differentiation that was reflected in the increased expression of elastin, lysyl oxidase, and fibulin-5 (Zhao et al., 2012).

Tendon cells were cocultured not only with stem cells but also with amniotic epithelial cells (AECs); in an attempt to provoke the differentiation of ovine AECs toward the tenogenic lineage, they were cocultured with primary sheep tenocytes (Barboni et al., 2012). As an outcome, the AECs were able to fully differentiate into tenocytes in the presence of an appropriate coculture with tenocytes. These findings are interesting because tenocytes are limited in terms of their availability, and the differentiation of AECs can therefore be another cell source for the generation of additional tenocytes after differentiation (Barboni et al., 2012). Interestingly, AECs express collagen 1 only when they are cultivated in aggregates, especially in tendon-like aggregates (Fig. 9.7).

## 9.2.6 Cellular therapy based on promoted tenogenesis of stem cells

Besides the key transcription factor in tenogenesis, SCX, another transcription factor has been shown to be an important regulator in tenogenesis (Liu et al., 2015). *Mohawk* (Mkx) is activated at certain stages in tendon development and is downregulated in

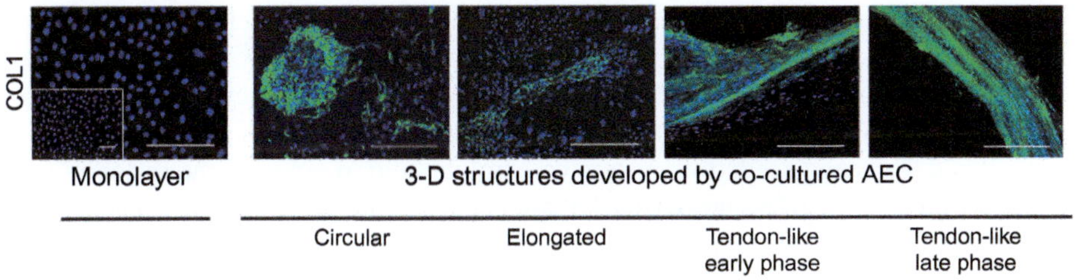


**Fig. 9.7** Collagen 1 (COL1) distribution in amniotic epithelial cells (AECs) aggregates developed in coculture with tenocytes. The images exemplify the COL1 protein distribution recorded by immunohistochemistry in the more representative typologies of cell aggregates obtained during AEC cocultures with primary tenocytes. The images show in *blue* the nuclei counterstained with 4′,6-diamidino-2-phenylindole (DAPI, a fluorescent stain) and in *green* the COL1 protein. The COL1 protein is undetectable when the AECs are organized in monolayer independently from the cultural conditions adopted (cocultured AEC, monolayer; AEC cultured alone, small inset). COL1 starts to appear in AEC forming cell aggregates and reaches its highest and widespread distribution within the tendon-like structures. In the early phase, COL1 is localized within the fusiform cells that start to be oriented along the longitudinal axis of the tendon-like structures. Later, the protein was either localized into the AEC or deposited within the extracellular matrix (tendon-like, late phase). Scale bar for all images = 100 μm.
From Barboni, B., Curini, V., Russo, V., Mauro, A., Di Giacinto, O., Marchisio, M., Alfonsi, M., Mattioli, M., 2012. Indirect co-culture with tendons or tenocytes can program amniotic epithelial cells towards stepwise tenogenic differentiation. PLoS ONE 7, e30974.

tendinopathy. When Mkx was compared with SCX in terms of MSC differentiation toward tenocytes, teratogenesis promoted by Mkx was more efficient than by SCX. Mechanistically, Mkx upregulated SCX, and both together accelerated tenogenesis. When Mkx-expressing MSCs (transduced C3H10T1/2 cells (mouse multipotent MSC line)) were transplanted to a murine full-size AT defect, it was shown that Mkx enhanced the regeneration of this tendon in various fashions compared to the control where enhanced green fluorescent protein (EGFP)-transduced cells were used (Fig. 9.8). First of all, histological examination showed that 4 weeks post operation, the regenerated tissue resembled natural tendon tissue better in the Mkx group—with bands of collagen fibrils having small crimped patterns and tenocytes with pronounced spindle-shaped morphology in contrast to the control group (with polycellular fibrils without the typical alignment and crimp). Moreover, the cross-sectional area of the fibrils in the Mkx group was larger than in the control group, indicating a higher progress in the regeneration process. Finally, also the biomechanics were enhanced; significantly higher failure load, failure stress, and absorbed energy at failure were detected compared to the control. In sum, Mkx-transduced MSCs added beneficially to the healing and regeneration of the lacerated tendon (Liu et al., 2015).

Not only transduction of (stem) cells but also external cues such as specifically prepared substrates with surfaces that trigger tenogenesis in stem cells seeded onto these materials have been examined to promote tenogenesis for cellular therapy purposes (Iannone et al., 2015).

For example, polydimethylsiloxane substrates with arrays of parallel channels (having 700 nm width and 1.4 μm pitch) guided human MSCs through the process of self-organization and differentiation, leading to the development of three-dimensional tissues with cellular and ECM organization, closely resembling that of an embryonic human tendon, while just flat surfaces of the same material did not (control) (Iannone et al., 2015). Hence, the specific nanopatterns on surfaces of materials implanted at tendon defects and seeded with stem cells might help to promote the differentiation process of the cells toward tenocytes that are needed at the wound site for regeneration.

Finally, also growth factor addition can promote tenogenesis in cellular therapy under the use of stem cells and therefore enhance tendon regeneration (Goncalves et al., 2013). In that sense, delivery devices may consist of peptides as has been presented by Crispim et al. (2015).

## 9.2.7 Platelet-rich plasma

A further option is to apply platelet-rich plasma (PRP) to enhance tendon healing. PRP is blood plasma that has been enriched with platelets (thrombocytes). As a concentrated source of autologous platelets, PRP contains (and releases through degranulation) several different growth factors and other cytokines that stimulate healing of tendons and ligaments. The efficacy of certain growth factors in healing various injuries and the concentrations of these growth factors found within PRP are the theoretical basis for the use of PRP in soft tissue repair. The platelets collected in PRP are activated by the addition of thrombin and calcium chloride, which induces the release of the mentioned factors from alpha granules. The growth factors and other cytokines present in PRP include platelet-derived growth factor (PDGF), TGF-β, FGF,

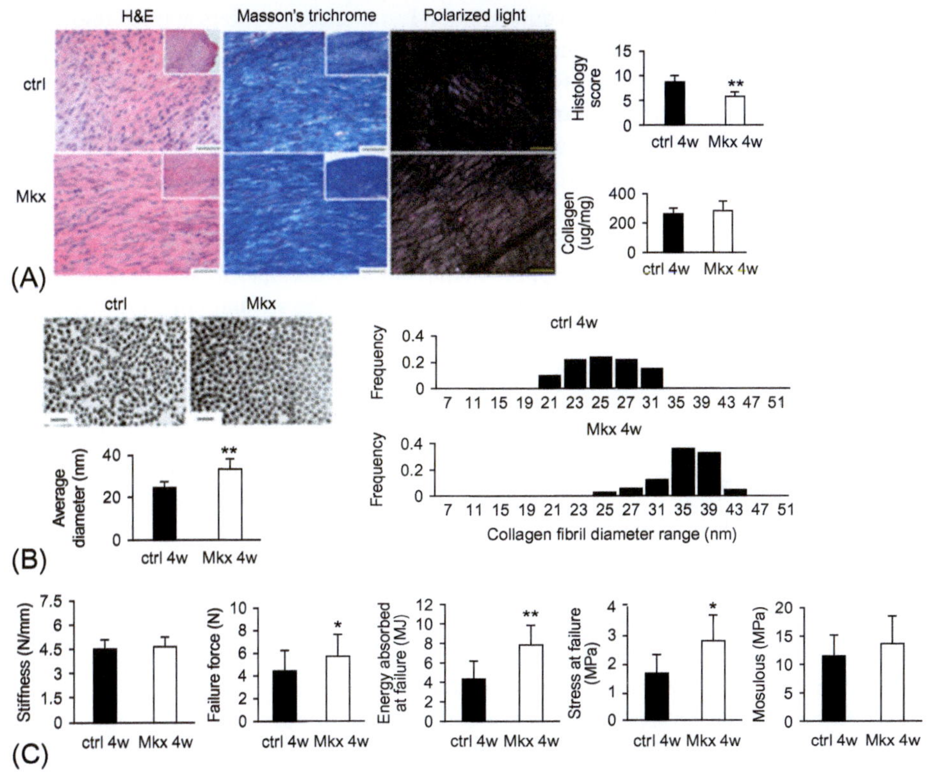


**Fig. 9.8** *Mohawk* (Mkx)-expressing MSCs enhance tendon regeneration in a mouse model of Achilles tendon defect. (A) Morphology of the repaired tissue site along the axis of the tendon in the Mkx and control groups (crtl, EGFP-transduced cells). H&E staining and polarized light microscopy showing the collagen fibrils at 4 weeks after implantation of MSCs. The maturation of repaired tendon was assessed by histology scoring. Masson's trichrome staining showing the deposited collagen at the repaired tissue site and quantitative analyses of the collagen content. (B) Transmission electron micrographs of cross sections of collagen fibrils in the control and Mkx groups from repaired tendons at 4 weeks after implantation. The average diameters and range of diameters of collagen fibrils in the Mkx group were significantly larger than in the control group. (C) The biomechanical properties of the repaired tendon (stiffness, failure force, energy absorbed at failure, stress at failure, and modulus) in the control and Mkx groups at 4 weeks after implantation. $^{*}p<0.05$, $^{**}p<0.01$ (Student's t-test). Scale bars 50 μm in (A), 100 nm in (B). Abbreviation: *H&E*, hematoxylin and eosin. From Liu, H., Zhang, C., Zhu, S., Lu, P., Zhu, T., Gong, X., Zhang, Z., Hu, J., Yin, Z., Heng, B.C., Chen, X., Ouyang, H.W., 2015. Mohawk promotes the tenogenesis of mesenchymal stem cells through activation of the TGF beta signaling pathway. Stem Cells 33, 443–455, © by Stem Cells, with permission from Wiley.

insulin-like growth factor 1 (IGF-1), IGF-2, vascular endothelial growth factor (VEGF), epidermal growth factor, interleukin 8, keratinocyte growth factor, and CTGF. The levels of these growth factors may vary among species as reported, for example, for dolphins where PRP had similar levels in PDGF as humans, however, lower levels in TGF-β and VEGF (Griffeth et al., 2014).

In a rabbit AT full-transection wound model, Lyras and coworkers applied one single shot of 0.5 mL of PRP into the tendon mass and examined the extracted ATs at 1, 2, 3, and 4 weeks post operation (Lyras et al., 2011). Histological and immunohistochemical findings showed that PRP accelerated tendon healing and increased the expression of IGF-1 in the early healing phase compared to the control group (no PRP). Additionally, there was a decrease in vessel density in combination with the better orientation of the fibroblasts in the PRP group during the last 2 weeks, suggesting that healing progressed more rapidly under the influence of PRP (Lyras et al., 2011). Therefore, it is not surprising that PRP is one cellular approach that has also find its way into clinical applications as reviewed for the PT (Jeong et al., 2014) and the medial collateral ligament (Andia and Maffulli, 2015) (see Section 9.4.2).

It was also proposed to combine several biological approaches that might synergistically support tendon and ligament regenerative healing (Isaac et al., 2012). Therefore, PRP application has also been combined with ASCs to improve tendon healing with a tendon hydrogel in vitro (Crowe et al., 2015) and in vivo in a rat AT model (Chiou et al., 2015). It was shown that the proliferation and the migration of ASCs that were exposed to culture medium with either plasma, plasma supplemented by growth factors or with PRP, and tendon hydrogel were stimulated most in the PRP group (Crowe et al., 2015). The tendon hydrogel was fabricated by human cadaver flexor tendons that were decellularized, milled, digested with pepsin, and mixed with phosphate-buffered solution to allow gelation at 37°C. In the in vivo study, five experimental groups were compared. A full-thickness defect created within the rat AT was either filled with saline (control), tendon hydrogel, hydrogel+PRP, hydrogel+ASCs, or hydrogel+PRP+ASCs. ASCs were green fluorescent protein positively stained. Bioluminescent, histological, and biomechanical analysis was performed 1, 2, 4, and 8 weeks post operation. in vivo results were confirming the in vitro data in terms of proliferation—showing that PRP significantly enhanced the proliferation of the ASCs. In terms of biomechanics, groups containing either PRP or ASCs or both were performing similarly, however, significantly better than the control or the tendon hydrogel alone at early time points (until 4 weeks post operation). At 8 weeks, the rat ATs achieved ultimate failure loads of normal tendons, and the differences between the experimental groups were small, with a tendency of the PRP group and the PRP+ASCs group having the highest ultimate loads (Chiou et al., 2015).

## 9.3 Application methods for cell therapies

After having chosen the appropriate cell type for tendon or ligaments regeneration (see Section 9.2), the question arises how to apply the cells so that they are most effective to support the healing process. Depending on the defect or disease of the corresponding tendon or ligament, several approaches have been investigated. For critical-size defects of tendons, tendon grafts that are seeded with cells may be one option; such grafts are made of either a natural or synthetic polymer to mimic the ECM and seeded with cells. Another option is to use natural tendons or ligament tissues that are decellularized and then reseeded with the corresponding cells. For smaller defects, also threads that are seeded with cells can be used to suture the lacerated site, or sheets that are seeded with cells can be wrapped around the repair site after being sutured conventionally. Furthermore, cells can also be injected intravenously and are then suggested to home to the injury site—or (more effectively) cells can be injected directly at the site of the lesion. As such, there are many ways to apply cells to promote the healing process. It has always to be taken into account that some procedures are more time-consuming and costly as, for example, decellularization of natural tissues and reseeding them compared to just injecting the corresponding cells at the site of injury. However, effectiveness and availability will turn the surgeon's interest to a particular application method—in research and in clinics.

To mimic nature closely, flexor tendons have been decellularized and then reseeded. The question was if such a treatment would alter the biocompatibility and the biomechanical properties of grafts. Trypsin and Triton-X treatment of rabbit flexor tendons led to their decellularization. The trypsinized cells were reseeded onto those constructs after expansion. Although the whole decellularized tendon tissues were incubated to a cell containing medium, only the surfaces of the graft were populated with the tenocytes (Fig. 9.9). Nevertheless, when these constructs were tested biomechanically and compared to the natural flexors, they had the same ultimate stress and elastic modulus (Chong et al., 2009b).

In order to treat tendinopathy in horses, stem cells were injected either intravenously or at the site of lesion. For that purpose, allogeneic bmMSCs were labeled with technetium-99m ($Tc^{99m}$), a radionuclide with a half-life of 6 h. In a tendon overstrain injury horse model, cell tracking after intravenous MSC application revealed, however, that all cells were trapped in the lung and did not home to the injury site. A more beneficial approach was the intralesional injection of the cells (Fig. 9.10) where the experiment revealed at least 24% of the cells had stayed at the injury site after 1 day (Becerra et al., 2013).

Another study examined the effect of intraarticular injection of GFP-labeled mesenchymal stromal cells to promote the healing process of a partially torn ACL in rats (Kanaya et al., 2007). After 2 and 4 weeks post injection, histological and biomechanical outcome revealed the beneficial impact of MSC application in ACL compared to cell-free treatment: ultimate failure loads were significantly higher at these time points (Fig. 9.11), and the histological scores including the degree of retraction, tissue coverage, predominant cell type, and collagen fiber organization were also obviously and significantly higher when MSCs were used (Fig. 9.12) (Kanaya et al., 2007).

## 9.4 Biological and mechanical outcome after cellular therapy

### 9.4.1 Animal models

What kind of outcome or improvement may be expected after cellular therapy? In the rabbit AT model, several studies are reported where the application of cells was beneficial. For example, Chong and coworkers investigated the effect of bmMSCs in a fibrin carrier when applied to a sharp complete transection through the midsubstance

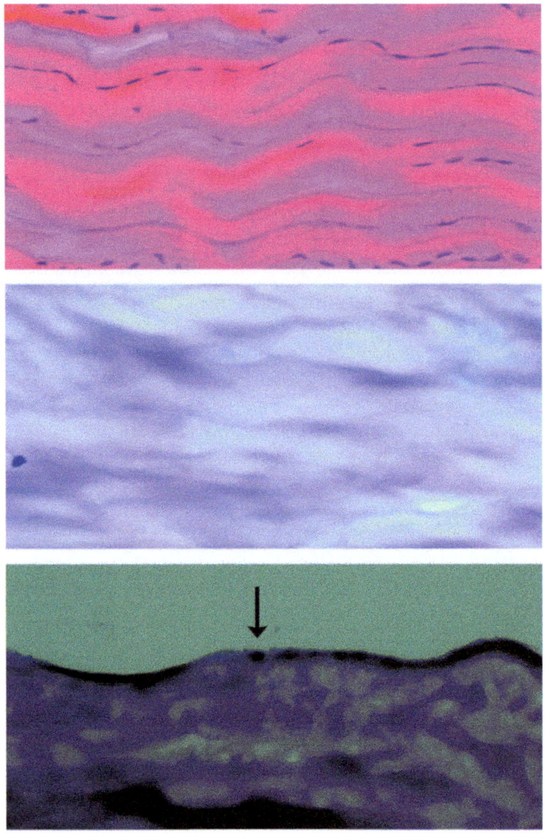

**Fig. 9.9** Representative sections of flexor tendons. (*Above*) Normal flexor tendon; note the flat tenocytes and crimp pattern. (*Center*) Acellularized tendon; note the absence of tenocytes and loss of crimp. (*Below*) Reseeded tendon construct; the arrow points to a single cell on the surface of tendon (hematoxylin and eosin; original magnification, ×40).
From Chong, A.K.S., Riboh, J., Smith, R.L., Lindsey, D.P., Pham, H.M., Chang, J., 2009b. Flexor tendon tissue engineering: acellularized and reseeded tendon constructs. Plast. Reconstr. Surg. 123, 1759–1766, with permission from the publisher Wolters Kluwer Health.

of the rabbit AT (control fibrin alone). The gross morphology of the tendons was the same whether stem cells were applied or not. Labeled bmMSCs remained viable and present in the intratendinous region for at least 6 weeks. At later time points, this became more diffuse. At early time points (3 weeks), the organization of the collagen fibers was better than in the cell-free fibrin. Biomechanical testing showed improved modulus in the treatment group as compared with the control group at 3 weeks, but not at 6 or 12 weeks. Hence, the application of bmMSCs may improve biological and mechanical outcome especially at early time point post operation (Chong et al., 2007). The presence and the paracrine action of the cells vanish over time together with the beneficial effects. Nevertheless, especially in early phases of the healing

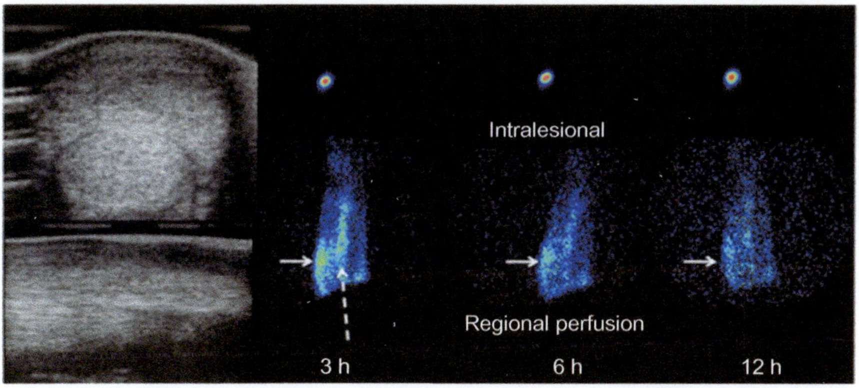


**Fig. 9.10** Gamma scintigraphs from a lesion on the distal superficial digital flexor tendon 22 cm distal to the accessory carpal bone (representative ultrasonographs shown on the *left*). They were performed at 3, 6, and 12 h after administration intralesionally (*top row*) and by intravenous regional perfusion (via the palmar digital vein in the pastern region; *bottom row*). Note the uptake of radiopharmaceutical over the distal superficial digital flexor tendon at the lesion site (*solid arrow*). The uptake denoted by the *dashed arrow* is persistence of uptake in the palmar digital vein.
From Becerra, P., Valdes Vazquez, M.A., Dudhia, J., Fiske-Jackson, A.R., Neves, F., Hartman, N.G., Smith, R.K.W., 2013. Distribution of injected technetium(99m)-labeled mesenchymal stem cells in horses with naturally occurring tendinopathy. J. Orthop. Res. 31, 1096–1102, © by Journal of Orthopaedic Research, with permission from Wiley.

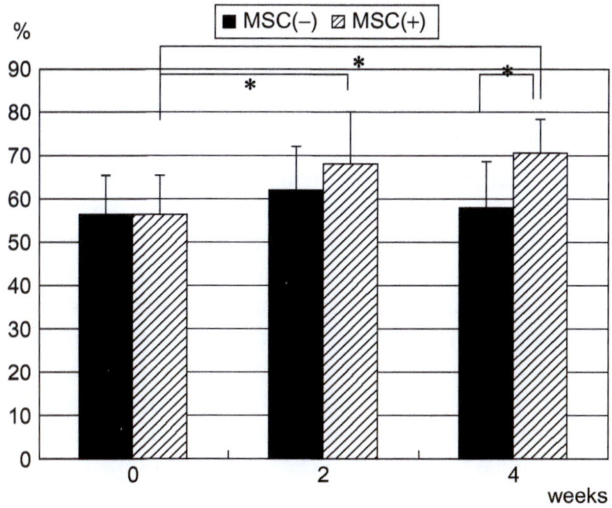


**Fig. 9.11** Injection of mesenchymal stromal cells (MSC) in anterior cruciate ligaments (ACL) in rat, ultimate failure load (percentage of sham side). The ultimate failure load in the MSC(+) group was significantly higher than that in the cell-free treatment group (MSC(−)) at 4 weeks after surgery ($P = 0.004$). In the MSC(+) group, there were significant differences between time zero and other periods ($P = 0.01$ at 2 weeks and $P = 0.002$ at 4 weeks). $*p < 0.05$.
From Kanaya, A., Deie, M., Adachi, N., Nishimori, M., Yanada, S., Ochi, M., 2007. Intra-articular injection of mesenchymal stromal cells in partially torn anterior cruciate ligaments in a rat model. Arthroscopy 23, 610–617, reprinted from The Journal of Arthroscopic & Related Surgery, Copyright 2007, with permission from Elsevier.

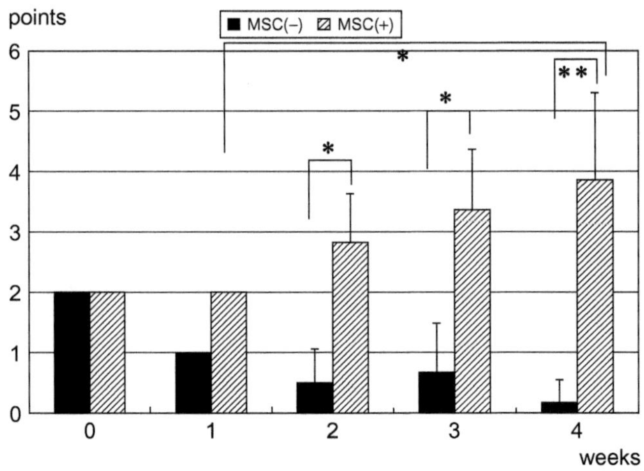


**Fig. 9.12** Injection of mesenchymal stromal cells (MSC) in anterior cruciate ligaments (ACL) in rat, histological scores (where the best score is 6 points). The score in the MSC(+) group was significantly better than that in the cell-free treatment group (MSC(−)) starting at 2 weeks after surgery ($P < 0.05$ at 2 and 3 weeks and $P < 0.0001$ at 4 weeks), and in the MSC(+) group, the score was significantly better at 4 weeks after surgery than that obtained 1 week after surgery ($P < 0.05$). $^{*}p < 0.05$, $^{**}p < 0.0001$.
From Kanaya, A., Deie, M., Adachi, N., Nishimori, M., Yanada, S., Ochi, M., 2007. Intra-articular injection of mesenchymal stromal cells in partially tom anterior cruciate ligaments in a rat model. Arthroscopy 23, 610–617, reprinted from The Journal of Arthroscopic & Related Surgery, Copyright 2007, with permission from Elsevier.

where rerupture is one big problem occurring, (stem) cell therapy may allow earlier active motion of the regenerating tendon or ligament, which is a big advantage because postoperative motion not only leads to fiber and tenocyte alignment but also strengthens the limb as such and enables the person to go back to daily activities earlier. Why the effectiveness of cellular therapy is often restricted to early phases during the healing has been discussed by addressing the various mechanisms of action (Chong et al., 2009a). Not only are MSCs effective through their autocrine and paracrine function and their obvious capability to differentiate (at least to a small extent according to (Fox et al., 2007)) into lineage-specific cells, but also are they capable of transferring mitochondria to cells with nonfunctional mitochondria (Spees et al., 2006). The latter are predominantly found in animal models for tendon lacerations and may therefore explain at least one part of the observed early support by the stem cells.

Nevertheless, not all cellular approaches have been shown to turn out beneficial as expected. When rabbit flexor tendons were decellularized by trypsin and Triton-X after a freeze-thaw cycle in an in vitro tissue engineering study and then reseeded with rabbit tenocytes, these constructs had lower ultimate stress values compared with the normal tendons and also compared with the decellularized tendons. Interestingly, the decellularized tendons were biomechanically not significantly different to normal tendons in terms of ultimate stress and elastic modulus. Hence, the reseeding with tenocytes did not enhance the biomechanical properties of the constructs, which does not mean, however, that in vivo outcome would not turn out better (biomechanically and with regard to tissue integration) (Chong et al., 2009b). Very often, cell-seeded constructs show a better in vivo performance compared with their cell-free equivalents. This was shown, for example, with a synthetic PLGA/silk fibroin and collagen matrix seeded with bone marrow stromal cells that turned out to have a maximum load of the regenerated rabbit AT of 60% of normal AT 4 months post operation, being significantly higher than the cell-free construct under otherwise equivalent conditions (Zhang et al., 2015).

### 9.4.2 Clinical setting

Cellular therapies to treat tendinopathy in the clinical setting are rather rarely chosen, and they are accompanied by some critical limitations. Often, only a small number of patients are enrolled and a control group having the otherwise same treatment without cells is lacking. Furthermore, long-term follow-ups with noninvasive assessments such as ultrasound and MRI are not always given, making clear comparisons among different therapies difficult, if not impossible (Jeong et al., 2014). Furthermore, if, for example, PRP is injected, which is an approach that found its way into clinics (Marques et al., 2015), comparison is made difficult by the fact that differently prepared PRPs may have completely different effects on cells and tissues as nicely demonstrated by an in vitro study where the migration of MSCs and human umbilical vein endothelial cells was investigated in the presence of different platelet-rich concentrates, among them being leukocyte- and platelet-rich fibrin, leukocyte- and PRP, or natural blood clot (Schaer et al., 2015). Nevertheless, several cell-based therapies have been tried out and compared to cell-free approaches. For example, to treat patellar tendinopathy, PRP injections were compared to shock wave therapy, and it was found that the PRP injections were much better compared to the shock wave therapy in terms of the Victorian Institute of Sports assessment-Patellar questionnaire, VISA-P, and the visual analog scale (VAS) scores for pain (Vetrano et al., 2013).

As the PT is ACL-compatible in shape and size, it is often used for ACL reconstruction. This graft is also ideal because earlier bone-to-bone healing was reported and because the tissue mechanics resemble that of ACL (Andia and Maffulli, 2015). What is left after grafting is the donor site where the PT defect has to be refilled. It was found that the application of PRP reduced donor site morbidity in PT; at 6 months post application, the patient group receiving PRP had a significantly smaller gap and a lower VAS (de Almeida et al., 2012).

Clinical trials using bone marrow mononuclear cells were also successful. Although these studies applied stem cells individually and with different procedures, an overall beneficial outcome was reported when the cells were injected in refractory chronic patellar tendinopathy (Pascual-Garrido et al., 2012; Clarke et al., 2011).

Hence, in contrast to the huge amount of preclinical reports dealing with cellular therapeutic approaches in tendon and ligament repair, clinical trials are still rare up to date. There is a need to generate high-quality evidences before recommending and applying cellular therapy to patients. Very often, standard protocols and reliable data are still missing, which makes clinical decisions difficult.

## 9.5 Summary

In this chapter, we presented various cell types that have been investigated in terms of cellular therapy of tendons and ligaments, including tenocytes, fibroblasts, synovial cells, (tendon) stem cells, and coculture systems. A special focus is laid on the promotion of tenogenesis when stem cells are used. Furthermore, different application methods are discussed, among them natural and synthetic materials seeded with cells and acting as grafts in critical-size defects, sheets wrapped around the corresponding limbs, or threads seeded with cells to apply them during suturing the lacerated tendons or ligaments. In addition, cells are injected, either intravenously or intralesionally. Finally, biological and biomechanical outcome after cellular therapeutic approaches are shown and discussed in terms of their effectiveness—in preclinical studies and in the clinical setting.

## References

Andia, I., Maffulli, N., 2015. Use of platelet-rich plasma for patellar tendon and medial collateral ligament injuries: best current clinical practice. J. Knee Surg. 28, 11–17.

Asai, S., Otsuru, S., Candela, M.E., Cantley, L., Uchibe, K., Hofmann, T.J., Zhang, K., Wapner, K.L., Soslowsky, L.J., Horwitz, E.M., Enomoto-Iwamoto, M., 2014. Tendon progenitor cells in injured tendons have strong chondrogenic potential: the CD105-negative subpopulation induces chondrogenic degeneration. Stem Cells 32, 3266–3277.

Barboni, B., Curini, V., Russo, V., Mauro, A., Di Giacinto, O., Marchisio, M., Alfonsi, M., Mattioli, M., 2012. Indirect co-culture with tendons or tenocytes can program amniotic epithelial cells towards stepwise tenogenic differentiation. PLoS ONE 7, e30974.

Baymurat, A.C., Ozturk, A.M., Yetkin, H., Ergun, M.A., Helvacioglu, F., Ozkızılcık, A., Tuzlakoğlu, K., Şener, E.E., Erdogan, D., 2015. Bio-engineered synovial membrane to prevent tendon adhesions in rabbit flexor tendon model. J. Biomed. Mater. Res. A 103, 84–90.

Becerra, P., Valdes Vazquez, M.A., Dudhia, J., Fiske-Jackson, A.R., Neves, F., Hartman, N.G., Smith, R.K.W., 2013. Distribution of injected technetium(99 m)-labeled mesenchymal stem cells in horses with naturally occurring tendinopathy. J. Orthop. Res. 31, 1096–1102.

Bi, Y., Ehirchiou, D., Kilts, T.M., Inkson, C.A., Embree, M.C., Sonoyama, W., Li, L., Leet, A.I., Seo, B.-M., Zhang, L., Shi, S., Young, M.F., 2007. Identification of tendon stem/progenitor cells and the role of the extracellular matrix in their niche. Nat. Med. 13, 1219–1227.

Bian, L., Zhai, D.Y., Mauck, R.L., Burdick, J.A., 2011. Coculture of human mesenchymal stem cells and articular chondrocytes reduces hypertrophy and enhances functional properties of engineered cartilage. Tissue Eng. A 17, 1137–1145.

Brown, J.P., Galassi, T.V., Stoppato, M., Schiele, N.R., Kuo, C.K., 2015. Comparative analysis of mesenchymal stem cell and embryonic tendon progenitor cell response to embryonic tendon biochemical and mechanical factors. Stem Cell Res. Ther. 6, 89.

Buckley, C.T., Vinardell, T., Thorpe, S.D., Haugh, M.G., Jones, E., Mcgonagle, D., Kelly, D.J., 2010. Functional properties of cartilaginous tissues engineered from infrapatellar fat pad-derived mesenchymal stem cells. J. Biomech. 43, 920–926.

Buschmann, J., Welti, M., Hemmi, S., Neuenschwander, P., Baltes, C., Rudin, M., Giovanoli, P., Calcagni, M., 2011. 3D co-cultures of osteoblasts and endothelial cells in

DegraPol foam: histological and high field MRI analyses of pre-engineered capillary networks in bone grafts. Tissue Eng. A 17, 291–299.

Buschmann, J., Gao, S., Härter, L., Hemmi, S., Welti, M., Werner, C.M.L., Calcagni, M., Cinelli, P., Wanner, G.A., 2013. Yield and proliferation rate of adipose-derived stem cells as a function of age, BMI and harvest site: increasing the yield by using adherent and supernatant fractions? Cytotherapy 15, 1098–1105.

Butler, D.L., Gooch, C., Kinneberg, K.R.C., Boivin, G.P., Galloway, M.T., Nirmalanandhan, V.S., Shearn, J.T., Dyment, N.A., Juncosa-Melvin, N., 2010. The use of mesenchymal stem cells in collagen-based scaffolds for tissue-engineered repair of tendons. Nat. Protoc. 5, 849–863.

Cadby, J.A., Buehler, E., Godbout, C., Van Weeren, P.R., Snedeker, J.G., 2014. Differences between the cell populations from the peritenon and the tendon core with regard to their potential implication in tendon repair. PLoS ONE 9, e92474.

Canseco, J.A., Kojima, K., Penvose, A.R., Ross, J.D., Obokata, H., Gomoll, A.H., Vacanti, C.A., 2012. Effect on ligament marker expression by direct-contact co-culture of mesenchymal stem cells and anterior cruciate ligament cells. Tissue Eng. A 18, 2549–2558.

Chen, J.L., Yin, Z., Shen, W.L.A., Chen, X.A., Heng, B.C., Zou, X.A.H., Ouyang, H.W., 2010. Efficacy of hESC-MSCs in knitted silk-collagen scaffold for tendon tissue engineering and their roles. Biomaterials 31, 9438–9451.

Chiou, G.J., Crowe, C., Mcgoldrick, R., Hui, K., Pham, H., Chang, J., 2015. Optimization of an injectable tendon hydrogel: the effects of platelet-rich plasma and adipose-derived stem cells on tendon healing in vivo. Tissue Eng. A 21, 1579–1586.

Chong, A.K.S., Ang, A.D., Goh, J.C.H., Hui, J.H.P., Lim, A.Y.T., Lee, E.H., Lim, B.H., 2007. Bone marrow-derived mesenchymal stem cells influence early tendon-healing in a rabbit Achilles tendon model. J. Bone Joint Surg. Am. 89A, 74–81.

Chong, A.K., Chang, J., Go, J.C.H., 2009a. Mesenchymal stem cells and tendon healing. Front. Biosci. 14, 4598–4605.

Chong, A.K.S., Riboh, J., Smith, R.L., Lindsey, D.P., Pham, H.M., Chang, J., 2009b. Flexor tendon tissue engineering: acellularized and reseeded tendon constructs. Plast. Reconstr. Surg. 123, 1759–1766.

Clarke, A.W., Alyas, F., Morris, T., Robertson, C.J., Bell, J., Connell, D.A., 2011. Skin-derived tenocyte-like cells for the treatment of patellar tendinopathy. Am. J. Sports Med. 39, 614–623.

Crispim, J., Fernandes, H., Jonkheijm, P., Saris, D., 2015. Use of synthetic peptides for the delivery of growth factors for tendon/ligament healing. Tissue Eng. A 21, S11.

Crowe, C.S., Chiou, G., Mcgoldrick, R., Hui, K., Pham, H., Chang, J., 2015. Tendon regeneration with a novel tendon hydrogel: in vitro effects of platelet-rich plasma on rat adipose-derived stem cells. Plast. Reconstr. Surg. 135, 981E–989E.

De Almeida, A.M., Demange, M.K., Sobrado, M.F., Rodrigues, M.B., Pedrinelli, A., Hernandez, A.J., 2012. Patellar tendon healing with platelet-rich plasma a prospective randomized controlled trial. Am. J. Sports Med. 40, 1282–1288.

Dhinsa, B.S., Mahapatra, A.N., Khan, W.S., 2015. Sources of adult mesenchymal stem cells for ligament and tendon tissue engineering. Curr. Stem Cell Res. Ther. 10, 26–30.

Discher, D.E., Janmey, P., Wang, Y.L., 2005. Tissue cells feel and respond to the stiffness of their substrate. Science 310, 1139–1143.

Elliot, D., Giesen, T., 2013. Primary flexor tendon surgery: the search for a perfect result. Hand Clin. 29, 191–206.

Engler, A.J., Sen, S., Sweeney, H.L., Discher, D.E., 2006. Matrix elasticity directs stem cell lineage specification. Cell 126, 677–689.
Fan, H., Liu, H., Toh, S.L., Goh, J.C.H., 2008. Enhanced differentiation of mesenchymal stem cells co-cultured with ligament fibroblasts on gelatin/silk fibroin hybrid scaffold. Biomaterials 29, 1017–1027.
Fox, J.M., Chamberlain, G., Ashton, B.A., Middleton, J., 2007. Recent advances into the understanding of mesenchymal stem cell trafficking. Br. J. Haematol. 137, 491–502.
Full, S.M., Delman, C., Gluck, J.M., Abdmaulen, R., Shemin, R.J., Heydarkhan-Hagvall, S., 2015. Effect of fiber orientation of collagen-based electrospun meshes on human fibroblasts for ligament tissue engineering applications. J. Biomed. Mater. Res. B Appl. Biomater. 103, 39–46.
Goncalves, A.I., Rodrigues, M.T., Lee, S.-J., Atala, A., Yoo, J.J., Reis, R.L., Gomes, M.E., 2013. Understanding the role of growth factors in modulating stem cell tenogenesis. PLoS ONE 8, e83734.
Griffeth, R.J., Garcia-Parraga, D., Mellado-Lopez, M., Luis Crespo-Picazo, J., Soriano-Navarro, M., Martinez-Romero, A., Moreno-Manzano, V., 2014. Platelet-rich plasma and adipose-derived mesenchymal stem cells for regenerative medicine-associated treatments in bottlenose dolphins (*Tursiops truncatus*). PLoS ONE 9, e108439.
Hirzinger, C., Tauber, M., Korntner, S., Quirchmayr, M., Bauer, H.-C., Traweger, A., Tempfer, H., 2014. ACL injuries and stem cell therapy. Arch. Orthop. Trauma Surg. 134, 1573–1578.
Iannone, M., Ventre, M., Formisano, L., Casalino, L., Patriarca, E.J., Netti, P.A., 2015. Nanoengineered surfaces for focal adhesion guidance trigger mesenchymal stem cell self-organization and tenogenesis. Nano Lett. 15, 1517–1525.
Isaac, C., Gharaibeh, B., Witt, M., Wright, V.J., Huard, J., 2012. Biologic approaches to enhance rotator cuff healing after injury. J. Shoulder Elb. Surg. 21, 181–190.
Jang, K.-M., Lim, H.C., Jung, W.Y., Moon, S.W., Wang, J.H., 2015. Efficacy and safety of human umbilical cord blood-derived mesenchymal stem cells in anterior cruciate ligament reconstruction of a rabbit model: new strategy to enhance tendon graft healing. Arthroscopy 31, 1530–1539.
Jeong, D.U., Lee, C.R., Lee, J.H., Pak, J., Kang, L.W., Jeong, B.C., Lee, S.H., 2014. Clinical applications of platelet-rich plasma in patellar tendinopathy. Biomed. Res. Int., 249498.
Jin, G.-Z., Han, C.-M., Kim, H.-W., 2015. In vitro co-culture strategies to prevascularization for bone regeneration: a brief update. Tissue Eng. Regen. Med. 12, 69–79.
Kajikawa, Y., Morihara, T., Watanabe, N., Sakamoto, H., Matsuda, K.-I., Kobayash, M., Oshima, Y., Yoshida, A., Kawata, M., Kubo, T., 2007. GFP chimeric models exhibited a biphasic pattern of mesenchymal cell invasion in tendon healing. J. Cell. Physiol. 210, 684–691.
Kanaya, A., Deie, M., Adachi, N., Nishimori, M., Yanada, S., Ochi, M., 2007. Intra-articular injection of mesenchymal stromal cells in partially tom anterior cruciate ligaments in a rat model. Arthroscopy 23, 610–617.
Lee, C.H., Lee, F.Y., Tarafder, S., Kao, K., Jun, Y., Yang, G., Mao, J.J., 2015. Harnessing endogenous stem/progenitor cells for tendon regeneration. J. Clin. Investig. 125, 2690–2701.
Lipner, J., Shen, H., Cavinatto, L., Liu, W., Havlioglu, N., Xia, Y., Galatz, L.M., Thomopoulos, S., 2015. In vivo evaluation of adipose-derived stromal cells delivered with a nanofiber scaffold for tendon-to-bone repair. Tissue Eng. A 21, 2766–2774.
Liu, H., Zhang, C., Zhu, S., Lu, P., Zhu, T., Gong, X., Zhang, Z., Hu, J., Yin, Z., Heng, B.C., Chen, X., Ouyang, H.W., 2015. Mohawk promotes the tenogenesis of mesenchymal stem cells through activation of the TGF beta signaling pathway. Stem Cells 33, 443–455.

Lohberger, B., Kaltenegger, H., Stuendl, N., Rinner, B., Leithner, A., Sadoghi, P., 2015. Impact of cyclic mechanical stimulation on the expression of extracellular matrix proteins in human primary rotator cuff fibroblasts. Knee Surg. Sports Traumatol. Arthrosc 1–8. http://dx.doi.org/10.1007/s00167-015-3790-6.

Lui, P.P.Y., Kong, S.K., Lau, P.M., Wong, Y.M., Lee, Y.W., Tan, C., Wong, O.T., 2014a. Allogeneic tendon-derived stem cells promote tendon healing and suppress immunoreactions in hosts: in vivo model. Tissue Eng. A 20, 2998–3009.

Lui, P.P.Y., Wong, O.T., Lee, Y.W., 2014b. Application of tendon-derived stem cell sheet for the promotion of graft healing in anterior cruciate ligament reconstruction. Am. J. Sports Med. 42, 681–689.

Luo, Q., Song, G., Song, Y., Xu, B., Qin, J., Shi, Y., 2009. Indirect co-culture with tenocytes promotes proliferation and mRNA expression of tendon/ligament related genes in rat bone marrow mesenchymal stem cells. Cytotechnology 61, 1–10.

Lyras, D.N., Kazakos, K., Georgiadis, G., Mazis, G., Middleton, R., Richards, S., O'Connor, D., Agrogiannis, G., 2011. Does a single application of PRP alter the expression of IGF-I in the early phase of tendon healing? J. Foot Ankle Surg. 50, 276–282.

Marques, L.F., Stessuk, T., Cherici Camargo, I.C., Sabeh Junior, N., Dos Santos, L., Ribeiro-Paes, J.T., 2015. Platelet-rich plasma (PRP): methodological aspects and clinical applications. Platelets 26, 101–113.

Meier Buergisser, G., Buschmann, J., 2015. History and performance of implant materials applied as peritendinous antiadhesives. J. Biomed. Mater. Res. B Appl. Biomater. 103, 212–228.

Ozturk, A.M., Yam, A., Chin, S.I., Heong, T.S., Helvacioglu, F., Tan, A., 2008. Synovial cell culture and tissue engineering of a tendon synovial cell biomembrane. J. Biomed. Mater. Res. A 84A, 1120–1126.

Pascual-Garrido, C., Rolon, A., Makino, A., 2012. Treatment of chronic patellar tendinopathy with autologous bone marrow stem cells: a 5-year-followup. Stem Cells Int., Unsp 953510.

Ramdass, B., Koka, P.S., 2015. Ligament and tendon repair through regeneration using mesenchymal stem cells. Curr. Stem Cell Res. Ther. 10, 84–88.

Schaer, M.O., Diaz-Romero, J., Kohl, S., Zumstein, M.A., Nesic, D., 2015. Platelet-rich concentrates differentially release growth factors and induce cell migration in vitro. Clin. Orthop. Relat. Res. 473, 1635–1643.

Scharf, A., Holmes, S., Thoresen, M., Mumaw, J., Stumpf, A., Peroni, J., 2015. Superparamagnetic iron oxide nanoparticles as a means to track mesenchymal stem cells in a large animal model of tendon injury. Contrast Media Mol. Imaging 10, 388–397.

Sharma, R.I., Snedeker, J.G., 2012. Paracrine interactions between mesenchymal stem cells affect substrate driven differentiation toward tendon and bone phenotypes. PLoS ONE 7, e31504.

Spees, J.L., Olson, S.D., Whitney, M.J., Prockop, D.J., 2006. Mitochondrial transfer between cells can rescue aerobic respiration. Proc. Natl. Acad. Sci. U. S. A. 103, 1283–1288.

Stanco, D., Vigano, M., Orfei, C.P., Di Giancamillo, A., Thiebat, G., Peretti, G., De Girolamo, L., 2014. In vitro characterization of stem/progenitor cells from semitendinosus and gracilis tendons as a possible new tool for cell-based therapy for tendon disorders. Joints 2, 159–168.

Stanco, D., Vigano, M., Orfei, C.P., Di Giancamillo, A., Peretti, G.M., De Girolamo, L., 2015. Characterization of tendon stem/progenitor cells and in vitro comparison with adipose derived stem cells. Tissue Eng. A 21, S174–S175.

Steinberg, P.J., Hodde, K.C., 1990. The morphology of synovial lining of various structures in several species as observed with scanning electron microscopy. Scanning Microsc. 4, 987–1019. discussion 1019–20.

Sun, H.B., Schaniel, C., Leong, D.J., Wang, J.H.C., 2015. Biology and mechano-response of tendon cells: progress overview and perspectives. J. Orthop. Res. 33, 785–792.

Tan, C., Lui, P.P.Y., Lee, Y.W., Wong, Y.M., 2014. Scx-transduced tendon-derived stem cells (TDSCs) promoted better tendon repair compared to mock-transduced cells in a rat patellar tendon window injury model. PLoS ONE 9, e97453.

Veronesi, F., Della Bella, E., Torricelli, P., Pagani, S., Fini, M., 2015. Effect of adipose-derived mesenchymal stromal cells on tendon healing in aging and estrogen deficiency: an in vitro co-culture model. Cytotherapy 17, 1536–1544.

Vetrano, M., Castorina, A., Vulpiani, M.C., Baldini, R., Pavan, A., Ferretti, A., 2013. Platelet-rich plasma versus focused shock waves in the treatment of jumper's knee in athletes. Am. J. Sports Med. 41, 795–803.

Wu, Y., Chen, L., Scott, P.G., Tredget, E.E., 2007. Mesenchymal stem cells enhance wound healing through differentiation and angiogenesis. Stem Cells 25, 2648–2659.

Wu, Y., Cao, H., Yang, Y., Zhou, Y., Gu, Y., Zhao, X., Zhang, Y., Zhao, Z., Zhang, L., Yin, J., 2013. Effects of vascular endothelial cells on osteogenic differentiation of noncontact co-cultured periodontal ligament stem cells under hypoxia. J. Periodontal Res. 48, 52–65.

Xing, Z., Xue, Y., Finne-Wistrand, A., Yang, Z.Q., Mustafa, K., 2013. Copolymer cell/scaffold constructs for bone tissue engineering: co-culture of low ratios of human endothelial and osteoblast-like cells in a dynamic culture system. J. Biomed. Mater. Res. A 101A, 1113–1120.

Zhang, M., Peng, Y.B., Zhou, Z., Zhou, J., Wang, Z., Lu, M.J., 2013. Differentiation of human adipose-derived stem cells co-cultured with urothelium cell line toward a urothelium-like phenotype in a nude murine model. Urology 81, WOS:000314686000081.

Zhang, W., Yang, Y., Zhang, K., Li, Y., Fang, G., 2015. Weft-knitted silk-poly(lactide-*co*-glycolide) mesh scaffold combined with collagen matrix and seeded with mesenchymal stem cells for rabbit Achilles tendon repair. Connect. Tissue Res. 56, 25–34.

Zhao, B., Liang, L., Yan, J., Ren, S., Ren, R., Zhang, X., 2012. Effect of mechanical stretch on the expressions of elastin, LOX and Fibulin-5 in rat BMSCs with ligament fibroblasts co-culture. Mol. Biol. Rep. 39, 6077–6085.

# In vitro–in vivo biomechanical performance of tissue-engineered constructs for tendon and ligament repair

**10**

## Abstract

It would save a lot of money and a lot of animal lives if the biomechanics of tissue-engineered constructs assessed in vitro could tell us about their in vivo performance. The question addressed in this chapter is if ultimate stress values of in vitro constructs can predict their in vivo outcome, the same for the elastic modulus, the ultimate load, and the stiffness. Hence, correlation plots of in vitro–in vivo data found in a series of peer-reviewed articles have been made and checked for correlations between the in vitro assessments and the in vivo outcome. As the data published for one and the same material in vitro and in vivo are rather sparse, however, only few examples are shown and discussed with a broad range of relations, having the same or less biomechanical outcome (group 1), having increased biomechanical outcome in vivo (group 2), or having an orders of magnitudes higher biomechanics compared to initial in vitro findings (group 3).

**Keywords:** Ultimate stress, Elastic modulus, Ultimate load, Stiffness, In vitro–in vivo correlation.

## Abbreviations

| | |
|---|---|
| **ACL** | anterior cruciate ligament (tendons and ligaments) |
| **CL** | crosslinked (treating methods of tissues in tissue engineering) |
| **FDA** | food and drug administration (authority in USA) |
| **FDP** | *flexor digitorum profundus* (tendons and ligaments) |
| **MCL** | medial collateral ligament (tendons and ligaments) |
| **MSCs** | mesenchymal stem cells (cells for seeding in tissue engineering) |
| **MSt** | mechanical stimulation (mechanical treatment in tissue engineering) |
| **Pa, MPa, GPa** | pascal, mega-pascal ($10^3$ Pa), giga-pascal ($10^6$ Pa) (unit of pressure, internal pressure, stress, Young's modulus and ultimate tensile strength, defined as one $N/m^2$) |
| **PT** | patellar tendon (tendons and ligaments) |
| **SIS** | small intestine submucosa (natural graft material) |
| **TEC(s)** | tissue-engineered construct(s) (tissue engineering method) |
| **TGF (−β*n*)** | transforming growth factor beta ($n = 1, 2, 3$) (protein in living cells) |

**Biomechanics of Tendons and Ligaments. http://dx.doi.org/10.1016/B978-0-08-100489-0.00010-7**

## 10.1 Introduction

Before a tissue-engineered tendon or ligament graft can be applied in the clinical setting, there is a long way to go including preclinical in vitro studies, subsequent in vivo animal studies, often from small animal models to large animal models. Only after generation of high-quality reliable data and FDA (US Food and Drug Administration) approval, clinical trials can be started. Hence, there are not only high costs for preclinical studies, but also a lot of animals have to be sacrificed. It was reported that even very promising medical devices may experience cessation just because no financial means were available to promote the necessary further research (Meier Buergisser and Buschmann, 2014). Furthermore, there is a high need to reduce animal experiments to a minimum in the sense of the 3Rs (Reduction, Refinement, and Replacement (Russel and Burch, 1959)). Therefore, if a reliable prediction of in vivo qualities based on the characterization of in vitro constructs is possible, this would not only reduce costs but may be ethically beneficial because the number of animals needed to create specific information could be reduced.

In this chapter, the main question was how in vitro biomechanical data correlate with in vivo findings. Natural and synthetic graft characteristics before implantation were therefore related to in vivo data. Graft manipulation such as cell seeding, mechanical stimulation in bioreactors, or addition of bioactive agents or chemicals was considered. in vitro and in vivo performance was compared with regard to ultimate tensile stress and elastic modulus as well as ultimate load and stiffness of the corresponding entities. The main limitation in this approach was that the data for a special tendon or ligament graft were often only acquired in vitro, but not in vivo, or vice versa. Hence, only reports are included where both assessments were performed or studies that dealt exactly with the same tendon or ligament graft reported in separate references focusing on either in vitro or in vivo characteristics.

## 10.2 Ultimate tensile stress

When mechanical properties of in vitro-constructed tendon or ligament grafts are compared to the properties of extracted specimen after in vivo implantation, different aspects may be considered. The material of the graft is fundamental for its mechanical properties. In addition, also cell seeding, mechanical stimulation, and/or chemical stimulation may have an impact on the resulting biomechanics. In Fig. 10.1, in vitro ultimate tensile stress was compared to in vivo ultimate tensile stress (see Table 10.2 for details and corresponding references). Roughly, three main groups can be formed: group 1 consisting of materials that exhibit a slightly lower or similar ultimate stress compared to preimplantation (in vitro), group 2 of materials with a better in vivo performance compared to in vitro ultimate stress, and group 3 of materials that exhibit an extremely better in vivo performance (better by orders of magnitudes) compared to in vitro (overview in Table 10.1).

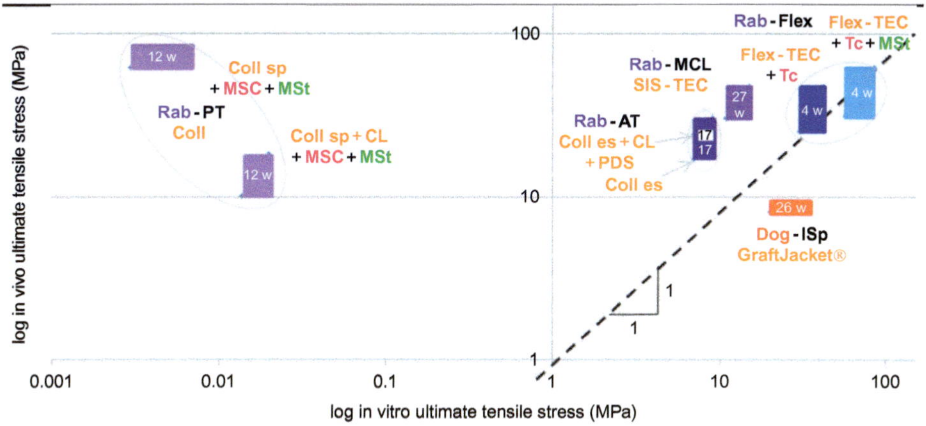


**Fig. 10.1** In vitro–in vivo comparison: ultimate tensile stress of in vitro grafts and reconstructed tendons & ligaments. Data boxes correspond means ± SD or SEM. Key see Table 10.2. Data from Saber, S., Zhang, A.Y., Ki, S.H., Lindsey, D.P., Smith, R.L., Riboh, J., Pham, H., Chang, J., 2010. Flexor tendon tissue engineering: bioreactor cyclic strain increases construct strength. Tissue Eng. A 16, 2085–2090; Thorfinn, J., Angelidis, I.K., Gigliello, L., Pham, H.M., Lindsey, D., Chang, J., 2012. Bioreactor optimization of tissue engineered rabbit flexor tendons in vivo. J. Hand Surg. Eur. 37, 109–114; Angelidis, I.K., Thorfinn, J., Connolly, I.D., Lindsey, D., Pham, H.M., Chang, J., 2010. Tissue engineering of flexor tendons: the effect of a tissue bioreactor on adipoderived stem cell-seeded and fibroblast-seeded tendon constructs. J. Hand Surg. 35, 1466–1472; Zhang, J., Wang, G.Y., Xiao, Y.P., Fan, L.Y., Wang, Q., 2011. The biomechanical behavior and host response to porcine-derived small intestine submucosa, pericardium and dermal matrix acellular grafts in a rat abdominal defect model. Biomaterials 32, 7086–7095; Musahl, V., Abramowitch, S.D., Gilbert, T.W., Tsuda, E., Wang, J.H.C., Badylak, S.F., Woo, S.L.Y., 2004. The use of porcine small intestinal submucosa to enhance the healing of the medial collateral ligament—a functional tissue engineering study in rabbits. J. Orthop. Res. 22, 214–220; Liang, R., Woo, S.L.Y., Takakura, Y., Moon, D.K., Jia, F., Abramowitch, S.D., 2006. Longterm effects of porcine small intestine submucosa on the healing of medial collateral ligament: a functional tissue engineering study. J. Orthop. Res. 24, 811–819; Adams, J.E., Zobitz, M.E., Reach Jr., J.S., An, K.-N.A., Teinmann, S.P., 2006. Rotator cuff repair using an acellular dermal matrix graft: an in vivo study in a canine model. Arthroscopy 22, 700–709; Chaudhury, S., Holland, C., Thompson, M.S., Vollrath, F., Carr, A.J., 2012. Tensile and shear mechanical properties of rotator cuff repair patches. J. Shoulder Elb. Surg. 21, 1168–1176; Juncosa-Melvin, N., Boivin, G.P., Gooch, C., Galloway, M.T., West, J.R., Dunn, M.G., Butler, D.L., 2006a. The effect of autologous mesenchymal stem cells on the biomechanics and histology of gel-collagen sponge constructs used for rabbit patellar tendon repair. Tissue Eng. 12, 369–379; Juncosa-Melvin, N., Shearn, J.T., Boivin, G.P., Gooch, C., Galloway, M.T., West, J.R., Nirmalanandhan, V.S., Bradica, G., Butler, D.L., 2006b. Effects of mechanical stimulation on the biomechanics and histology of stem cell-collagen sponge constructs for rabbit patellar tendon repair. Tissue Eng. 12, 2291–2300; Nirmalanandhan, V.S., Juncosa-Melvin, N., Shearn, J.T., Boivin, G.P., Galloway, M.T., Gooch, C., Bradica, G., Butler, D.L., 2009. Combined effects of scaffold stiffening and mechanical preconditioning cycles on construct biomechanics, gene expression, and tendon repair biomechanics. Tissue Eng. A 15, 2103–2111; Moshiri, A., Oryan, A., Meimandi-Parizi, A., 2013a. Role of tissue-engineered artificial tendon in healing of a large Achilles tendon defect

**Table 10.1 In vitro–in vivo comparison: graft material groups with different impacts on resulting ultimate tensile stress**

| Group | Ultimate tensile stress: impact in vivo, compared to in vitro | |
|---|---|---|
| 1 | Lower | 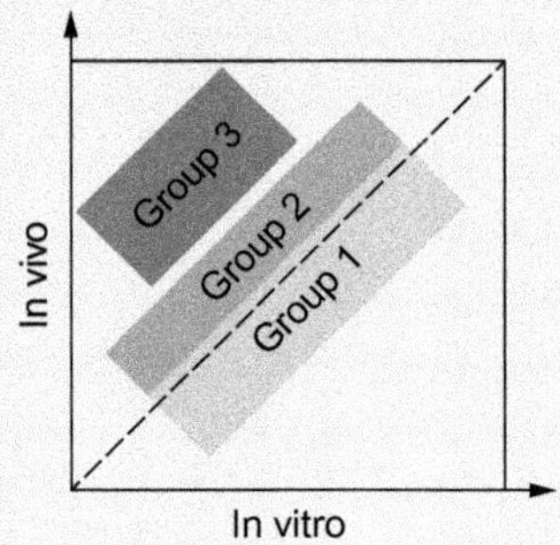  |
| 2 | Higher | |
| 3 | Extremely higher | |

Starting with group 1, there are materials used as tendon or ligament grafts that exhibit a similar in vivo biomechanical performance as determined before implantation. For example, flexor tendon tissue-engineered constructs (TECs) based on decellularized rabbit hind-paw *flexor digitorum profundus* (FDP) that were reseeded with tenocytes had the same ultimate stress in vitro as in vivo 4 weeks postoperation (Saber et al., 2010). When the constructs were cultivated in a bioreactor with cyclic strain, the ultimate tensile stress was significantly increased in vitro and in vivo; however, the in vitro increase was higher compared to the in vivo increase (from 36–71 MPa in vitro, and from 36–46 MPa in vivo) (Thorfinn et al., 2012). In the in vivo–in vitro plot, such data lie below the 1:1 line (Fig. 10.1). It might be speculated about the reason of why a loss of material property is found (here the ultimate stress decreasing from 71 to 46 MPa when moving from in vitro to in vivo). Probably, the cells receiving a mechanical stimulation in vitro were stimulated for higher collagen production, among others, while this predominant effect might be faded in vivo caused by cell migration into the surrounding tissue as well as cell immigration from the surroundings (like macrophages among others), leading to another balance of soft tissue composition.

Similar findings for in vitro–in vivo performance were reported for GraftJacket, a commercially available decellularized human dermis collagen matrix (Adams et al., 2006). GraftJacket exhibited an ultimate tensile stress of $27 \pm 7$ MPa in vitro. When

model in rabbits. J. Am. Coll. Surg. 217, 421–441; Moshiri, A., Oryan, A., Meimandiparizi, A., 2013c. A novel application of biosynthetic tissue engineered tridimensional implant on large tendon defects: a comprehensive detailed in vivo investigation with significant clinical value. Connect. Tissue Res. 54 (3), 227–243; Meimandi-Parizi, A., Oryan, A., Moshiri, A., Silver, I.A., 2013b. Novel application of a tissue-engineered collagen-based three-dimensional bio-implant in a large tendon defect model: a broad-based study with high value in translational medicine. Tissue Cell 45, 282–294; Oryan, A., Moshiri, A., Meimandi-Parizi, A., 2013a. Graft selection in ACL reconstructive surgery: past, present, and future. Curr. Orthop. Pract. 24, 321–333.

**Table 10.2 In vitro–in vivo comparison: mechanical properties of reconstructed tendon and ligaments (ultimate stress [MPa], elastic modulus [MPa])**

| Experimental model | | | | | Tissue engineering | | | | Ultimate stress (MPa) | | | Elastic modulus (MPa) | | | Source | |
|---|---|---|---|---|---|---|---|---|---|---|---|---|---|---|---|---|
| M | T/L | Rec | HT (w) | Sc | Sc tr | C | Me St | Ch St | In vitro | In vivo | Native | In vitro | In vivo | Native | Literature | Tab. |
| ***Tendon auto-/allo-graft, TEC*** | | | | | | | | | | | | | | | | |
| Rab | Flx | R | 4 | Flx | | Tc | | | 36 ± 6 | 36 ± 11 | 64–73 | 632 ± 86 | 505 ± 148 | 980–1166 | Saber et al. (2010), Thorfinn et al. (2012), Angelidis et al. (2010) | 6.1a |
| | | | | Flx | | | | | 71 ± 14 | 46 ± 15 | | 1091 ±169 | 699 ± 197 | | Saber et al. (2010), Thorfinn et al. (2012), Angelidis et al. (2010) | 6.1a+b |
| ***Small intestinal submucosa (SIS-TEC)*** | | | | | | | | | | | | | | | | |
| Rab | MCL | R | 27 | SIS | | | | | 9–15 | 36 ± 11 | 76–84 | 2200 ±500 | 387 ± 99 | 1142 ±412 | Zhang et al. (2011), Musahl et al. (2004), Liang et al. (2006) | 6.2a |
| ***Fascia lata (ECM-TEC)*** | | | | | | | | | | | | | | | | |
| n.d. | RC | – | | ECM | AB | | | | | | | 304–532 | | | Aurora et al. (2007), Derwin et al. (2008) | 6.2a |

*Continued*

**Table 10.2 Continued**

| Experimental model | | | | | Tissue engineering | | | | Ultimate stress (MPa) | | | Elastic modulus (MPa) | | | Source | |
|---|---|---|---|---|---|---|---|---|---|---|---|---|---|---|---|---|
| M | T/L | Rec | HT (w) | Sc | Sc tr | C | Me St | Ch St | In vitro | In vivo | Native | In vitro | In vivo | Native | Literature | Tab. |
| ***Human umbilical vein (HUV)*** | | | | | | | | | | | | | | | | |
| n.d. | T/L | – | | HUV | | MSC | | | 1.7 ± 0.2 | | | 9 ± 2 | | | Abousleiman et al. (2008), Issa et al. (2011) | 6.2a |
| | | | | HUV | | MSC | dyn | | 2.7 ± 0.8 | | | 10 ± 2 | | | | |
| ***Silk-TEC*** | | | | | | | | | | | | | | | | |
| Rab | AT | W | 4 | Silk-nf | | | | | 0.9 ± 0.7 | 4 ± 1 | 10 ± 2 | 30 ± 21 | 8 ± 3 | 19 ± 3 | Ni et al. (2012) | 6.2a |
| ***Chitin/Chitosan-based polymer*** | | | | | | | | | | | | | | | | |
| Rab | ISp | R | 12 | Chito-san fi | HA | | | | | 9 ± *1* | 54 ± *13* | | 89 ± *7*[a] | 583 ± 93[a] | Funakoshi et al. (2005) | 6.2a |

| | | | | | | | | | | | | | | | | |
|---|---|---|---|---|---|---|---|---|---|---|---|---|---|---|---|---|
| ***Veins*** | | | | | | | | | | | | | | | | |
| n.d. | Flx | – | | Vein | | | | | 3±1 | | | | | | Rodger et al. (2015) | 6.2a |
| ***Collagen based scaffolds*** | | | | | | | | | | | | | | | | |
| Dog | ISp | I<br>– | 26 | GJ | | | | | 27±7 | ~8.5 | | 48±50 | | | Adams et al. (2006), Chaudhury et al. (2012) | 7.1a+7.2a |
| Rab | PT | R | 12 | Coll sp | | MSC | dyn | | 0.005 ±0.002 | 72±*11* | 101±*16* | 0.020 ±0.010 | 441±*26* | 861±*106* | Juncosa-Melvin et al. (2006a,b) | 7.1b+7.2a |
| | | | | Coll sp | CL | MSC | dyn | | 0.017 ±*0.003* | 14±*4* | | 0.046 ±*0.010* | 51±12 | | Juncosa-Melvin et al. (2006a), Nirmalanandhan et al. (2009) | |
| | | | | Coll g/sp | | MSC | dyn | | | 62±14 | | 0.14±0.04 | 341±47 | | Juncosa-Melvin et al. (2006a, 2007), Shearn et al. (2007) | |
| Rab | AT | R | 17 | Coll es | | | | | 8±1 | 19±2 | 20–34 | 1.0±0.1 | 1.4±0.2 | 1.5–3.2 | Meimandi-Parizi et al. (2013a), Oryan et al. (2013b), Moshiri et al. (2013a) | 7.2a |
| | | | | Coll es | CL+PDS | | | | 9±2 | 27±3 | | 0.610 ±0.057 | 2.1±0.3 | | Moshiri et al. (2013a) | |
| ***Synthetic polymeric scaffolds*** | | | | | | | | | | | | | | | | |
| Chi | FDP | W | 3 | PELA | Ibp | | | | 2.9–3.4 | 2±1 | | | | | Liu et al. (2013) | 8.2a |

[a]Tangent modulus.

*Experimental Model*: *M*, in vivo model (*Rab*, rabbit; *Chi*, chicken; *n.d.*, in vivo experiment not done); *T/L*, tendon & ligament (*Target organ*: *Flx*, flexors of the hand/foot/paw; *MCL*, medial collateral ligament; *RC*, rotor cuff; *AT*, Achilles tendon; *ISp*, infraspinatus; *PT*, patellar tendon; *FDP*, *flexor digitorum profundus*); *Rec*, reconstruction: *R*, replacement; *A*, augmentation; *I*, interposition; *W*, wrapping; *HT*, healing time (*w*, weeks). *Tissue engineering*: *Sc*, scaffold (*TEC*, tissue engineering construct; *SIS*, small intestinal submucosa; *ECM*, extra cellular matrix; *HUV*, human umbilical vein; *nf*, nanofiber; *fi*, fiber; *un*, unwoven; *GJ*, GraftJacket; *Coll*, collagen based scaffold; *sp*, sponge; *g*, gel; *g/sp*, gel/sponge; *es*, electrospun; *PELA*, poly(L-lactic acid)–polyethylene glycol); *Sc tr*, scaffold treatment (*AB*, antibiotic; *HA*, hyaluronic acid; *CL*, cross linking; *PDS*, polydioxanon; *Ibp*, ibuprofen); *C*, seeding cell type (*Tc*, tenocytes; *MSC*, bone marrow mesenchymal stem cells); *MSt*, mechanical stimulation (*dyn*, dynamic stimulation); *CSt*, chemical stimulation.

Data: *Mean*±*SD*=Standard deviation; *SEM*, *Standard Error of Mean (italic)*.

used as an interposition in a large rotator cuff tear model in dogs, however, the ultimate stress was significantly lower at 6 months postoperation, being only 8.5 MPa, although the gross appearance of the extracted specimen as well as the histology mimicked natural appearance quite well (Adams et al., 2006). Again, for such findings, the corresponding data are situated below the 1:1 line (Fig. 10.1), indicating an in vivo ultimate stress that is lower than the in vitro performance might suggest. Obviously, rotator cuff regeneration led to a less strong bridge consisting of the collagen matrix repopulated by invading cells from the surroundings.

In group 2, a higher ultimate stress was found in vivo when compared to the in vitro data. Examples for this group are TECs based on natural small intestinal submucosa (SIS) used for rabbit MCL (medial collateral ligament) reconstruction (Zhang et al., 2011; Musahl et al., 2004; Liang et al., 2006) or synthetic electrospun collagen meshes, treated by crosslinking and added polydioxanone for rabbit Achilles tendon (AT) reconstruction (Meimandi-Parizi et al., 2013a; Oryan et al., 2013a; Moshiri et al., 2013a,c).

As for the SIS used for MCL reconstruction, a 6 mm gap was reported to be closed with a strip of SIS. Compared to nontreated controls, the ultimate stress at 26 weeks postoperation was increased by 50%. Here, the in vitro biomechanics of the cell-free SIS with 10–15 MPa ultimate stress (Zhang et al., 2011), underestimated the beneficial long term in vivo outcome with a resulting ultimate stress of around 40 MPa (Liang et al., 2006). This better in vivo performance can be attributed to several additional findings reported in this study: collagen fibrils with larger diameters in the SIS-treated ligaments were found in comparison with uniformly small fibrils for the non-SIS-treated group, implying that the SIS has a positive effect on fibril diameter. Moreover, a higher cell density compared to the non-SIS-treated group was found in the SIS-group. Obviously, the biological scaffold SIS leads to enhanced biomechanical properties after implantation because it acts bioactive during the healing process of the ligament tissue, stimulating the formation of larger collagen fibrils and higher cell density, among others (Liang et al., 2006).

On the other hand, collagen fiber meshes produced by electrospinning exhibited an ultimate stress of $0.40 \pm 0.11$ MPa and after crosslinking $2.69 \pm 0.47$ MPa (Meimandi-Parizi et al., 2013a). When used as a defect filler in rabbit AT reconstruction, the ultimate stress increased significantly to $19 \pm 2$ MPa and for the cross-linked scaffold to $27 \pm 3$ MPa 17 weeks postsurgery (Moshiri et al., 2013c). Like in the SIS examples mentioned earlier, the collagen fiber mesh was populated by invading cells from the surrounding which led to the production of more collagen fibers in an aligned orientation, which caused the higher ultimate stress of the material in vivo compared to its cell-free random spun fiber mesh.

Finally, a third group of materials shows extreme improvements in terms of ultimate stress in vivo when compared to their previous in vitro properties. Examples are collagen sponges that are seeded with mesenchymal stem cells (MSCs) cultivated under dynamic conditions in a bioreactor (Juncosa-Melvin et al., 2006a). Before being used for rabbit patellar tendon (PT) repair, such grafts exhibited an ultimate stress of $0.005 \pm 0.002$ MPa, while after 12 weeks in vivo, it was increased to $72 \pm 11$ MPa. Consultation of the natural healthy reference value $101 \pm 16$ MPa reveals that the regenerated tendon had almost recovered to a material with the natural ultimate stress

(Juncosa-Melvin et al., 2006a). This is impressive, as the ultimate stress increased by four orders of magnitudes. The explanation for this improvement cannot primarily lie in the repopulation of the graft because it had been cell-seeded before implantation. However, in this case, MSCs have been used that may be properly differentiated towards the tenocyte cell type in vivo (Sharma and Snedeker, 2012). Moreover, stem cells are known to exhibit paracrine actions (Shen et al., 2014), among which the attraction of resident cells accelerating the regeneration might lead to an overall more stable material (Geburek et al., 2016). Finally, also antiinflammatory actions of stem cells were reported (Shang et al., 2015), which might have the beneficial effect of suppressing the inflammatory reaction towards the collagen sponges at least to some extent.

Similarly, also cross-linked collagen sponges seeded with MSCs and cultivated under dynamic conditions exhibited an extremely better in vivo performance in terms of ultimate stress when compared to previous in vitro ultimate stress data. It was reported that the cross-linked collagen sponges exhibited an ultimate stress of $0.017 \pm 0.003$ MPa, while 12 weeks postsurgery the extracted rabbit PT had an ultimate stress of $14 \pm 4$ MPa—three orders of magnitudes higher (Nirmalanandhan et al., 2009). Interestingly, although in vitro data of non-cross-linked and cross-linked collagen sponges imply the cross-linked version to be a stronger graft material because the ultimate stress is threefold higher, the in vivo outcome is the opposite; used in the same animal model for the same time, the ultimate stress of the cross-linked collagen sponge acting as a bridge in the rabbit PT defect exhibited a fivefold lower ultimate stress. This can be nicely seen in Fig. 10.1, as there is a shift from the non-cross-linked data (denoted by Coll sp+MSC+MSt, 12 w) to the cross-linked data (denoted by Coll sp+CL+MSC+MSt, 12 w) in the south-west direction (Nirmalanandhan et al., 2009). Here, in vitro assessments of this biomechanical parameter would not have been conclusive and predictive of the in vivo outcome. Usually, a stronger in vitro tested material is assumed to reveal a better final in vivo performance; nevertheless, both materials, the non-cross-linked and the cross-linked collagen sponges, were strongly improving in terms of ultimate stress after in vivo implantation, therefore placed highly above the 1:1 line (Fig. 10.1).

## 10.3 Elastic modulus

Like for the ultimate stress, an in vitro–in vivo correlation figure is presented for the elastic modulus (Fig. 10.2). Obviously, the studies dealing with flexor tendon TECs based on decellularized rabbit hind-paw FDP reseeded with tenocytes had similar elastic moduli in vitro as in vivo 4 weeks postsurgery, similar to the findings for the ultimate stress (group 1). The in vivo performance with regard to the elastic modulus was only a little less that the in vitro data, in other words while the TECs exhibited an elastic modulus of $632 \pm 86$ and $1091 \pm 169$ MPa in vitro, for the static and dynamically cultivated tenocytes, respectively, after 4 weeks in vivo it was $505 \pm 148$ and $699 \pm 197$ MPa, respectively (Saber et al., 2010; Thorfinn et al., 2012). The in vitro dynamical stimulation lost some of its beneficial effects after implantation.

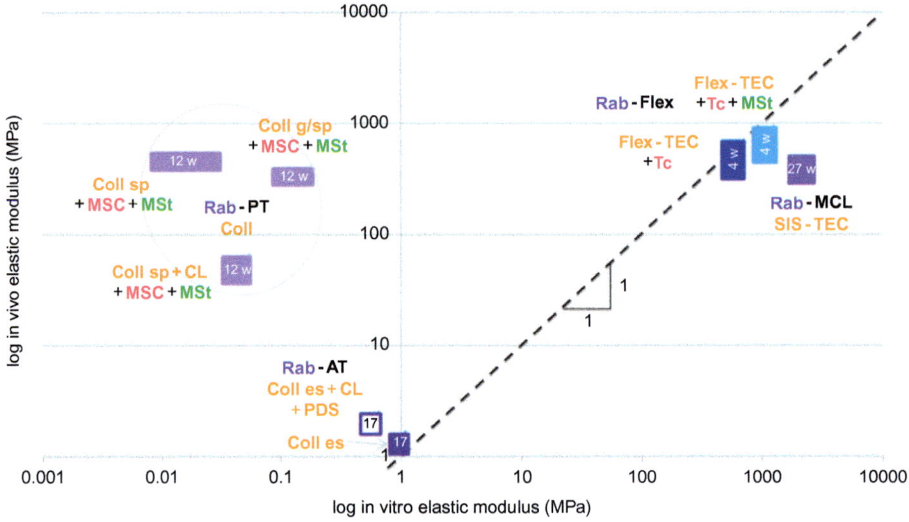


**Fig. 10.2** In vitro–in vivo comparison: elastic modulus of in vitro grafts and reconstructed tendons & ligaments. Data boxes correspond means ± SD or SEM. Key: see Table 10.2. Data from Saber, S., Zhang, A.Y., Ki, S.H., Lindsey, D.P., Smith, R.L., Riboh, J., Pham, H., Chang, J., 2010. Flexor tendon tissue engineering: bioreactor cyclic strain increases construct strength. Tissue Eng. A 16, 2085–2090; Thorfinn, J., Angelidis, I.K., Gigliello, L., Pham, H.M., Lindsey, D., Chang, J., 2012. Bioreactor optimization of tissue engineered rabbit flexor tendons in vivo. J. Hand Surg. Eur. 37, 109–114; Angelidis, I.K., Thorfinn, J., Connolly, I.D., Lindsey, D., Pham, H.M., Chang, J., 2010. Tissue engineering of flexor tendons: the effect of a tissue bioreactor on adipoderived stem cell-seeded and fibroblast-seeded tendon constructs. J. Hand Surg. 35, 1466–1472; Zhang, J., Wang, G.Y., Xiao, Y.P., Fan, L.Y., Wang, Q., 2011. The biomechanical behavior and host response to porcine-derived small intestine submucosa, pericardium and dermal matrix acellular grafts in a rat abdominal defect model. Biomaterials 32, 7086–7095; Musahl, V., Abramowitch, S.D., Gilbert, T.W., Tsuda, E., Wang, J.H.C., Badylak, S.F., Woo, S.L.Y., 2004. The use of porcine small intestinal submucosa to enhance the healing of the medial collateral ligament—a functional tissue engineering study in rabbits. J. Orthop. Res. 22, 214–220; Liang, R., Woo, S.L.Y., Takakura, Y., Moon, D.K., Jia, F., Abramowitch, S.D., 2006. Long-term effects of porcine small intestine submucosa on the healing of medial collateral ligament: a functional tissue engineering study. J. Orthop. Res. 24, 811–819; Adams, J.E., Zobitz, M.E., Reach Jr., J.S., An, K.-N.A., Teinmann, S.P., 2006. Rotator cuff repair using an acellular dermal matrix graft: an in vivo study in a canine model. Arthroscopy 22, 700–709; Chaudhury, S., Holland, C., Thompson, M.S., Vollrath, F., Carr, A.J., 2012. Tensile and shear mechanical properties of rotator cuff repair patches. J. Shoulder Elb. Surg. 21, 1168–1176; Juncosa-Melvin, N., Boivin, G.P., Gooch, C., Galloway, M.T., West, J.R., Dunn, M.G., Butler, D.L., 2006a. The effect of autologous mesenchymal stem cells on the biomechanics and histology of gel-collagen sponge constructs used for rabbit patellar tendon repair. Tissue Eng. 12, 369–379; Juncosa-Melvin, N., Shearn, J.T., Boivin, G.P., Gooch, C., Galloway, M.T., West, J.R., Nirmalanandhan, V.S., Bradica, G., Butler, D.L., 2006b. Effects of mechanical stimulation on the biomechanics and histology of stem cell-collagen sponge constructs for rabbit patellar tendon repair. Tissue Eng. 12, 2291–2300; Nirmalanandhan, V.S., Juncosa-Melvin, N., Shearn, J.T., Boivin, G.P., Galloway, M.T., Gooch, C., Bradica, G., Butler,

For group 2 with respect to the ultimate stress, the findings for the elastic modulus differ depending on the material investigated. While the SIS TEC clearly showed a significant decrease in elastic modulus from 2200 ± 500 MPa in vitro to only 387 ± 99 MPa in vivo when used for rabbit MCL reconstruction (Liang et al., 2006) and is therefore situated below the 1:1 dashed line (Fig. 10.2), the electrospun collagen meshes (either crosslinked and treated with PDS or not) exhibited a slightly better in vivo elastic modulus than in vitro—the same as reported for the ultimate stress (Meimandi-Parizi et al., 2013a,b).

For group 3, the findings for the elastic modulus were similar like for the ultimate stress: an orders of magnitude higher elastic modulus was found for the in vivo specimen compared to the in vitro constructs. While collagen gel/collagen sponge constructs increased their in vitro elastic modulus of 0.14 ± 0.04 MPa to 341 ± 47 in a rabbit PT model 12 weeks postsurgery (Juncosa-Melvin et al., 2006a), it was found for the collagen sponges without crosslinking a corresponding increase from 0.020 ± 0.010 to 441 ± 26 Pa (Juncosa-Melvin et al., 2006b), and for the cross-linked collagen sponges from 0.046 ± 0.010 to 51 ± 12 MPa (Nirmalanandhan et al., 2009).

## 10.4 Ultimate load

### 10.4.1 Grafts

Ultimate in vivo loads may not well be predicted based on in vitro testing as is shown in Fig. 10.3 (for data see Table 10.3). Obviously, there are materials like SIS exhibiting a certain in vitro ultimate load, but upon implantation they gain ultimate loads in a range of two orders of magnitude (red circle)—depending on where they are applied (like *infraspinatus* interposition or ACL reconstruction—ACL = anterior cruciate ligament) and also depending on the extraction time postsurgery. For example, while SIS used as an *infraspinatus* interposition in a dog model increased its in vivo ultimate load after 27 weeks by a factor of 3 (Dejardin et al., 2001), the in vivo performance when used as an ACL graft for reconstruction in a goat model was 30 times better 52 weeks postoperation in terms of ultimate load (Badylak et al., 1999).

D.L., 2009. Combined effects of scaffold stiffening and mechanical preconditioning cycles on construct biomechanics, gene expression, and tendon repair biomechanics. Tissue Eng. A 15, 2103–2111; Meimandi-Parizi, A., Moshiri, A., Oryan, A., 2013a. Role of tissue-engineered artificial tendon in healing of a large Achilles tendon defect model in rabbits. J. Am. Coll. Surg. 217, 421–441; Moshiri, A., Oryan, A., Meimandiparizi, A., 2013c. A novel application of biosynthetic tissue engineered tridimensional implant on large tendon defects: a comprehensive detailed in vivo investigation with significant clinical value. Connect. Tissue Res. 54 (3), 227–243; Meimandi-Parizi, A., Oryan, A., Moshiri, A., Silver, I.A., 2013b. Novel application of a tissue-engineered collagen-based three-dimensional bio-implant in a large tendon defect model: a broad-based study with high value in translational medicine. Tissue Cell 45, 282–294; Oryan, A., Moshiri, A., Meimandi-Parizi, A., 2013a. Graft selection in ACL reconstructive surgery: past, present, and future. Curr. Orthop. Pract. 24, 321–333.

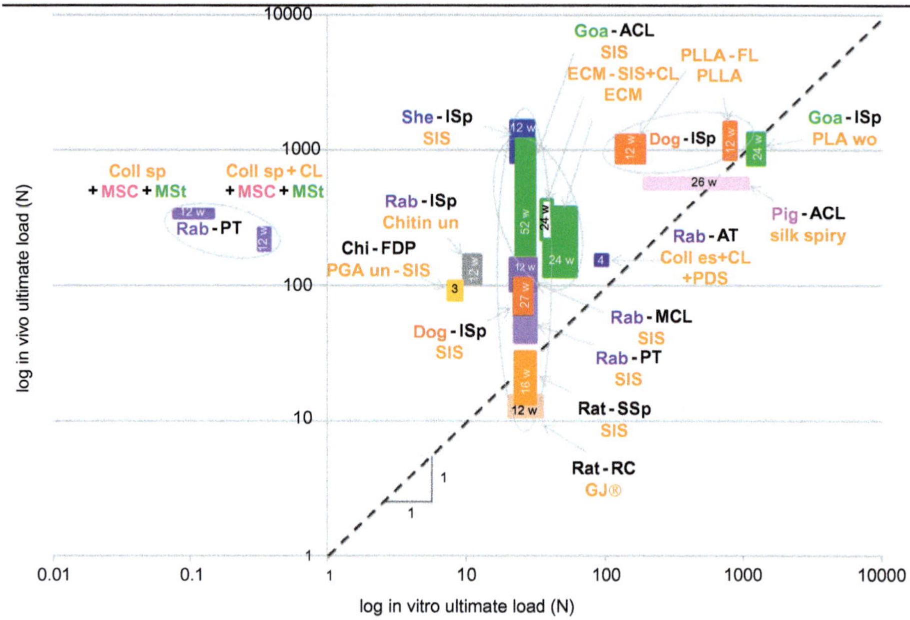


**Fig. 10.3** In vitro–in vivo comparison: ultimate load of in vitro grafts and reconstructed tendons & ligaments. Data boxes correspond means ± SD or SEM. Key: see Table 10.3. Data from Zhang, J., Wang, G.Y., Xiao, Y.P., Fan, L.Y., Wang, Q., 2011. The biomechanical behavior and host response to porcine-derived small intestine submucosa, pericardium and dermal matrix acellular grafts in a rat abdominal defect model. Biomaterials 32, 7086–7095; Dejardin, L.M., Arnoczky, S.P., Ewers, B.J., Haut, R.C., Clarke, R.B., 2001. Tissue-engineered rotator cuff tendon using porcine small intestine submucosa. Histologic and mechanical evaluation in dogs. Am. J. Sports Med. 29, 175–184; Funakoshi, T.,Majima, T., Suenaga, N., Iwasaki, N., Yamane, S., Minami, A., 2006. Rotator cuff regeneration using chitin fabric as an acellular matrix. J. Shoulder Elb. Surg. 15, 112–118; Juncosa-Melvin, N., Boivin, G.P., Gooch, C., Galloway, M.T., West, J.R., Dunn, M.G., Butler, D.L., 2006a. The effect of autologous mesenchymal stem cells on the biomechanics and histology of gel-collagen sponge constructs used for rabbit patellar tendon repair. Tissue Eng. 12, 369–379; Juncosa-Melvin, N., Shearn, J.T., Boivin, G.P., Gooch, C., Galloway, M. T., West, J.R., Nirmalanandhan, V.S., Bradica, G., Butler, D.L., 2006b. Effects of mechanical stimulation on the biomechanics and histology of stem cell-collagen sponge constructs for rabbit patellar tendon repair. Tissue Eng. 12, 2291–2300; Nirmalanandhan, V.S., Juncosa-Melvin, N., Shearn, J.T., Boivin, G.P., Galloway, M.T., Gooch, C., Bradica, G., Butler, D.L., 2009. Combined effects of scaffold stiffening and mechanical preconditioning cycles on construct biomechanics, gene expression, and tendon repair biomechanics. Tissue Eng. A 15, 2103–2111; Meimandi-Parizi, A., Oryan, A., Moshiri, A., 2013a. Role of tissue engineered collagen based tridimensional implant on the healing response of the experimentally induced large Achilles tendon defect model in rabbits: a long term study with high clinical relevance. J. Biomed. Sci. 20, 28; Meimandi-Parizi, A., Oryan, A., Moshiri, A., Silver, I.A., 2013b. Novel application of a tissue-engineered collagen-based three-dimensional bio-implant in a large tendon defect model: a broad-based study with high value in translational medicine. Tissue Cell 45, 282–294; Oryan, A., Moshiri, A., Parizi Meimandi, A., Silver, I.A., 2013b. A long-term in vivo investigation on the effects of xenogenous based,

Other materials like collagen sponges (either with or without crosslinking) show much higher ultimate loads compared to in vitro findings, which is according to the situation found for the ultimate stress, denoted as group 3 (Fig. 10.1). Again, as found for the ultimate stress, the cross-linked collagen sponges were stronger in vitro than the non-cross-linked collagen sponges; when implanted; however, the 12-week outcome was that the non-cross-linked specimens were stronger than the cross-linked ones. More precisely, while the in vivo performance of the non-cross-linked collagen sponge acting as a rabbit PT reconstruction was 3000 times stronger than the previous in vitro construct, it was "only" 600 times stronger in the cross-linked version (Juncosa-Melvin et al., 2006a,b; Nirmalanandhan et al., 2009).

### 10.4.2 Augmentations

Furthermore, prediction of the in vivo outcome is not possible at all if, for example, SIS is used as an augmentation device, because the most part of the load bearing tissue is the reconstructed tendon or ligament and the SIS makes up only a part of the further stability found in vivo. Nevertheless, reference values to estimate approximately the

electrospun, collagen implants on the healing of experimentally-induced large tendon defects. J. Musculoskelet. Neuronal Interact. 13, 353–367; Moshiri, A., Oryan, A., Meimandi-Parizi, A., 2013a. Role of tissue-engineered artificial tendon in healing of a large Achilles tendon defect model in rabbits. J. Am. Coll. Surg. 217, 421–441; Moshiri, A., Oryan, A., Meimandiparizi, A., 2013c. A novel application of biosynthetic tissue engineered tridimensional implant on large tendon defects: a comprehensive detailed in vivo investigation with significant clinical value. Connect. Tissue Res. 54 (3), 227–243; Chaudhury, S., Holland, C., Thompson, M.S., Vollrath, F., Carr, A.J., 2012. Tensile and shear mechanical properties of rotator cuff repair patches. J. Shoulder Elb. Surg. 21, 1168–1176; Ide, J., Kikukawa, K., Hirose, J., Iyama, K.-I., Sakamoto, H., Mizuta, H., 2009. Reconstruction of large rotator-cuff tears with acellular dermal matrix grafts in rats. J. Shoulder Elb. Surg. 18, 288–295; Fisher, M.B., Liang, R., Jung, H.-J., Kim, K.E., Zamarra, G., Almarza, A.J., Mcmahon, P.J., Woo, S.L.Y., 2012. Potential of healing a transected anterior cruciate ligament with genetically modified extracellular matrix bioscaffolds in a goat model. Knee Surg. Sports Traumatol. Arthrosc. 20, 1357–1365; Schlegel, T.F., Hawkins, R., Lewis, C.W., Motta, T., Turner, A.S., 2006. The effects of augmentation with swine small intestine submucosa on tendon healing under tension: histologic and mechanical evaluations in sheep. Am. J. Sports Med. 34, 275–280; Santoni, B.G., Mcgilvray, K.C., Lyons, A.S., Bansal, M., Turner, A.S., Macgillivray, J.D., Coleman, S.H., Puttlitz, C.M., 2010. Biomechanical analysis of an ovine rotator cuff repair via porous patch augmentation in a chronic rupture model. Am. J. Sports Med. 38, 679–686; Nicholson, G.P., Breur, G.J., Van Sickle, D., Yao, J.Q., Kim, J., Blanchard, C.R., 2007. Evaluation of a cross-linked acellular porcine dermal patch for rotator cuff repair augmentation in an ovine model. J. Shoulder Elb. Surg. 16, 184S–190S; Derwin, K.A., Codsi, M.J., Milks, R.A., Baker, A.R., Mccarron, J.A., Iannotti, J.P., 2009. Rotator cuff repair augmentation in a canine model with use of a woven poly-L-lactide device. J. Bone Joint Surg. Am. 91, 1159–1171; Baker, A.R., Mccarron, J.A., Tan, C.D., Iannotti, J.P., Derwin, K.A., 2012. Does augmentation with a reinforced fascia patch improve rotator cuff repair outcomes? Clin. Orthop. Relat. Res. 470, 2513–2521; Macgillivray, J. D., Fealy, S., Terry, M.A., Koh, J.L., Nixon, A.J., Warren, R.F., 2006. Biomechanical evaluation of a rotator cuff defect model augmented with a bioresorbable scaffold in goats. J. Shoulder Elb.

**Table 10.3 In vitro–in vivo comparison: structural properties of reconstructed tendon and ligaments (ultimate load [N], stiffness [N/mm])**

| Experimental model | | | | Tissue engineering | | | | | Ultimate load (N) | | | Stiffness (N/mm) | | | Source | |
|---|---|---|---|---|---|---|---|---|---|---|---|---|---|---|---|---|
| M | T/L | Rec | HT (w) | Sc | Sc tr. | C | Me St | Ch St | In vitro | In vivo | Native | In vitro | In vivo | Native | Literature | Tab. |
| ***Tendon auto-/allo-graft, TEC*** | | | | | | | | | | | | | | | | |
| Dog | ISp | I | 27 | SIS | | | | | 23–30 | 85 ± 24 | 187 ± 24 | | | | Zhang et al. (2011), Dejardin et al. (2001) | 6.2b |
| Rab | MCL | A | 12 | SIS | | | | | | 117 ± 35 | 332 ± 51 | | 46 ± 13 | 90 ± 15 | Zhang et al. (2011), Musahl et al. (2004) | |
| Rab | PT | A | 12 | SIS | | | | | | 78 ± 39 | 200–262 | | 39 ± 16 | 128–150 | Zhang et al. (2011), Karaoglu et al. (2008) | |
| Goa | ACL | R | 52 | SIS | | | | | | 706 ± 544 | 3151 ± 407 | | 180 ± 113 | 480 ± 42 | Zhang et al. (2011), Badylak et al. (1999) | |
| She | ISp | A | 12 | SIS | | | | | | 1252 ± 402 | 5533 ± 736 | | 215 ± 44 | 545 ± 47 | Zhang et al. (2011), Schlegel et al. (2006) | |
| Rat | SSp | A | 16 | SIS | | | | | | 23 ± 9 | 29–42 | | 18 ± 8 | 24–64 | Zhang et al. (2011), Zalavras et al. (2006) | |

| | | | | | | | | | | | | | | | | |
|---|---|---|---|---|---|---|---|---|---|---|---|---|---|---|---|---|
| ***Fascia lata (ECM*** | | | | | | | | | | | | | | | | |
| n.d. | T/L | – | | FL | | | | | | | | 85 ± 15 | | | Aurora et al. (2007), Derwin et al. (2008) | 6.2b |
| | | | | ECM | | | | | | | | 71–98 | | | | |
| ***Muscle (M)*** | | | | | | | | | | | | | | | | |
| Rat | AT | A | 8 | M fl | | | | | | ~100 | 60–80 | | ~53 | 36–86 | Majewski et al. (2008) | 6.2b |
| | | | | M | | | | GF | | ~181 | | | ~68 | | Majewski et al. (2012) | |
| ***Silk-TEC*** | | | | | | | | | | | | | | | | |
| Pig | ACL | R | 26 | Silk y | | | | | 9.4 ± 0.3 | | | 2.0 ± 0.1 | | | Woo et al. (1991), Li et al. (2013, 2014), Li and Snedeker (2013) | 6.2b |
| | | | | Silk w | | | | | 1550 ± 150 | | 2160 ± 157 | 285 ± 25 | | 242 ± 28 | | |
| | | | | Silk w | | Fb | dyn | | 480 ± 50 | | | 485 ± 15 | | | | |
| | | | | Silksp | | | | | 800 ± 200 | 566 ± 29 | 1749 ± 284 | 285 ± 20 | 183 ± 10 | 259 ± 15 | | |
| ***Chitin/Chitosan-based polymer*** | | | | | | | | | | | | | | | | |
| Rab | ISp | R | 12 | Chito-san fi | HA | | | | 16 ± 2 | | | 7 ± *1* | | | Funakoshi et al. (2005) | 6.2b |
| | | | | Chitin un | | | | | 9–12 | 135 ± 28 | | | 47 ± 14 | | Funakoshi et al. (2006) | |
| ***Collagen based scaffolds*** | | | | | | | | | | | | | | | | |
| Rab | PT | R | 12 | Coll sp | | MSC | dyn | | 0.11 ± 0.03 | 339 ± *11* | 471 ± *67* | 0.05 ± 0.02 | 142 ± *11* | 160 ± *11* | Juncosa-Melvin et al. (2006a,b) Juncosa 2006 # 2291/Juncosa, Butler 2006 #369 | 7.1b |
| | | | | Coll sp | CL | MSC | dyn | | 0.35 ± *0.01* | 218 ± *35* | 471 ± *67* | 0.07 ± *0.01* | 60 ± 10 | 160 ± *11* | Nirmalanandhan et al. (2009), Juncosa-Melvin et al. (2006a) | |

*Continued*

**Table 10.3 Continued**

| Experimental model | | | | Tissue engineering | | | | | Ultimate load (N) | | | Stiffness (N/mm) | | | Source | |
|---|---|---|---|---|---|---|---|---|---|---|---|---|---|---|---|---|
| M | T/L | Rec | HT (w) | Sc | Sc tr. | C | Me St | Ch St | In vitro | In vivo | Native | In vitro | In vivo | Native | Literature | Tab. |
| | | | | Coll g/sp | | MSC | dyn | | | 308 ± 13 | 471 ± *67* | 0.15 ± 0.10 | 125 ± 11 | 160 ± *11* | Juncosa-Melvin et al. (2006a), Shearn et al. (2007) | |
| Rab | AT | R | 17 | Coll es | | | | | | 119 ± 16 | 212–351 | 5 ± 1 | 59 ± 5 | 57–212 | Meimandi-Parizi et al. (2013a), Oryan et al. (2013b), Moshiri et al. (2013a) | 7.2b |
| | | | | Coll es | CL+PDS | | | | 95 ± 5 | 153 ± 8 | 212–351 | | 78 ± 4 | 57–212 | Meimandi-Parizi et al. (2013a), Oryan et al. (2013b), Moshiri et al. (2013a,b) | |
| Dog | ISp | A | 26 | GJ | | | | | 47 ± 11 | | | 7 ± 1 | 58 ± 23 | 100 ± 30 | Smith et al. (2012) | 6.2b |
| Rat | RC | R | 12 | GJ | | | | | 28 ± 7 | 13 ± 2 | 18 ± 3 | | | | Chaudhury et al. (2012), Ide et al. (2009) | 7.2b |
| Goa | ACL | W | 12 | ECM | | | | | 50 ± 12 | 249 ± 129 | 1385–1634 | | 53 ± 19 | 112–125 | Chaudhury et al. (2012), Fisher et al. (2012) | |
| She | ISp | A | 24 | ECM-SIS | CL | | | | 38 ± 3 | 316 ± 97 | 3516–5233 | 7 ± 2 | | 244–545 | Schlegel et al. (2006), Santoni et al. (2010), Nicholson et al. (2007) | 7.2b +2.2 |

| | | | | | | | | | | | | | | | | |
|---|---|---|---|---|---|---|---|---|---|---|---|---|---|---|---|---|
| ***Synthetic polymeric scaffolds*** | | | | | | | | | | | | | | | | |
| Chi | FDP | A | 3 | PGAun/ SIS | | *Tc* | | | 9±1 | 92±12 | 110±9 | | | | Liu et al. (2013) | 8.2a |
| Dog | ISp | A | 12 | PLLA | | | | | 796±34 | 1223±316 | 1595±285 | 195±2 | 243±45 | 515±141 | Derwin et al. (2009) | .2b |
| | | | | PLLA/ FL | Sut fi | | | | 153±27 | 1037±189 | 1595±285 | | 248±43 | 515±141 | Derwin et al. (2009), Baker et al. (2012) | 8.2b |
| Goa | ISp | A | 24 | PLA wo | | | | | 1214±93 | 1046±248 | | | | | MacGillivray et al. (2006) | 8.2b |

*Experimental Model*: *M*, in vivo model (*Rab*, rabbit; *Goa*, goat; *She*, sheep; *Chi*, chicken; *n.d.*, in vivo experiment not done); *T & L*, tendon & ligament (*Target organ*: *MCL*, medial collateral ligament; *RC*, rotor cuff; *ISp*, infraspinatus; *SSp*, supraspinatus; *AT*, Achilles tendon; *PT*, patellar tendon; *FDP*, *flexor digitorum profundus*); *Rec*, reconstruction; *R*, replacement; *A*, augmentation; *I*, interposition; *W*, wrapping; *HT*, healing time (*w*, weeks). *Tissue engineering*: *Sc*, scaffold (*TEC*, tissue engineering construct; *SIS*, small intestinal submucosa; *FL*, fascia lata; *ECM*, extra cellular matrix; *M*, muscle; *M fl*, muscle flap; *HUV*, human umbilical vein; *nf*, nanofiber; *fi*, fiber *w*, woven *un*, unwoven; *GJ*, GraftJacket; *Coll*, collagen based scaffold; *sp*, sponge; *g*, gel; *g/sp*, gel/sponge; *es*, electrospun; *PELA*, poly(L-lactic acid)–polyethylene glycol); *Sc tr* scaffold treatment (*AB*, antibiotic; *HA*, hyaluronic acid; *CL*, cross linking; *PDS*, polydioxanon; *Ibp*, ibuprofen); *C*, seeding cell type (*Fb*, fibroblasts; *Tc*, tenocytes; *MSC*, bone marrow mesenchymal stem cells); *MSt*, mechanical stimulation (*dyn*, dynamic stimulation); *CSt*, chemical stimulation (*BMP*, bone morphogenetic proteins (BMP-12 cDNA del); *GF*, growth factor (TGF-β1).
Data: Mean ± SD = Standard deviation; *SEM, Standard Error of Mean (italic)*.

expectable strength after implantation can brought together; both a rabbit MCL and a PT augmentation by SIS lead to around three times lower ultimate loads compared to natural MCL or PT, respectively, at 12 weeks postsurgery (Musahl et al., 2004; Karaoglu et al., 2008). When SIS was used as an infraspinatus augmentation in sheep, ultimate loads at 12 weeks were reported to be around one-fourth of natural values (Schlegel et al., 2006). Hence, although no in vitro–in vivo predictability can be deduced, the approximate in vivo ultimate load can be estimated to be around one-third to one-fourth of the ultimate loads exhibited by native tendons or ligaments.

## 10.5 Stiffness

An in vitro–in vivo plot for the stiffness of a series of tendon or ligament TECs is shown in Fig. 10.4 (for data, see Table 10.3). There are examples for the three groups defined earlier. While the silk used for porcine ACL reconstruction exhibited a stiffness of only 70% of the in vitro stiffness (Li and Snedeker, 2013; Li et al., 2014), being attributed to group 1, the electrospun collagen mesh used for rabbit AT reconstruction, in accordance to group 2-behaviour, exhibited a 12-fold higher stiffness in vivo after 17 weeks compared to its initial stiffness in vitro (Meimandi-Parizi et al., 2013a; Oryan et al., 2013b; Moshiri et al., 2013c). Finally, the collagen sponges, either cross-linked or not, exhibited the typical group 3-behaviour, as they had an orders of magnitude higher stiffness when implanted as rabbit PT constructs for 12 weeks (Juncosa-Melvin et al., 2006a,b).

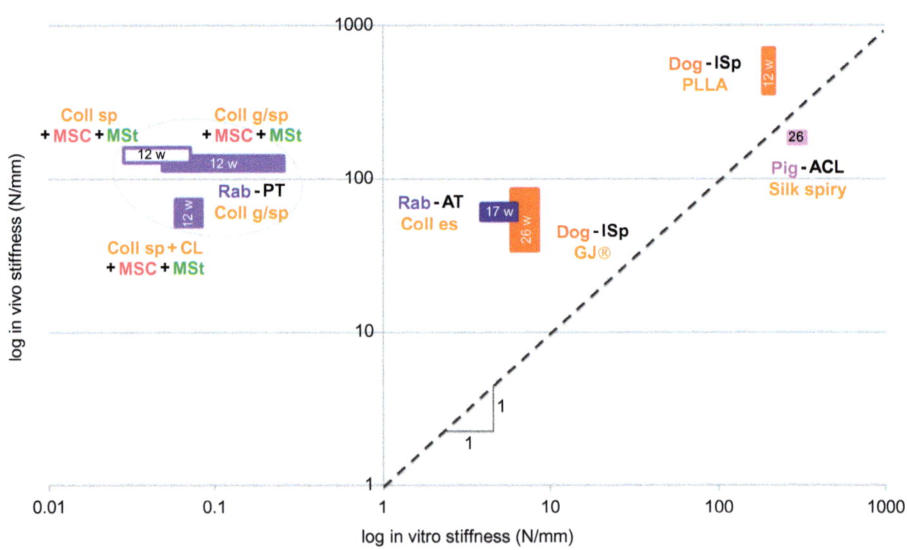


Fig. 10.4 In vitro–in vivo comparison: Stiffness of in vitro grafts and reconstructed tendons & ligaments. Data boxes correspond means ± SD or SEM. Key: see Table 10.3.

Data from Li, X., He, J., Bian, W., Li, Z., Zhang, W., Li, D., Snedeker, J.G., 2014. A novel silk-based artificial ligament and tricalcium phosphate/polyether ether ketone anchor for anterior cruciate ligament reconstruction—safety and efficacy in a porcine model. Acta Biomater. 10, 3696–3704; Juncosa-Melvin, N., Boivin, G.P., Gooch, C., Galloway, M.T., West, J.R., Dunn, M.G., Butler, D.L., 2006a. The effect of autologous mesenchymal stem cells on the biomechanics and histology of gel-collagen sponge constructs used for rabbit patellar tendon repair. Tissue Eng. 12, 369–379; Juncosa-Melvin, N., Shearn, J.T., Boivin, G.P., Gooch, C., Galloway, M.T., West, J.R., Nirmalanandhan, V.S., Bradica, G., Butler, D.L., 2006b. Effects of mechanical stimulation on the biomechanics and histology of stem cell-collagen sponge constructs for rabbit patellar tendon repair. Tissue Eng. 12, 2291–2300; Nirmalanandhan, V.S., Juncosa-Melvin, N., Shearn, J.T., Boivin, G.P., Galloway, M.T., Gooch, C., Bradica, G., Butler, D.L., 2009. Combined effects of scaffold stiffening and mechanical preconditioning cycles on construct biomechanics, gene expression, and tendon repair biomechanics. Tissue Eng. A 15, 2103–2111; Shearn, J.T., Juncosa-Melvin, N., Boivin, G.P., Galloway, M.T., Goodwin, W., Gooch, C., Dunn, M.G., Butler, D.L., 2007. Mechanical stimulation of tendon tissue engineered constructs: effects on construct stiffness, repair biomechanics, and their correlation. J. Biomech. Eng. 129, 848–854; Meimandi-Parizi, A., Oryan, A., Moshiri, A., Silver, I.A., 2013b. Novel application of a tissue-engineered collagen-based three-dimensional bio-implant in a large tendon defect model: a broad-based study with high value in translational medicine. Tissue Cell 45, 282–294; Oryan, A., Moshiri, A., Meimandi-Parizi, A., 2013a. Graft selection in ACL reconstructive surgery: past, present, and future. Curr. Orthop. Pract. 24, 321–333; Moshiri, A., Oryan, A., Meimandiparizi, A., 2013c. A novel application of biosynthetic tissue engineered tridimensional implant on large tendon defects: a comprehensive detailed in vivo investigation with significant clinical value. Connect. Tissue Res. 54 (3), 227–243; Fini, M., Bondioli, E., Castagna, A., Torricelli, P., Giavaresi, G., Rotini, R., Marinelli, A., Guerra, E., Orlandi, C., Carboni, A., Aiti, A., Benedettini, E., Giardino, R., Melandri, D., 2012. Decellularized human dermis to treat massive rotator cuff tears: in vitro evaluations. Connect. Tissue Res. 53, 298–306; Chaudhury, S., Holland, C., Thompson, M.S., Vollrath, F., Carr, A.J., 2012. Tensile and shear mechanical properties of rotator cuff repair patches. J. Shoulder Elb. Surg. 21, 1168–1176; Fisher, M.B., Liang, R., Jung, H.-J., Kim, K.E., Zamarra, G., Almarza, A.J., Mcmahon, P.J., Woo, S.L.Y., 2012. Potential of healing a transected anterior cruciate ligament with genetically modified extracellular matrix bioscaffolds in a goat model. Knee Surg. Sports Traumatol. Arthrosc. 20, 1357–1365; Schloegl, W., Marschall, V., Witting, M.Y., Volkmer, E., Drosse, I., Leicht, U., Schieker, M., Wiggenhorn, M., Schaubhut, F., Zahler, S., Friess, W., 2012. Porosity and mechanically optimized PLGA based in situ hardening systems. Eur. J. Pharm. Biopharm. 82, 554–562; Santoni, B.G., Mcgilvray, K.C., Lyons, A.S., Bansal, M., Turner, A.S., Macgillivray, J.D., Coleman, S.H., Puttlitz, C.M., 2010. Biomechanical analysis of an ovine rotator cuff repair via porous patch augmentation in a chronic rupture model. Am. J. Sports Med. 38, 679–686; Nicholson, G.P., Breur, G.J., Van Sickle, D., Yao, J.Q., Kim, J., Blanchard, C.R., 2007. Evaluation of a cross-linked acellular porcine dermal patch for rotator cuff repair augmentation in an ovine model. J. Shoulder Elb. Surg. 16, 184S–190S; Derwin, K.A., Codsi, M.J., Milks, R.A., Baker, A.R., Mccarron, J.A., Iannotti, J.P., 2009. Rotator cuff repair augmentation in a canine model with use of a woven poly-L-lactide device. J. Bone Joint Surg. Am. 91, 1159–1171.

## 10.6 Predictability in general; appropriate mathematical model

Originally, the aim of this book chapter was to generate a (linear) equation correlating in vitro biomechanical parameters to their corresponding in vivo values. This equation was intended to be modified with increments changing either the slope and/or the intercept for the material under use, the cell seeding and mechanical stimulation. It was planned to correlate individual in vitro–in vivo raw data pairs and analyse them with linear regression, as for example reported for stiffness and elastic modulus of stem cell-collagen sponge constructs for rabbit PT repair (Juncosa-Melvin et al., 2006b). However, the realization of such an equation was hampered by the lack of sufficient data, especially for individual raw data sets. Traditionally, data are presented and reported as means and standard deviations or errors, but not as single determined data for each specimen under view. In order to have data sets of a number of diverse materials, either cell-free or cell-seeded, and cell-seeded under static or dynamic conditions, a benevolent attitude of many authors publishing data of interest would have been the premise to achieve the necessary data, because such authors would have to be contacted and asked to provide all original raw data used for their articles of interest.

Hence, we reduced our intentions to the mere presentation of the data (as means and standard deviations) in an in vitro–in vivo plot and discussed deviations from the 1:1 line. Nevertheless, even with this reduced aim, it was still very difficult to get enough data for these in vitro–in vivo plots, because most of the papers published in this field either deal with in vitro experiments or with in vivo tests alone, but do not present both at the same time. In order to overcome this problem, we were looking for the same materials or tissue-engineered constructs that were investigated in vitro and in vivo in two different articles; but also this was sometimes a problem as not exactly the same material had been used (e.g. in one case cell-seeded in the other cell-free, respectively).

Therefore, the predictability of in vivo biomechanics based on in vitro measurements in general cannot yet be fully and detailed discussed or even summarized in an equation, because of the lack of sufficient data and because of the difficult availability of the raw data standing behind the published mean values.

## 10.7 Summary

In this chapter, four in vitro–in vivo correlation plots are presented and discussed; for the ultimate stress, the elastic modulus, the ultimate load as well as the stiffness. It was clearly found that in vitro assessment of biomechanical properties do not predict the in vivo outcome, because the time point of extraction, the way the corresponding grafts are used as well as the cell seeding, type of cells used, cultivation techniques may influence the outcome. It has to be emphasized, however, that the lack of a sufficient amount of raw data pairs (in vitro of the exact same material as in vivo) hampers a detailed analysis in this regard. In the future, authors experimenting in this field should be encouraged to share raw data in order to get more insight into the relationship of in vitro and in vivo biomechanical characteristics.

## References

Abousleiman, R.I., Reyes, Y., Mcfetridge, P., Sikavitsas, V., 2008. The human umbilical vein: a novel scaffold for musculoskeletal soft tissue regeneration. Artif. Organs 32, 735–742.

Adams, J.E., Zobitz, M.E., Reach Jr., J.S., An, K.-N.A., Teinmann, S.P., 2006. Rotator cuff repair using an acellular dermal matrix graft: an in vivo study in a canine model. Arthroscopy 22, 700–709.

Angelidis, I.K., Thorfinn, J., Connolly, I.D., Lindsey, D., Pham, H.M., Chang, J., 2010. Tissue engineering of flexor tendons: the effect of a tissue bioreactor on adipoderived stem cell-seeded and fibroblast-seeded tendon constructs. J. Hand Surg. 35, 1466–1472.

Aurora, A., Mccarron, J., Iannotti, J.P., Derwin, K., 2007. Commercially available extracellular matrix materials for rotator cuff repairs: state of the art and future trends. J. Shoulder Elb. Surg. 16, S171–S178.

Badylak, S., Arnoczky, S., Plouhar, P., Haut, R., Mendenhall, V., Clarke, R., Horvath, C., 1999. Naturally occurring extracellular matrix as a scaffold for musculoskeletal repair. Clin. Orthop. Relat. Res. (Suppl. 367), S333–S343.

Baker, A.R., Mccarron, J.A., Tan, C.D., Iannotti, J.P., Derwin, K.A., 2012. Does augmentation with a reinforced fascia patch improve rotator cuff repair outcomes? Clin. Orthop. Relat. Res. 470, 2513–2521.

Chaudhury, S., Holland, C., Thompson, M.S., Vollrath, F., Carr, A.J., 2012. Tensile and shear mechanical properties of rotator cuff repair patches. J. Shoulder Elb. Surg. 21, 1168–1176.

Dejardin, L.M., Arnoczky, S.P., Ewers, B.J., Haut, R.C., Clarke, R.B., 2001. Tissue-engineered rotator cuff tendon using porcine small intestine submucosa. Histologic and mechanical evaluation in dogs. Am. J. Sports Med. 29, 175–184.

Derwin, K.A., Baker, A.R., Spragg, R.K., Leigh, D.R., Farhat, W., Iannotti, J.P., 2008. Regional variability, processing methods, and biophysical properties of human fascia lata extracellular matrix. J. Biomed. Mater. Res. Part A 84, 500–507.

Derwin, K.A., Codsi, M.J., Milks, R.A., Baker, A.R., Mccarron, J.A., Iannotti, J.P., 2009. Rotator cuff repair augmentation in a canine model with use of a woven poly-L-lactide device. J. Bone Joint Surg. Am. 91, 1159–1171.

Fisher, M.B., Liang, R., Jung, H.-J., Kim, K.E., Zamarra, G., Almarza, A.J., Mcmahon, P.J., Woo, S.L.Y., 2012. Potential of healing a transected anterior cruciate ligament with genetically modified extracellular matrix bioscaffolds in a goat model. Knee Surg. Sports Traumatol. Arthrosc. 20, 1357–1365.

Funakoshi, T., Majima, T., Iwasaki, N., Suenaga, N., Sawaguchi, N., Shimode, K., Minami, A., Harada, K., Nishimura, S.-I., 2005. Application of tissue engineering techniques for rotator cuff regeneration using a chitosan-based hyaluronan hybrid fiber scaffold. Am. J. Sports Med. 33, 1193–1201.

Funakoshi, T., Majima, T., Suenaga, N., Iwasaki, N., Yamane, S., Minami, A., 2006. Rotator cuff regeneration using chitin fabric as an acellular matrix. J. Shoulder Elb. Surg. 15, 112–118.

Geburek, F., Mundle, K., Conrad, S., Hellige, M., Walliser, U., Van Schie, H.T.M., Van Weeren, R., Skutella, T., Stadler, P.M., 2016. Tracking of autologous adipose tissue-derived mesenchymal stromal cells with in vivo magnetic resonance imaging and histology after intralesional treatment of artificial equine tendon lesions—a pilot study. Stem Cell Res. Ther. 7, 21.

Ide, J., Kikukawa, K., Hirose, J., Iyama, K.-I., Sakamoto, H., Mizuta, H., 2009. Reconstruction of large rotator-cuff tears with acellular dermal matrix grafts in rats. J. Shoulder Elb. Surg. 18, 288–295.

Issa, R.I., Engebretson, B., Rustom, L., Mcfetridge, P.S., Sikavitsas, V.I., 2011. The effect of cell seeding density on the cellular and mechanical properties of a mechanostimulated tissue-engineered tendon. Tissue Eng. A 17, 1479–1487.

Juncosa-Melvin, N., Boivin, G.P., Gooch, C., Galloway, M.T., West, J.R., Dunn, M.G., Butler, D.L., 2006a. The effect of autologous mesenchymal stem cells on the biomechanics and histology of gel-collagen sponge constructs used for rabbit patellar tendon repair. Tissue Eng. 12, 369–379.

Juncosa-Melvin, N., Shearn, J.T., Boivin, G.P., Gooch, C., Galloway, M.T., West, J.R., Nirmalanandhan, V.S., Bradica, G., Butler, D.L., 2006b. Effects of mechanical stimulation on the biomechanics and histology of stem cell-collagen sponge constructs for rabbit patellar tendon repair. Tissue Eng. 12, 2291–2300.

Juncosa-Melvin, N., Matlin, K.S., Holdcraft, R.W., Nirmalanandhan, V.S., Butler, D.L., 2007. Mechanical stimulation increases collagen type I and collagen type III gene expression of stem cell—collagen sponge constructs for patellar tendon repair. In: ASME Summer Bioengineering Conference, Jun 24–28, 2007 Keystone, CO. Amer Soc Mechanical Engineers, pp. 981–982.

Karaoglu, S., Fisher, M.B., Woo, S.L.Y., Fu, Y.-C., Liang, R., Abramowitch, S.D., 2008. Use of a bioscaffold to improve healing of a patellar tendon defect after graft harvest for ACL reconstruction: a study in rabbits. J. Orthop. Res. 26, 255–263.

Li, X., Snedeker, J.G., 2013. Wired silk architectures provide a biomimetic ACL tissue engineering scaffold. J. Mech. Behav. Biomed. Mater. 22, 30–40.

Li, Y., Fessel, G., Georgiadis, M., Snedeker, J.G., 2013. Advanced glycation end-products diminish tendon collagen fiber sliding. Matrix Biol. 32, 169–177.

Li, X., He, J., Bian, W., Li, Z., Zhang, W., Li, D., Snedeker, J.G., 2014. A novel silk-based artificial ligament and tricalcium phosphate/polyether ether ketone anchor for anterior cruciate ligament reconstruction—safety and efficacy in a porcine model. Acta Biomater. 10, 3696–3704.

Liang, R., Woo, S.L.Y., Takakura, Y., Moon, D.K., Jia, F., Abramowitch, S.D., 2006. Long-term effects of porcine small intestine submucosa on the healing of medial collateral ligament: a functional tissue engineering study. J. Orthop. Res. 24, 811–819.

Liu, S., Hu, C.M., Li, F.F., Li, X.J., Cui, W.G., Fan, C.Y., 2013. Prevention of peritendinous adhesions with electrospun ibuprofen-loaded poly(L-lactic acid)-polyethylene glycol fibrous membranes. Tissue Eng. A 19, 529–537.

Macgillivray, J.D., Fealy, S., Terry, M.A., Koh, J.L., Nixon, A.J., Warren, R.F., 2006. Biomechanical evaluation of a rotator cuff defect model augmented with a bioresorbable scaffold in goats. J. Shoulder Elb. Surg. 15, 639–644.

Majewski, M., Betz, O., Ochsner, P.E., Liu, F., Porter, R.M., Evans, C.H., 2008. Ex vivo adenoviral transfer of bone morphogenetic protein 12 (BMP-12) cDNA improves Achilles tendon healing in a rat model. Gene Ther. 15, 1139–1146.

Majewski, M., Porter, R.M., Betz, O.B., Betz, V.M., Clahsen, H., Fluckiger, R., Evans, C.H., 2012. Improvement of tendon repair using muscle grafts transduced with TGF-beta1 cDNA. Eur. Cell. Mater. 23, 94–101. discussion 101–2.

Meier Buergisser, G., Buschmann, J., 2014. History and performance of implant materials applied as peritendinous antiadhesives. J. Biomed. Mater. Res. B Appl. Biomater. 103 (1), 212–228.

Meimandi-Parizi, A., Oryan, A., Moshiri, A., 2013a. Role of tissue engineered collagen based tridimensional implant on the healing response of the experimentally induced large Achilles tendon defect model in rabbits: a long term study with high clinical relevance. J. Biomed. Sci. 20, 28.

Meimandi-Parizi, A., Oryan, A., Moshiri, A., Silver, I.A., 2013b. Novel application of a tissue-engineered collagen-based three-dimensional bio-implant in a large tendon defect model: a broad-based study with high value in translational medicine. Tissue Cell 45, 282–294.

Moshiri, A., Oryan, A., Meimandi-Parizi, A., 2013a. Role of tissue-engineered artificial tendon in healing of a large Achilles tendon defect model in rabbits. J. Am. Coll. Surg. 217, 421–441.

Moshiri, A., Oryan, A., Meimandi-Parizi, A., Silver, I.A., Tanideh, N., Golestani, N., 2013b. Effectiveness of hybridized nano- and microstructure biodegradable, biocompatible, collagen-based, three-dimensional bioimplants in repair of a large tendon-defect model in rabbits. J. Tissue Eng. Regen. Med. 10 (6), 451–465.

Moshiri, A., Oryan, A., Meimandiparizi, A., 2013c. A novel application of biosynthetic tissue engineered tridimensional implant on large tendon defects: a comprehensive detailed in vivo investigation with significant clinical value. Connect. Tissue Res. 54 (3), 227–243.

Musahl, V., Abramowitch, S.D., Gilbert, T.W., Tsuda, E., Wang, J.H.C., Badylak, S.F., Woo, S.L.Y., 2004. The use of porcine small intestinal submucosa to enhance the healing of the medial collateral ligament—a functional tissue engineering study in rabbits. J. Orthop. Res. 22, 214–220.

Ni, T., Senthil-Kumar, P., Dubbin, K., Aznar-Cervantes, S.D., Datta, N., Randolph, M.A., Cenis, J.L., Rutledge, G.C., Kochevar, I.E., Redmond, R.W., 2012. A photoactivated nanofiber graft material for augmented Achilles tendon repair. Lasers Surg. Med. 44, 645–652.

Nicholson, G.P., Breur, G.J., Van Sickle, D., Yao, J.Q., Kim, J., Blanchard, C.R., 2007. Evaluation of a cross-linked acellular porcine dermal patch for rotator cuff repair augmentation in an ovine model. J. Shoulder Elb. Surg. 16, 184S–190S.

Nirmalanandhan, V.S., Juncosa-Melvin, N., Shearn, J.T., Boivin, G.P., Galloway, M.T., Gooch, C., Bradica, G., Butler, D.L., 2009. Combined effects of scaffold stiffening and mechanical preconditioning cycles on construct biomechanics, gene expression, and tendon repair biomechanics. Tissue Eng. A 15, 2103–2111.

Oryan, A., Moshiri, A., Meimandi-Parizi, A., 2013a. Graft selection in ACL reconstructive surgery: past, present, and future. Curr. Orthop. Pract. 24, 321–333.

Oryan, A., Moshiri, A., Parizi Meimandi, A., Silver, I.A., 2013b. A long-term in vivo investigation on the effects of xenogenous based, electrospun, collagen implants on the healing of experimentally-induced large tendon defects. J. Musculoskelet. Neuronal Interact. 13, 353–367.

Rodger, M.P., Theobald, P., Giddins, G., 2015. Vein grafts to augment flexor tendon repairs: a biomechanical study on strength and gap resistance. J. Hand Surg. 40, 695–699.

Russel, W.M.S., Burch, R.L., 1959. The Principle of Humane Experimental Technique. Universities Federation for Animal Welfare, London.

Saber, S., Zhang, A.Y., Ki, S.H., Lindsey, D.P., Smith, R.L., Riboh, J., Pham, H., Chang, J., 2010. Flexor tendon tissue engineering: bioreactor cyclic strain increases construct strength. Tissue Eng. A 16, 2085–2090.

Santoni, B.G., Mcgilvray, K.C., Lyons, A.S., Bansal, M., Turner, A.S., Macgillivray, J.D., Coleman, S.H., Puttlitz, C.M., 2010. Biomechanical analysis of an ovine rotator cuff repair via porous patch augmentation in a chronic rupture model. Am. J. Sports Med. 38, 679–686.

Schlegel, T.F., Hawkins, R., Lewis, C.W., Motta, T., Turner, A.S., 2006. The effects of augmentation with swine small intestine submucosa on tendon healing under tension: histologic and mechanical evaluations in sheep. Am. J. Sports Med. 34, 275–280.

Shang, Q., Bai, Y., Wang, G., Song, Q., Guo, C., Zhang, L., Wang, Q., 2015. Delivery of adipose-derived stem cells attenuates adipose tissue inflammation and insulin resistance in obese mice through remodeling macrophage phenotypes. Stem Cells Dev. 24, 2052–2064.

Sharma, R.I., Snedeker, J.G., 2012. Paracrine interactions between mesenchymal stem cells affect substrate driven differentiation toward tendon and bone phenotypes. PLoS One. 7 (2)e31504.

Shearn, J.T., Juncosa-Melvin, N., Boivin, G.P., Galloway, M.T., Goodwin, W., Gooch, C., Dunn, M.G., Butler, D.L., 2007. Mechanical stimulation of tendon tissue engineered constructs: effects on construct stiffness, repair biomechanics, and their correlation. J. Biomech. Eng. 129, 848–854.

Shen, W., Chen, J., Zhu, T., Chen, L., Zhang, W., Fang, Z., Heng, B.C., Yin, Z., Chen, X., Ji, J., Chen, W., Ouyang, H.-W., 2014. Intra-articular injection of human meniscus stem/progenitor cells promotes meniscus regeneration and ameliorates osteoarthritis through stromal cell-derived factor-1/CXCR4-mediated homing. Stem Cells Transl. Med. 3, 387–394.

Smith, M.J., Cook, J.L., Kuroki, K., Jayabalan, P.S., Cook, C.R., Pfeiffer, F.M., Waters, N.P., 2012. Comparison of a novel bone-tendon allograft with a human dermis-derived patch for repair of chronic large rotator cuff tears using a canine model. Arthroscopy 28, 169–177.

Thorfinn, J., Angelidis, I.K., Gigliello, L., Pham, H.M., Lindsey, D., Chang, J., 2012. Bioreactor optimization of tissue engineered rabbit flexor tendons in vivo. J. Hand Surg. Eur. 37, 109–114.

Woo, S.L., Hollis, J.M., Adams, D.L., Lyon, R.M., Takai, S., 1991. Tensile properties of the human femur-anterior cruciate ligament-tibia complex. The effects of specimen age and orientation. Am. J. Sports Med. 19, 217–225.

Zalavras, C.G., Gardocki, R., Huang, E., Stevanovic, M., Hedman, T., Tibone, J., 2006. Reconstruction of large rotator cuff tendon defects with porcine small intestinal submucosa in an animal model. J. Shoulder Elb. Surg. 15, 224–231.

Zhang, J., Wang, G.Y., Xiao, Y.P., Fan, L.Y., Wang, Q., 2011. The biomechanical behavior and host response to porcine-derived small intestine submucosa, pericardium and dermal matrix acellular grafts in a rat abdominal defect model. Biomaterials 32, 7086–7095.

# Role of cellular response in the healing process of tendons and ligaments

# 11

## Abstract

Cellular responses occurring as cascades of inflammatory cells and tenocytes immigrating the wound site orchestrate the healing process in lacerated tendons and ligaments. Time-dependent concentrations of cytokines accompany this very well-balanced sequence of inflammation, remodelling, and maturation in the healing tissue. Extrinsic healing (cells migrating from synovium or tendon sheath towards the wound site) and intrinsic healing (cells primarily from endotenon and epitenon contribute to wound healing) lead to regeneration. However, unwanted scar-like tissue may be formed, associated with an insufficient resolution of inflammation and excessive remodelling. Scar tissue exhibits lower biomechanical properties and is prone to rerupture. Ways to promote scarless healing are discussed. In addition, gene expression and growth factor levels as a function of time after injury are presented.

**Keywords:** Intrinsic and extrinsic healing, Immunoreaction, Growth factors, Remodelling, Consolidation.

## Abbreviation

| | |
|---|---|
| **ACL** | anterior cruciate ligament (tendons and ligaments) |
| **AT** | Achilles tendon (tendons and ligaments) |
| **bFGF** | basic fibroblastic growth factor (protein in living cells) |
| **BMP** | bone morphogenetic protein (element in tissues) |
| **CD** | cluster of differentiation (cell development) |
| **COMP** | cartilage oligomeric matrix protein (protein in living cells, also known as thrombospondin-5, an extracellular matrix protein primarily present in cartilage) |
| **COX** | cyclooxygenase (enzyme in living cells) |
| **CTGF** | connective tissue growth factor (protein in living cells) |
| **ECM** | extra cellular matrix (matrix in living cells) |
| **EGF** | epidermal growth factor (protein in living cells) |
| **FDP** | *flexor digitorum profundus* (tendons and ligaments) |
| **FPR2/ ALX** | *N*-formyl peptide receptor 2/A lipoxin (G-protein coupled receptor, interacting with lipoxin) |
| **IGF (−*n*)** | insulin-like growth factor ($n=1$, 2, 3) (protein in living cells) |
| **IL** | interleukin-*n* ($n=1$–17) (protein in living cells) |
| **LX** | lipoxin (bioactive product with immunomodulatory and anti-inflammatory action) |

**Biomechanics of Tendons and Ligaments. http://dx.doi.org/10.1016/B978-0-08-100489-0.00011-9**

**MMP (−*n*)** matrix metalloproteinase-*n* ($n = 1, 2, 3$) (enzymes in living cells)
**mRNA** messenger ribonucleic acid (element of living cells)
**NF-κB** nuclear factor kappa B (protein complex controlling transcription of DNA, cytokine production, and cell survival)
**NT** no treatment (no treated tendon or tissue, mainly contralateral part of the body)
**PCR** polymerase chain reaction (technical feature of gene technology lab)
**PDGF** platelet-derived growth factor (protein in living cells)
**PGE** prostaglandins (physiologically active lipid compounds having diverse hormone-like effects)
**PRP** platelet rich plasma (enriched blood plasma)
**RNA** ribonucleic acid (element of living cells)
**TGF (−β*n*)** transforming growth factor beta ($n = 1, 2, 3$) (protein in living cells)
**TNF** tumor necrosis factors (family) (group of cytokines that can cause cell death (apoptosis))
**VEGF** vascular endothelial growth factor (protein in living cells)

## 11.1 Introduction

In contrast to other limbs of the musculoskeletal system, the healing of lacerated tendons and ligaments usually takes a long time because of the inherent characteristics of these connective tissues. Not only is there a lack of an inclusive vascular network but also have the cells a low metabolism, which adds to the poor intrinsic healing capacities and the little regenerative potential. Often, healing is accompanied by the development of a scar tissue in addition to truly regenerated tissue or in worst case as a sole final tissue; this ends up in poor mechanical properties which leads to one of the two most common problems occurring after tendon injuries—the rerupture as a consequence of overuse (Elliot and Giesen, 2013).

After acute injuries and after surgeries, tendon healing occurs in three main phases (Table 11.1) (Sharma and Maffulli, 2006). There is an initial cellular phase, where the

**Table 11.1 Overview and duration of the three phases in tendon healing according to (Sharma and Maffulli, 2006)**

| Phase | Process | | Time | Repair tissue |
|---|---|---|---|---|
| 1 | Initial phase with immigration of erythrocytes and inflammatory cells | | <day 1 | Cellular phase |
| 2 | Remodelling stage | | Day 2–6 weeks | |
| 3a | Modelling stage | Consolidation stage | 6–10 weeks | Fibrous phase |
| 3b | | Maturation stage | 10 weeks–1 year | Scar-like tissue |

otherwise cell-poor tendon tissue is populated by immigrating erythrocytes and inflammatory cells such as neutrophils, monocytes, and macrophages. An inflammatory cascade is induced where inflammatory cytokines such as Interleukin-1 (IL-1) and tumor necrosis factor-α (TNF-α) are released (Huegle et al., 2012). Moreover, vasoactive factors lead to an increased vascular permeability and initiate angiogenesis. Tenocytes proliferate and migrate to the wound site where they predominantly produce collagen type III (Sharma and Maffulli, 2006). For the following 6 weeks, during the remodelling stage, collagen type III synthesis is predominant and the water and glycosaminoglycans' content of the healing tissue are higher than in healthy tissue (Fig. 11.1). Besides monocytes and macrophages occurring immediately after injury, other inflammatory cell types such as lymphocytes come later, during the remodelling stage, and are part of the adaptive immune system, providing the anti-inflammatory cytokine IL-10 and TNF-receptor 2 which neutralizes TNF-α and as such initiate the end of the inflammatory reaction (Schulze-Tanzil et al., 2011). In the third-phase (modelling stage), cellular density decreases with a change from a cellular tissue to a fibrous tissue; fibers get more aligned during this phase, whereas they were not well oriented before. Moreover, from around 10 weeks postlaceration, a scar-like tissue is most often the predominant tissue found. Until a tendon or ligament tissue has fully recovered, 1 year may be needed

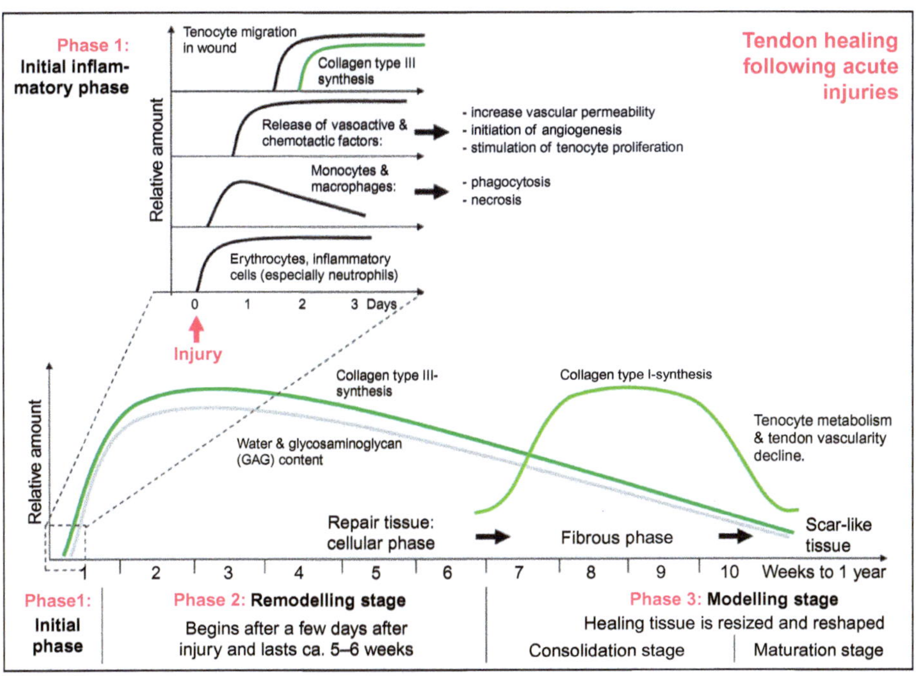


**Fig. 11.1** Overview of the three phases in tendon and ligament healing according to (Sharma and Maffulli, 2006) cell immigration, collagen synthesis and formation of the regenerated tendon tissue.

(maturation stage). In this last phase of regeneration, tenocyte metabolism and vascularity decline—nevertheless ending up in a fibrovascular scar tissue that never reaches the characteristics of normal healthy tendon or ligament tissue.

## 11.2 Intrinsic versus extrinsic healing

When cellular responses to tendon and ligament injuries are under view, intrinsic and extrinsic healing patterns are discussed (Sharma and Maffulli, 2006). Intrinsic tendon healing involves tenoblasts and tenocytes from within the tendon tissue itself, predominantly originating from endotenon and epitenon and migrating towards the wound site. On the other hand, extrinsic healing refers to inflammatory cells and fibroblasts that are migrating from outside of the tendon tissue to the repair site, for example, from the synovium or from the tendon sheath. It has often been discussed between specialists which of the two mechanisms might be predominant and whether one mechanism alone would lead to a proper regenerated tissue. Generally, intrinsic healing is associated with an improved biomechanical outcome, while extrinsic healing is often regarded to be responsible for adhesion formation to the surrounding tissues.

The relative contribution of intrinsic and extrinsic healing patterns, however, was concluded to be highly dependent on the type of trauma, its anatomical position, the presence of a synovial sheath, and the extent of force and stress acting upon the repair site and induced by early postoperative motion (Koob, 2002). Nevertheless, these two healing patterns were investigated in-depth and compared, for example, in terms of collagen production or cell proliferation kinetics depending on a certain type of tendon or ligament laceration. For example, canine flexor tendon cells from tendon core tissue produced more mature and larger collagen molecules compared to tenoblasts originating from the epitenon (Fujita et al., 1992). Cells that migrate from outside to the tendon repair site such as cells from the synovial sheath or the peritenon produce even less collagen compared to cells from epitenon or endotenon; however, their proliferation and migration rate is higher compared to epitenon, endotenon or internal cells from the tendon core tissue. In addition, in the peritenon, a subpopulation of tendon stem cells exist which contribute to the extrinsic healing—being significantly higher expressed when compared to the core tendon cells (Cadby et al., 2014).

The question if one healing pattern might be enough for an adequate restoration of tendons or ligaments has also been addressed. For example, it has been reported that intrinsic healing may be sufficient, and there is no need for vascularity or cellular support from outside to achieve a properly restored tendon. These findings were reported for a canine flexor tendon model 12 weeks postoperation (Gelberman et al., 1986). Newer studies, however, emphasize the interplay of cells originating from multiple sites, in other words from the tendon core as well as from outside (Thomopoulos et al., 2015). It is reported that flexor tendon healing starts with fibroblasts migrating from the epitenon to the wound site. Simultaneously, vascularity increases (Manning et al., 2014). Only later internal tenocytes proliferate and migrate to the repair site. In contrast, if the rotator cuff is torn, fibroblasts from the tendon core as well as from the surrounding start to migrate immediately to the repair site which is often the tendon-to-bone interface. There, they produce a disorganized collagen structure resulting in a biomechanically weak fibrous scar tissue (Carpenter et al., 1998).

The optimization of intrinsic and extrinsic healing has been addressed by various approaches; among them by the application of an electrospun core–shell fiber mesh having incorporated hyaluronan particles from which mitomycin-C was sustainably released (Zhao et al., 2015). Mitomycin-C inhibits collagen production in fibroblasts and induces fibroblast apoptosis. Applied as a sheath around the tendon, the hypothesis was that fibroblasts outside the sheath were led to apoptosis (suppression of the extrinsic healing), while intrinsic healing was the only mechanism of action. The authors checked the antiadhesive effect in two animal models, the rat AT (Achilles tendon) and the rabbit flexor(s). It was found that the adhesion was significantly reduced, while the biomechanics remained unchanged. Western blotting for adhesion-associated proteins revealed significantly lower levels in mitomycin-C tendons (Zhao et al., 2015).

## 11.3 Scarless healing

As mentioned earlier, tendon and ligament regeneration most often ends up in poor fibrous scar-like tissue lacking the collagen organization seen in normal healthy limbs. A main question is if and how the healing could be supported to avoid scar formation. Are there measures and ways to direct a proper tendon healing leading to normal tendon tissue with all its tissue-specific and mechanical characteristics? It is well known that the foetus of mammals has the capacity to heal skin wounds without any scar formation. This is attributed to the fact that there is no inflammatory response due to the absence of the immune system. Only in the presence of a functional immune system, an inflammation occurs and leads to a scar-mediated way of healing. Although other factors in foetal scarless wound healing may play a role, the absence of an inflammatory reaction seems to be a crucial factor (Galatz et al., 2015).

Therefore, animal models for foetal tendon wound healing have been developed in order to study wound healing without any inflammatory response. For example, the lateral extensor tendons were partially lacerated in foetal sheep and simultaneously in maternal limbs (Beredjiklian et al., 2003). While at the wound site of the adult sheep, a gap filled with typical granulation tissue and inflammatory cells was found 1 week after injury; it looked completely different in the foetal tendon. There, a reconstitution of the parallel aligned collagen fibers was found. Going along with this, only a minimum expression of transforming growth factor TGF-β was found at the foetal wound site, while a clear upregulation of this growth factor was detected in the adult sheep (Beredjiklian et al., 2003).

A subsequent question in terms of regenerative capacity is whether the foetal tendon as such is capable of scarless healing or if it is the environment in the foetus that makes this beneficial healing process possible. Hence, injured foetal sheep tendons were transplanted into adult sheep, and the healing was studied in this environment (Favata et al., 2006). It was found that the injured foetal tendon in the adult sheep was still able to regenerate without scar formation—indicating that the scarless repair is intrinsic to the foetal tendon itself. Interestingly, increased levels of TGF-β1, bFGF (basic fibroblastic growth factor), and CD44 (cluster of differentiation) were only found at the adults' specimen wound sites, while no such changes were recognized in the foetal specimen.

Nevertheless, an inflammatory reaction following any injury is not disadvantageous *per se*—it depends on the site of injury. Inflammation has been reported to be positive in axonal and neuronal regeneration (Yin et al., 2006). Furthermore, when two mouse strains were compared, which were bred for maximum and for minimum inflammatory reaction, respectively, it was found that the strain with maximum inflammatory response showed a complete ear hole closure with the replacement of cartilage and without scar formation after puncture. This was attributed to a macrophage-enhanced wound healing process (Koh and DiPietro, 2011). However, in tendon and ligament healing, the mechanisms of true regeneration, in other words, the mechanisms underlying scarless healing, have still to be elucidated in order to be able to offer better clinical therapies.

## 11.4 Inflammatory reaction

The inflammatory reaction that occurs after tendon or ligament injury in adults is necessary for the debridement; however, a persistent inflammation has adverse consequences resulting in a fibrous tissue that lacks the properties of a truly regenerated tendon or ligament. Understanding inflammatory response in detail and finding ways of resolution of inflammation at an appropriate time point are therefore important in this field. Specifically, the time-dependent occurrence and concentrations of cytokines, prostaglandins, and proresolving mediators such as lipoxins have to be in a concentration-dependent balance to provide a way back to the normal tendon and ligament homeostasis in the regenerated tissue after acute injury. Dakin and coworkers (2014) propose that an insufficient resolution of the inflammatory reaction occurring in acute tendon and ligament injuries may lead to a persistent low-level inflammation, which increases the propensity for fibrotic rather than regenerative healing and in consequence rerupture or reinjury (Fig. 11.2) (Dakin et al., 2014).

### *11.4.1 Key players in regulating inflammation: Cytokines, prostaglandins, lipoxins*

In tendon and ligament homeostasis, cytokines play an important role in terms of cellular activity and differentiation. These small proteins are involved in matrix synthesis and exert their effects already at very low concentrations ($10^{-12}$ molar). After an acute tendon or ligament injury, proinflammatory cytokines are released by proinflammatory macrophages (M1) that are otherwise not found in healthy tissue.

Typically, interleukin-1-α (IL-1α) and IL-1β, TNF-α, and interferon gamma (IFN-γ) are detected at elevated (Dakin et al., 2014). Furthermore, there is a change

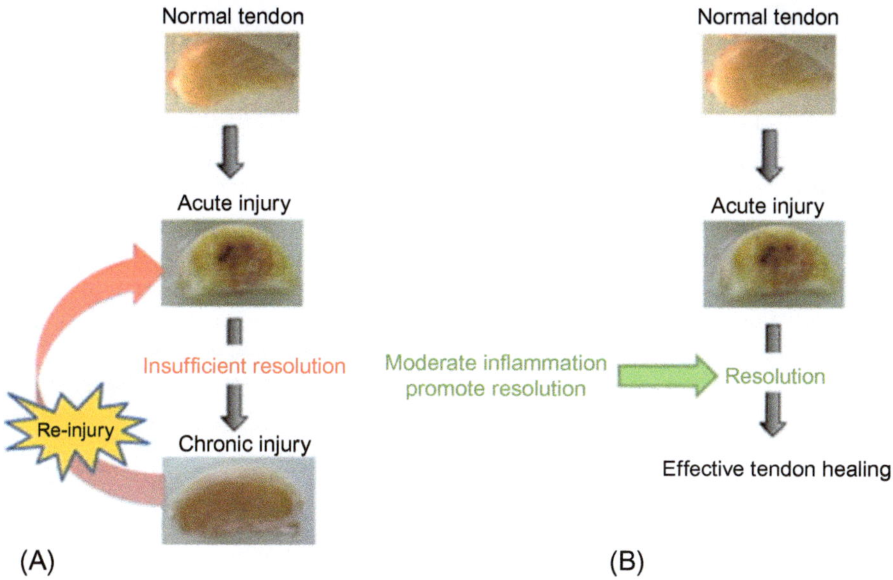


**Fig. 11.2** Schematic to propose the relationship between inflammation and resolution in the development of tendon injury. (A) In early stage injury, inflammation triggers a tendon resolution response, which appears to be transient and reduces with age and time after injury. During the later stages of healing (chronic injury), we propose that insufficient or dysregulated resolution allows low-level inflammation to persist, increasing the propensity for fibrotic healing and reinjury. (B) To improve the healing response of tendon, a potential therapeutic strategy is to moderate inflammation whilst simultaneously enhancing the tendons resolution response. From Dakin, S.G., Dudhia, J., Smith, R.K.W., 2014. Resolving an inflammatory concept: the importance of inflammation and resolution in tendinopathy. Vet. Immunol. Immunopathol. 158, 121–127, reprinted from Veterinary Immunology and Immunopathology, Copyright 2014, with permission from Elsevier.

in prostaglandin levels as a response to injury. Prostaglandins are arachidonic acid-based molecules. Arachidonic acid is derived from the cell membrane phospholipid bilayer by phospholipase after an injury and also in response to proinflammatory IL-1β (Tsuzaki et al., 2003; Yang et al., 2005). Cyclooxygenase-1 (COX-1) and COX-2 are involved in the synthesis of prostaglandins from arachidonic acid. From the three existing series of prostaglandins, series 2 prostaglandin, $PGE_2$, is more biologically active than others. The release of $PGE_2$ during tendon healing has adverse as well as positive effects; it has degradative effects via increased expression of metalloproteinases MMP-1 and MMP-3 (Thampatty et al., 2007; Manning et al., 2014). However, when injected into a rat patellar tendon compared to saline-injected tendons, it improved biomechanical properties of the regenerated tendon (Ferry et al., 2007). The normal prostaglandin production by tenocytes is induced by repetitive motion (Zhang and Wang, 2010; Dakin et al., 2012). Therefore, higher mechanical strength associated with higher prostaglandin levels is not surprising.

### 11.4.2 The end of inflammation

It has to be noted that returning back to the normal homeostatic state of tendons and ligaments after injury is not just a cessation of inflammatory cells and a decline of their secreted cytokines; however, it is a highly regulated process. The resolution of inflammation is induced by proresolving mediators such as lipoxins. For example, lipoxin A4 ($LXA_4$) is acting upon receptor FPR2/ALX (Ye et al., 2009). This receptor is expressed by inflammatory cells, especially monocytes and macrophages. As such, expression of $LXA_4$ is important to control the duration and also magnitude of the inflammatory cellular response. If the inflammatory response is not adequately switched off by such proresolving actors, chronic inflammation may develop. The formation of a fibrogenic scar tissue is an unwanted consequence thereof. As it has been shown that receptor FPR2/ALX is reduced with ageing, development of reruptures after tendon injuries in elderly persons is found because the resolution of inflammatory response cannot be performed sufficiently (Dakin et al., 2012).

## 11.5 Healing patterns of specific tendons and ligaments

In order to apply growth factors for biological tendon and ligament therapy, challenging questions have to be answered. Which growth factor is the most effective during natural wound healing; is it a single one or a combination of growth factors that direct an orchestrate the proper healing? Moreover, as there is a time-dependent occurrence of growth factors in different concentrations, upregulations of growth factors on the gene- as well as on the protein-level as a function of time postlaceration have to be exactly elucidated (Gulotta and Rodeo, 2009). Finally, the question about the choice of the appropriate delivery device to have a sustained release and an adequate release kinetics of the according proteins (Lavin et al., 2013) remains to be discussed and has to be tuned and matched according to the desired healing pattern.

### 11.5.1 Flexor tendon

In a study with chickens, the *flexor digitorum profundi* were completely transected and harvested at early time points after surgical repair. Gene expression of growth factors by polymerase chain reaction and production of the corresponding proteins by immunohistochemical staining were studied in addition to healing status examination by histology (Fig. 11.3) (Chen et al., 2008).

It was shown that gene expression of TGF-β was high or moderate during the first 3 weeks postsurgery which was confirmed on the protein level (Chen et al., 2008). Flexor tendon adhesion formation has been linked to TGF-β; with TGF-β1 being expressed early and vanishing after 3 weeks and TGF-β3 coming only around 3 weeks postinjury and remaining thereafter (Juneja et al., 2013). While CTGF, VEGF, and insulin-like growth factor-I (IGF-1) were also expressed moderately high during the first 3 weeks like TGF-β, this was not the case for platelet-derived growth factor B (PDGF-B) and neither for bFGF, which was only found to be minimally

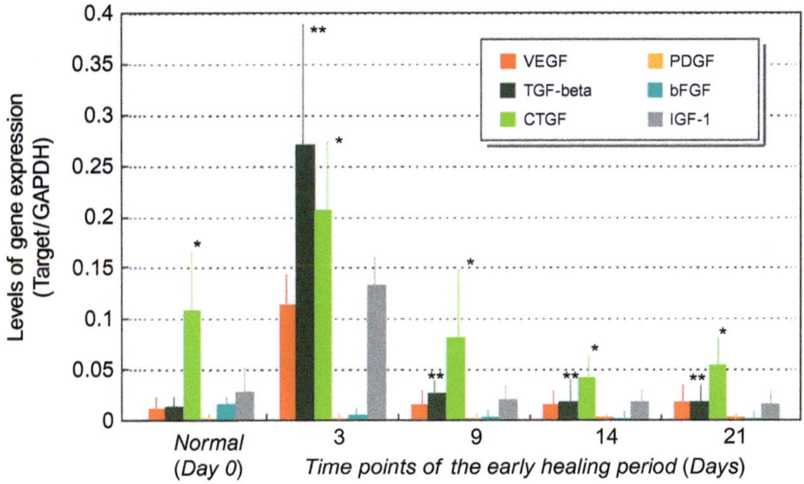


**Fig. 11.3** Levels of expression of 6 growth factor genes in normal and healing flexor tendons in chicken. Among 6 growth factors, the connective tissue growth factor (CTGF) gene expressed at the highest level in normal tendons and in healing tendons. The transforming growth factor TGF-gene expressed at the highest levels at day 3 and expressed at high or moderately high levels at other healing time points. Asterisks (* or **) indicate CTGF and TGF-gene having expression levels significantly higher than platelet-derived growth factor B (PDGF-B) and basic fibroblastic growth factor (bFGF) genes in the same days. The PDGF-B gene expressed at the lowest level in normal tendons and at postsurgical days 3 and 9. At days 14 and 21, bFGF gene expression was the lowest of these growth factors. Expression of genes of all growth factors except bFGF increased significantly at 3 days compared with normal tendons.
From Chen, C.H., Cao, Y., Wu, Y.F., Bais, A.J., GAO, J.S., Tang, J.B., 2008. Tendon healing in vivo: gene expression and production of multiple growth factors in early tendon healing period. J. Hand Surg. 33, 1834–1842, reprinted from The Journal of Hand Surgery, Copyright 2008, with permission from Elsevier.

increased at 3 days postsurgery (Chen et al., 2008). The minimum gene expression of PDGF-B, however, stands in contrast with other flexor healing studies. For example, Duffy and coworkers reported that in a canine tendon model, where the flexor tendons were transected between the second and fourth annular pulleys and subsequently repaired, at 3, 10, and 17 days postsurgery, PDGF protein (as well as epidermal growth factor EGF protein) was expressed in larger levels than in healthy tendons (Duffy et al., 1995). As for bFGF, which was also reported to be expressed in low levels at early time points postoperation (Chen et al., 2008), this growth factor nevertheless has been shown to increase rabbit synovial tenocytes' proliferation and enhance collagen production by activating the nuclear factor kappa B (NF-κB) pathway including a significant upregulation of NF-κB-inducing kinase (NIK), inhibitor of kappa B kinase alpha (IKKα) and beta (IKKβ), and NF-kB on the gene and protein level (Tang et al., 2004).

The pattern of upregulated genes and expressed proteins at different time points, however, may also depend on the animal model chosen. In a horse model, the temporal expression of IGF-I, TGF-β, and collagen types I and III in flexor lesions was investigated (Dahlgren et al., 2005). In accordance with the chicken flexor study mentioned earlier, the moderately increased levels of IGF-1 were also detected. It was found that the temporal changes in IGF-1 peaked at 4 weeks postsurgery at the RNA-level as well as on the protein-level; as for the TGF-β1, RNA-expression was highest 1 week postlaceration, while protein expression was highest at 4 weeks (Dahlgren et al., 2005).

The roles and function of those timely induced growth factor cascades have been investigated widely (Hsu and Chang, 2004). Vascular endothelial growth factor (VEGF) has been found to be negligibly expressed in human healthy ATs; however, when they were injured, higher levels of VEGF have been detected (Pufe et al., 2001). During canine flexor tendon healing, the levels of VEGF mRNA expression peaked at 7 until 14 days and then returned to baseline values (Boyer et al., 2001); interestingly, predominantly tenocytes of the core tendon tissue expressed VEGF (around 70%) and not tenocytes migrating from the epitenon to the wound site, which implicates that the core cells are responsible for inducing early angiogenesis at the repair site (Bidder et al., 2000). Insulin-like growth factor (IGF) has been reported to be expressed in healthy flexor tendons as well as in injured ones, the main role being based on its mitotic activity and collagen synthesis (Abrahamsson, 1997). Furthermore, IGF was found to stimulate synthesis of DNA, and proteoglycan, as well as cell proliferation in healthy rabbit flexor tendons (Abrahamsson et al., 1991). IGF is induced by PDGF, the platelet-derived growth factor, which is upregulated in injured tendons compared to healthy tendons. Its main function in early tendon healing relies on its capacity to enhance fibroblast proliferation and differentiation, collagen production, and angiogenesis (Deuel et al., 1991).

### 11.5.2 Achilles tendon

As found for many other tendons, the healing AT often turns out in a fibrovascular tissue that is biomechanically weaker compared to normal healthy tissues—having consequences as pain and rerupture. There are many attempts to face this problem by applying biological therapy, among them platelet rich plasma, bone marrow aspirate, bone morphogenic proteins BMP-2, -12, and -14 as well as stem cells, sometimes supported by scaffold materials (Shapiro et al., 2015). Besides these biological approaches, also mechanical therapies such as early mechanical loading have been reported to lead to an approved AT healing in the rabbit model (Wang et al., 2015). It was shown that early mechanical loading (the rabbits were allowed to move after operation and reattachment of the detached AT-bone interface) resulted in a significantly higher ultimate stress 4 weeks postoperation compared to immobilized ATs. Moreover, also from a histological point of view, collagen fiber were much better organized in the nonimmobilized group (Wang et al., 2015).

The healing of fully transected rabbit ATs was also investigated by ultrasound and correlated to the histological findings (Buschmann et al., 2014). It was shown that cellular composition in the healing tendon tissue between 3 and 6 weeks postoperation (a conventional 4-strand Becker suture) changed significantly; while in healthy intact tendons, the cell density in the endo- and epitenon was similar and significantly higher compared to the tendon tissue in between which is generally very low populated by tenoblasts and tenocytes, the ratio between the cell densities in the endo-to-epitenon was reduced to around 1/3 at 3 weeks and to 1/6 at 6 weeks postsurgery, respectively. In addition, while the endotenon cell densities were significantly lower compared to healthy tissue, the epitenon cell densities were significantly higher when compared to intact tissues. This implies that especially cells from the endotenon migrate to the wound site in order to promote healing—with the result of a transient depletion of cells in the endotenon. Also the collagen structure was much more disorganized compared to intact tendon tissue both at 3 and 6 weeks postsurgery, respectively, and this was pronounced especially in the granulation tissue.

Otherwise, also scoring systems to describe the status of tendon tissue during the healing were developed (Stoll et al., 2011) and applied to the healing rabbit AT. This system included inspection of extracellular matrix (ECM) organization of the whole repaired tendon (scores 0, 1, and 2), cellularity (scores 0, 1, and 2), cell alignment (scores 0, 1, and 2), cell nucleus morphology (scores 0, 1, and 2), cell distribution (scores 0 and 1), organization of the tendon callus (scores 0, 1, and 2), configuration of callus (scores 0 and 1 based on macroscopic observation of the specimen), integration of constructs to the normal tissue (scores 0, 1, and 2), vascularization (scores 0 and 1), metaplasia (scores 0, 1, 2, and 3), and features of inflammation (scores 0 and 1). The scores were summed up (maximum 20), with 0 scores standing for damaged tissue and 20 scores for healthy tissue. As can be seen in Fig. 11.4, there is a gradual increase in scores during a period up to 3 months; after this time, the tendon tissue still has not

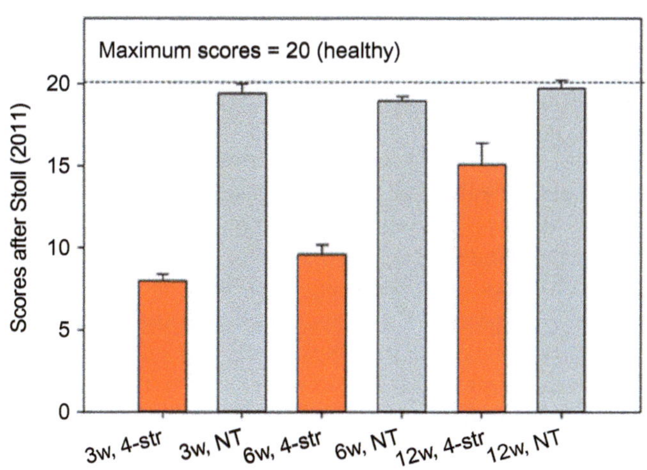


**Fig. 11.4** Rabbit Achilles tendon (AT) histopathological analysis. Scores determined after a system developed by Stoll and coworkers (2011). Key *4-str*, 4-strand Becker sutured ATs after full transection; *NT*, no treatment (healthy counter legs).

fully regenerated; however, when looking at which aspects are not reaching the full set of points, these are aspects concerning primarily the cells—such as the nucleus morphology, the tenocyte distribution, the cell alignment, and the cell-to-matrix ratio as well as ECM organization did often reach only 1 point from a range of 0–2 points, thereby lowering the sum of scores to around 15 points (from a maximum of 20) even 12 weeks postoperation.

### 11.5.3 Anterior cruciate ligament

The anterior cruciate ligament is one of the most vulnerable ligaments of the knee. Therefore, many attempts such as autografts, xenografts, tissue-engineered constructs (Peach et al., 2012), with or without cells (Mifune et al., 2012), and growth factors (Kim et al., 2011), among others, have been investigated and reviewed (Shirazi et al., 2015; Oryan et al., 2013).

### 11.5.4 Rotator cuff

There are many approaches to address rotator cuff tears, such as application of decellularized tendon grafts (Pan et al., 2014), bioreactor-based ligament and tendon tissue engineering (Wang et al., 2013), or growth factor application to enhance biomechanical properties at early time points as investigated for PDGF-BB in a rat supraspinatus model (Kovacevic et al., 2015). To support the healing of rotator cuff tears by growth factor application, it is important to know the time-dependent concentrations during the healing process. Therefore, temporal changes in growth factors were studied in a rat model undergoing bilateral supraspinatus detachment and repair (Wuergler-Hauri et al., 2007). It was found for the midsubstance that several growth factors significantly increased compared to normal healthy tendon, among them bFGF, BMP-12, -13, and -14, COMP, CTGF, PDGF-B, and TGF-β1 (Fig. 11.5 according to Table II, scores for intensity of staining).

Also the rabbit supraspinatus has been explored in terms of growth factor expression in the early healing phase (Kobayashi et al., 2006). At 1, 3, 5, 7, 9, 11, 14, 21, and 28 days postoperation, immunohistochemical assessment of four growth factors was evaluated (Fig. 11.6). It was found that bFGF peaked in the midsubstance at 7–9 days postoperation; for IGF-1, it was between days 3 and 5; however, at later time points (7–14 days), there was still an upregulation towards the proximal side. Moreover, PDGF was found to be highest between day 7 and 9, with high levels remaining towards the proximal side up to 14 days. Finally, TGF-β was expressed throughout the observed time frame, although the midsubstance was clearly stained between 7 and 14 days; before and after that time, it was more towards the proximal end of the tendon where TGF-β was expressed (Kobayashi et al., 2006).

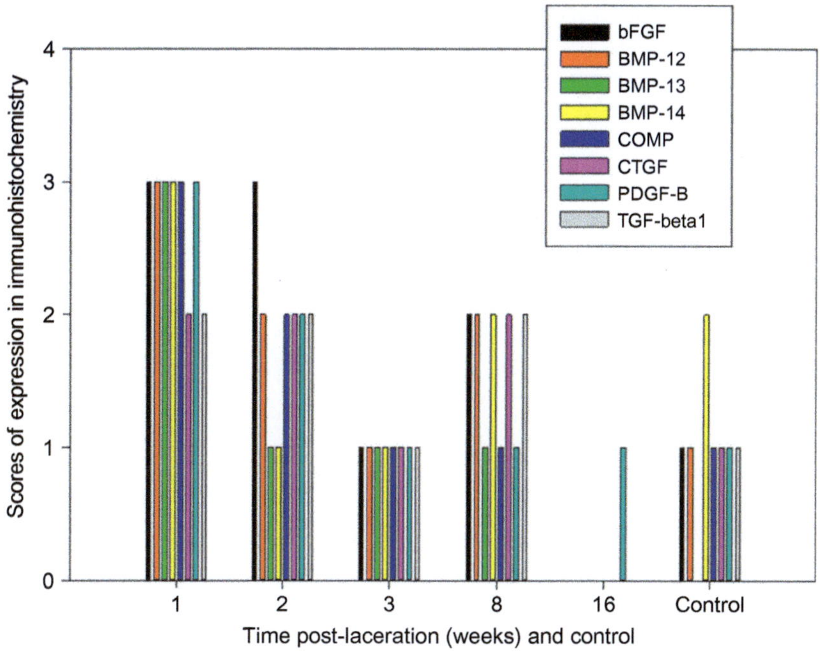


**Fig. 11.5** Growth factors in the midsubstance of repaired supraspinatus tendon in comparison with healthy tendon in a rat model. bFGF, basic fibroblastic growth factor; *BMP*, bone morphogenic protein; *COMP*, cartilage oligomeric matrix protein; *CTGF*, connective tissue growth factor; *PDGF-B*, platelet-derived growth factor B; TGF-β1, transforming growth factor beta1.
Data from Wuergler-Hauri, C.C., Dourte, L.M., Baradet, T.C., Williams, G.R., Soslowsky, L.J., 2007. Temporal expression of 8 growth factors in tendon-to-bone healing in a rat supraspinatus model. J. Shoulder Elb. Surg. 16, 198S–203S.

## 11.6 Summary

In this chapter, tendon and ligament healing processes are described in terms of cellular infiltration and action at the wound site. After acute injuries, an inflammatory response starts immediately, including proinflammatory macrophages that release IL-1 and TNF-α. Tenoblast and tenocyte migration follow and lymphocytes initiating the end of the inflammatory reaction do also occur. During the remodelling phase, collagen III levels, glycosaminoglycans, and water are usually high and decline after around 6 weeks, where the consolidation stage with collagen fiber alignment and a decrease in cellularity are characteristic. Fibrous scar-like tissue may result instead of truly regenerated tendon tissue, which is associated with an insufficient termination of the inflammatory cellular response—often ending up in reruptures caused by overuse of the biomechanically weaker tendons and ligaments.

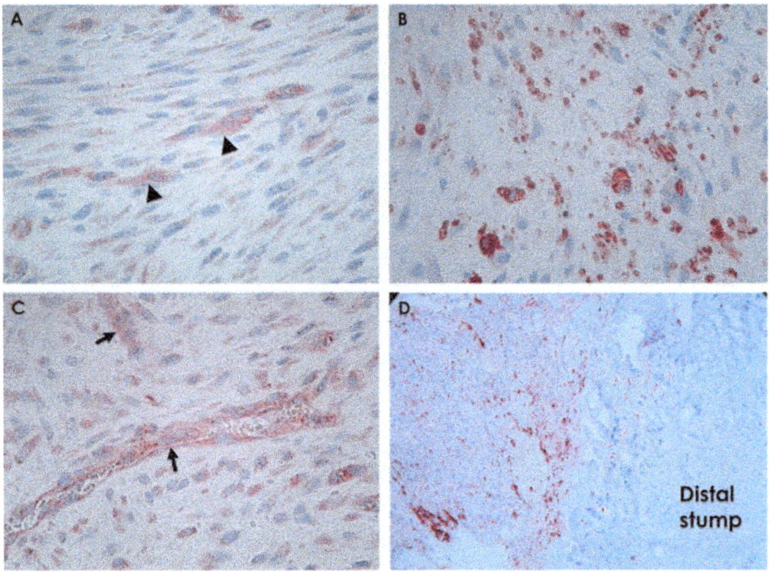


**Fig. 11.6** Growth factors-expression in supraspinatus tendon in a rabbit model. Immunohistochemical staining: (A) Basic fibroblast growth factor (bFGF). Positive-staining fibroblast-like cells are shown as red staining in cytoplasm (arrowheads) (original magnification: ×200). (B) IGF-1. Positive-staining blood cells are shown as dark red staining (original magnification: ×200). (C) PDGF. Positive-staining vascular endothelial cells are shown as red staining (arrows) (original magnification: ×200). (D) TGF-β. Positive-staining blood cells are shown as red staining at distal area of defect. Negative-staining area is the distal stump of the tendon.
Original magnification: ×20, from Kobayashi, M., Itoi, E., Minagawa, H., Miyakoshi, N., Takahashi, S., Tuoheti, Y., Okada, K., Shimada, Y., 2006. Expression of growth factors in the early phase of supraspinatus tendon healing in rabbits. J. Shoulder Elb. Surg. 15, 371–377, © Reprinted from Journal of Shoulder and Elbow Surgery, with permission from Elsevier.

# References

Abrahamsson, S.O., 1997. Similar effects of recombinant human insulin-like growth factor-I and II on cellular activities in flexor tendons of young rabbits: experimental studies in vitro. J. Orthop. Res. 15, 256–262.

Abrahamsson, S.O., Lundborg, G., Lohmander, L.S., 1991. Recombinant human insulin-like growth factor-I stimulates in vitro matrix synthesis and cell-proliferation in rabbit flexor tendon. J. Orthop. Res. 9, 495–502.

Beredjiklian, P.K., Favata, M., Cartmell, J.S., Flanagan, C.L., Crombleholme, T.M., Soslowsky, L.J., 2003. Regenerative versus reparative healing in tendon: a study of biomechanical and histological properties in fetal sheep. Ann. Biomed. Eng. 31, 1143–1152.

Bidder, M., Towler, D.A., Gelberman, R.H., Boyer, M.I., 2000. Expression of mRNA for vascular endothelial growth factor at the repair site of healing canine flexor tendon. J. Orthop. Res. 18, 247–252.

Boyer, M.I., Watson, J.T., Lou, J.R., Manske, P.R., Gelberman, R.H., Cai, S.R., 2001. Quantitative variation in vascular endothelial growth factor mRNA expression during early flexor tendon healing: an investigation in a canine model. J. Orthop. Res. 19, 869–872.

Buschmann, J., Puippe, G., Meier Buergisser, G., Bonavoglia, E., V, P., Calcagni, M., 2014. Correspondence of high-frequency ultrasound and histomorphometry of healing rabbit Achilles tendon tissue. Connect. Tissue Res. 55, 123–131.

Cadby, J.A., Buehler, E., Godbout, C., Van Weeren, P.R., Snedeker, J.G., 2014. Differences between the cell populations from the peritenon and the tendon core with regard to their potential implication in tendon repair. PLoS One. 9(3). e92474.

Carpenter, J.E., Thomopoulos, S., Flanagan, C.L., Debano, C.M., Soslowsky, L.J., 1998. Rotator cuff defect healing: a biomechanical and histologic analysis in an animal model. J. Shoulder Elb. Surg. 7, 599–605.

Chen, C.H., Cao, Y., Wu, Y.F., Bais, A.J., Gao, J.S., Tang, J.B., 2008. Tendon healing in vivo: gene expression and production of multiple growth factors in early tendon healing period. J. Hand Surg. 33, 1834–1842.

Dahlgren, L.A., Mohammed, H.O., Nixon, A.J., 2005. Temporal expression of growth factors and matrix molecules in healing tendon lesions. J. Orthop. Res. 23, 84–92.

Dakin, S.G., Dudhia, J., Werling, N.J., Werling, D., Abayasekara, D.R.E., Smith, R.K.W., 2012. Inflamm-aging and arachidonic acid metabolite differences with stage of tendon disease. PLoS One. 7(11). e48978.

Dakin, S.G., Dudhia, J., Smith, R.K.W., 2014. Resolving an inflammatory concept: the importance of inflammation and resolution in tendinopathy. Vet. Immunol. Immunopathol. 158, 121–127.

Deuel, T.F., Kawahara, R.S., Mustoe, T.A., Pierce, G.F., 1991. Growth-factors and wound-healing—platelet-derived growth-factor as a model cytokine. Annu. Rev. Med. 42, 567–584.

Duffy Jr., F.J., Seiler, J.G., Gelberman, R.H., Hergrueter, C.A., 1995. Growth factors and canine flexor tendon healing: Initial studies in uninjured and repair models. J. Hand Surg. 20, 645–649.

Elliot, D., Giesen, T., 2013. Primary flexor tendon surgery: the search for a perfect result. Hand Clin. 29, 191–206.

Favata, M., Beredjiklian, P.K., Zgonis, M.H., Beason, D.P., Crombleholme, T.M., Jawad, A.F., Soslowsky, L.J., 2006. Regenerative properties of fetal sheep tendon are not adversely affected by transplantation into an adult environment. J. Orthop. Res. 24, 2124–2132.

Ferry, S.T., Dahners, L.E., Afshari, H.M., Weinhold, P.S., 2007. The effects of common anti-inflammatory drugs on the healing rat patellar tendon. Am. J. Sports Med. 35, 1326–1333.

Fujita, M., Hukuda, S., Doida, Y., 1992. Experimental study of intrinsic healing of the flexor tendon collagen synthesis of the cultured flexor tendon cells of the canine. J. Jpn. Orthop. Assoc. 66, 326–333.

Galatz, L.M., Gerstenfeld, L., Heber-Katz, E., Rodeo, S.A., 2015. Tendon regeneration and scar formation: the concept of scarless healing. J. Orthop. Res. 33, 823–831.

Gelberman, R.H., Manske, P.R., Akeson, W.H., Woo, S.L.Y., Lundborg, G., Amiel, D., 1986. Kappa-delta award paper—flexor tendon repair. J. Orthop. Res. 4, 119–128.

Gulotta, L.V., Rodeo, S.A., 2009. Growth factors for rotator cuff repair. Clin. Sports Med. 28, 13–23.

Hsu, C., Chang, J., 2004. Clinical implications of growth factors in flexor tendon wound healing. J. Hand Surg. 29, 551–563.

Huegle, T., Majewski, M., Valderrabano, V., 2012. Zell- und molekularbiologische Abläufe bei der Heilung von Sehnenverletzungen. Schweizerische Zeitschrift für "Sportmedizin und Sporttraumatologie" 60, 49–51.

Juneja, S.C., Schwarz, E.M., O'Keefe, R.J., Awad, H.A., 2013. Cellular and molecular factors in flexor tendon repair and adhesions: a histological and gene expression analysis. Connect. Tissue Res. 54, 218–226.

Kim, H.M., Galatz, L.M., Das, R., Havlioglu, N., Rothermich, S.Y., Thomopoulos, S., 2011. The role of transforming growth factor beta isoforms in tendon-to-bone healing. Connect. Tissue Res. 52, 87–98.

Kobayashi, M., Itoi, E., Minagawa, H., Miyakoshi, N., Takahashi, S., Tuoheti, Y., Okada, K., Shimada, Y., 2006. Expression of growth factors in the early phase of supraspinatus tendon healing in rabbits. J. Shoulder Elb. Surg. 15, 371–377.

Koh, T.J., Dipietro, L.A., 2011. Inflammation and wound healing: the role of the macrophage. Expert Rev. Mol. Med. 13e23.

Koob, T.J., 2002. Biomimetic approaches to tendon repair. In: Annual Meeting of the Society-for-Integrative-and-Comparative-Biology, Jan 04 2002 Anaheim, California. Elsevier Science Inc., pp. 1171–1192.

Kovacevic, D., Gulotta, L.V., Ying, L., Ehteshami, J.R., Deng, X.-H., Rodeo, S.A., 2015. rhPDGF-BB promotes early healing in a rat rotator cuff repair model. Clin. Orthop. Relat. Res. 473, 1644–1654.

Lavin, D.M., Zhang, L., Furtado, S., Hopkins, R.A., Mathiowitz, E., 2013. Effects of protein molecular weight on the intrinsic material properties and release kinetics of wet spun polymeric microfiber delivery systems. Acta Biomater. 9, 4569–4578.

Manning, C.N., Havlioglu, N., Knutsen, E., Sakiyama-Elbert, S.E., Silva, M.J., Thomopoulos, S., Gelberman, R.H., 2014. The early inflammatory response after flexor tendon healing: a gene expression and histological analysis. J. Orthop. Res. 32, 645–652.

Mifune, Y., Matsumoto, T., Ota, S., Nishimori, M., Usas, A., Kopf, S., Kuroda, R., Kurosaka, M., Fu, F.H., Huard, J., 2012. Therapeutic potential of anterior cruciate ligament-derived stem cells for anterior cruciate ligament reconstruction. Cell Transplant. 21, 1651–1665.

Oryan, A., Moshiri, A., Meimandi-Parizi, A., 2013. Graft selection in ACL reconstructive surgery: past, present, and future. Curr. Orthop. Pract. 24, 321–333.

Pan, J., Liu, G.-M., Ning, L.-J., Zhang, Y., Luo, J.-C., Huang, F.-G., Qin, T.-W., 2014. Rotator cuff repair using a decellularized tendon slices graft: an in vivo study in a rabbit model. Knee Surg. Sports Traumatol. Arthrosc. 23 (5), 1524–1535.

Peach, M.S., James, R., Toti, U.S., Deng, M., Morozowich, N.L., Allcock, H.R., Laurencin, C.T., Kumbar, S.G., 2012. Polyphosphazene functionalized polyester fiber matrices for tendon tissue engineering: in vitro evaluation with human mesenchymal stem cells. Biomed. Mater. 7 (4), 045016.

Pufe, T., Petersen, W., Tillmann, B., Mentlein, R., 2001. The angiogenic peptide vascular endothelial growth factor is expressed in foetal and ruptured tendons. Virchows Arch. 439, 579–585.

Schulze-Tanzil, G., Al-Sadi, O., Wiegand, E., Ertel, W., Busch, C., Kohl, B., Pufe, T., 2011. The role of pro-inflammatory and immunoregulatory cytokines in tendon healing and rupture: new insights. Scand. J. Med. Sci. Sports 21, 337–351.

Shapiro, E., Grande, D., Drakos, M., 2015. Biologics in Achilles tendon healing and repair: a review. Curr. Rev. Musculoskelet. Med. 8, 9–17.

Sharma, P., Maffulli, N., 2006. Biology of tendon injury: healing, modeling and remodeling. J. Musculoskelet. Neuronal Interact. 6, 181–190.

Shirazi, A.N., Chrzanowski, W., Khademhosseini, A., Dehghani, F., 2015. Anterior cruciate ligament: structure, injuries and regenerative treatments. In: Bertassoni, L.E., Coelho, P.G. (Eds.), Engineering Mineralized and Load Bearing Tissues. Springer International Publishing, Switzerland.

Stoll, C., John, T., Conrad, C., Lohan, A., Hondke, S., Ertel, W., Kaps, C., Endres, M., Sittinger, M., Ringe, J., Schulze-Tanzil, G., 2011. Healing parameters in a rabbit partial tendon defect following tenocyte/biomaterial implantation. Biomaterials 32, 4806–4815.

Tang, J.B., Xu, Y., Wang, X.T., 2004. Tendon healing in vitro: activation of NIK, IKK alpha, IKK beta, and NF-kappa B genes in signal pathway and proliferation of tenocytes. Plast. Reconstr. Surg. 113, 1703–1711.

Thampatty, B.P., Li, H., Im, H.-J., Wang, J.H.C., 2007. EP4 receptor regulates collagen type-I, MMP-1, and MMP-3 gene expression in human tendon fibroblasts in response to IL-1β treatment. Gene 386, 154–161.

Thomopoulos, S., Parks, W.C., Rifkin, D.B., Derwin, K.A., 2015. Mechanisms of tendon injury and repair. J. Orthop. Res. 33, 832–839.

Tsuzaki, M., Guyton, G., Garrett, W., Archambault, J.M., Herzog, W., Almekinders, L., Bynum, D., Yang, X., Banes, A.J., 2003. IL-1β induces COX2, MMP-1, -3 and -13, ADAMTS-4, IL-1β and IL-6 in human tendon cells. J. Orthop. Res. 21, 256–264.

Wang, T., Gardiner, B.S., Lin, Z., Rubenson, J., Kirk, T.B., Wang, A., Xu, J.K., Smith, D.W., Lloyd, D.G., Zheng, M.H., 2013. Bioreactor design for tendon/ligament engineering. Tissue Eng. Part B Rev. 19, 133–146.

Wang, J., Jiang, D., Wen, S., Jing, S., Fan, D., Hao, Z., Han, C., 2015. Improved Achilles tendon healing by early mechanical loading in a rabbit model. Int. J. Clin. Exp. Med. 8, 1144–1149.

Wuergler-Hauri, C.C., Dourte, L.M., Baradet, T.C., Williams, G.R., Soslowsky, L.J., 2007. Temporal expression of 8 growth factors in tendon-to-bone healing in a rat supraspinatus model. J. Shoulder Elb. Surg. 16, 198S–203S.

Yang, G., Im, H.-J., Wang, J.H.C., 2005. Repetitive mechanical stretching modulates IL-1β induced COX-2, MMP-1 expression, and PGE2 production in human patellar tendon fibroblasts. Gene 363, 166–172.

Ye, R.D., Boulay, F., Wang, J.M., Dahlgren, C., Gerard, C., Parmentier, M., Serhan, C.N., Murphy, P.M., 2009. International union of basic and clinical pharmacology. LXXIII. Nomenclature for the formyl peptide receptor (FPR) family. Pharmacol. Rev. 61, 119–161.

Yin, Y.Q., Henzl, M.T., Lorber, B., Nakazawa, T., Thomas, T.T., Jiang, F., Langer, R., Benowitz, L.I., 2006. Oncomodulin is a macrophage-derived signal for axon regeneration in retinal ganglion cells. Nat. Neurosci. 9, 843–852.

Zhang, J., Wang, J.H.C., 2010. Mechanobiological response of tendon stem cells: implications of tendon homeostasis and pathogenesis of tendinopathy. J. Orthop. Res. 28, 639–643.

Zhao, X., Jiang, S., Liu, S., Chen, S., Lin, Z.Y., Pan, G., He, F., Li, F., Fan, C., Cui, W., 2015. Optimization of intrinsic and extrinsic tendon healing through controllable water-soluble mitomycin-C release from electrospun fibers by mediating adhesion-related gene expression. Biomaterials 61, 61–74.

# Evolving treatments and emerging strategies for tendon and ligament reconstruction

# 12

## Abstract

The topic of emerging trends in a certain field of research is always bound to the actual situation. As books usually have quite a long production process, emerging trends in tendon and ligament healing in this chapter refer mostly to the situation of April 2016. At that time, it was found that reports and publications mostly dealt with combinations of earlier findings to improve tendon and ligament healing by the application of especially fabricated grafts with several features addressing antiadhesion, acceleration of healing processes, and stimulation of neo-angiogenesis, among others. Moreover, cellular approaches were being investigated where transduction approaches as well as the application of cell mixtures, platelet rich plasma, or whole blood were used and shown to have beneficial effects in terms of biomechanical properties of the regenerated tendons or ligaments.

**Keywords:** Coating, Core electrospinning, Transduction, Early active motion, Synergism.

## Abbreviations

| | |
|---|---|
| **3D** | three dimensions (technical feature of imaging techniques) |
| **ACL** | anterior cruciate ligament (tendons and ligaments) |
| **Ag** | silver (elemental metal) |
| **ASCs** | adipose-derived stem cells (cells for seeding in tissue engineering) |
| **AT** | Achilles tendon (tendons and ligaments) |
| **bFGF** | basic fibroblast growth factor (protein in living cells) |
| **BMP** | bone morphogenetic protein (element in tissues) |
| **BMSCs** | bone-marrow-derived stromal cells (cells for seeding in tissue engineering) |
| **CD** | cluster of differentiation (cell development) |
| **DiI** | fluorescent lipophilic cationic indocarbocyanine dye (dye used in biosciences) |
| **ECM** | extracellular matrix (matrix in living cells, natural graft material) |
| **FACS** | fluorescence-activated cell sorting (method in biotechnology) |
| **FDP** | flexor *digitorum profundus* (tendons and ligaments) |
| **FIB** | fibroblast (cells in humans and animals) |
| **FLT (-*n*)** | soluble fms-like tyrosine kinase-1, antagonist of VEGF (protein in living cells) |
| **G-CSF** | granulocyte-colony stimulation growth factor (protein in living cells) |
| **GDF (-*n*)** | growth/differentiation factor-5 (protein in living cells) |
| **GFP** | green fluorescent protein (labeling in imaging technique) |

**Biomechanics of Tendons and Ligaments. http://dx.doi.org/10.1016/B978-0-08-100489-0.00012-0**

| | |
|---|---|
| **HA** | hyaluronic acid (natural agent in tissue engineering) |
| **HFFs** | human foreskin fibroblasts (cells for seeding in tissue engineering) |
| **kDa** | kilo dalton (unity for protein size) |
| **MDCs** | muscle-derived cells (cells for seeding in tissue engineering) |
| **miR-218** | mikroRNA = ribonucleic acid-218 (element of living cells) |
| **MRL** | Murphy Roths Large (mouse strain) |
| **MSCs** | mesenchymal stem cells (cells for seeding in tissue engineering) |
| **NaCl** | natrium cloride (salt) |
| **p21** | cyclin-dependent kinase inhibitor (protein in living cells) |
| **PBS** | phosphate-buffered saline (physiological conditioned solution) |
| **PCL** | poly(ε-caprolactone) (synthetic scaffold material) |
| **PE** | polyethylene (synthetic scaffold material) |
| **PET** | polyethylene terephthalate (synthetic scaffold material) |
| **PGA** | poly(glycolide) acid (synthetic scaffold material) |
| **PRP** | platelet rich plasma (enriched blood plasma) |
| **PT** | patellar tendon (tendons and ligaments) |
| **ROM** | range of motion (measurement in biomechanics) |
| **SLRPs** | small leucine-rich proteoglycans (element of living cells) |
| **T&L** | tendon & ligaments (tendon and ligaments) |
| **TENO** | tenocytes (cells in tendon and ligaments) |
| **TOB** | transducer of ERbB2 (protein in living cells) |
| **VEGF** | vascular endothelial growth factor (protein in living cells) |

## 12.1 Introduction

A search for literature in PubMed in April 2016 using the keywords "tendon or ligament," "rec* or aug*," "biomech*" and restricted to the years 2015/2016 revealed a series of 592 publications (including books and full original papers). Interestingly, references dealing with the knee's tendons and ligaments were abundant and significantly higher in number compared to publications referring to shoulder, hand, wrist, or foot ankle/foot (Fig. 12.1).

However, preclinical as well as clinical literature was actually more a continuing cascade of research referring to former experiments or insights than truly novel ideas. We found that tissue engineering of tendon and ligaments was getting more specialized in terms of combinations of previous beneficial findings. For example, while cell seeding on different materials used as scaffolds had been widely examined before, an emerging trend was found to use transfected cells in combination with growth factors being released: a sophisticated scaffold material like a polymer coated with bioactive or at least biocompatible layers is fabricated, sometimes in gradients, releasing bioactive substances in a sustained way. Each component or cue had been established in a series of papers before, however, the "new" findings mostly revealed emerging synergistic features caused by a specific combination of these approaches. Moreover, superior animal models were getting established exhibiting better healing processes, what should help to understand the mechanisms standing behind successful tendon and ligament regeneration (Lalley et al., 2015) in order to circumvent the unwanted

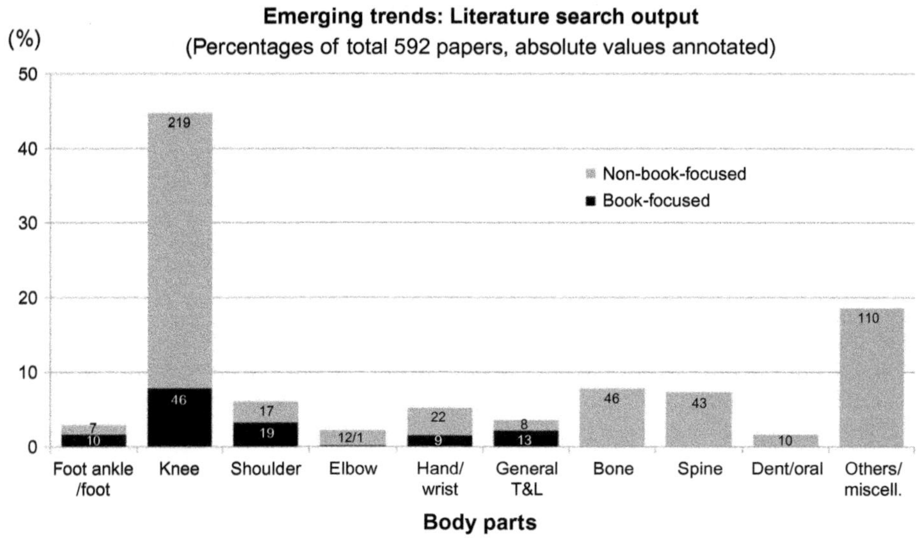


**Fig. 12.1** Results of literature search with percentage and absolute numbers, sorted by organ addressed (keywords used: "tendon or ligament," "rec* or aug*," "biomech*" for the years 2015/2016. *Non-book-focused*, no biomechanical measurements presented; *T&L*, tendon and ligaments; *miscell.*, miscellaneous.

scar formation which often goes along with poor biomechanics, in worst case leading to reruptures (Galatz et al., 2015). Finally, the best candidates to act as allografts are elicited with a special focus on time zero biomechanics (Qu et al., 2015) because the trend in using such grafts was to provide a full load-bearing capacity from the beginning, meaning immediately postoperation (Nau et al., 2015).

Nevertheless, although combinations of cues seem currently to be trendy, still studies on single components of agents adding to the benefit in tendon or ligament regeneration were being investigated, like taurine improving strength and reducing adhesion (Akdemir et al., 2015), granulocyte-colony stimulation growth factor to enhance regeneration (Buchmann et al., 2015), patch augmentation in the repair of rotator cuff tears (Flury, 2016), the preconditioning effect of grafts, and many more.

## 12.2 Coating, formulation, fabrication

When dealing with new trends in tendon and ligament repair approaches, there is often a combination of several features that have been established and shown to be beneficial when applied as single cues in earlier studies. This is also found with regard to the material used for tendon and ligament grafting. For example, artificial ligaments based on polyethylene which is as such an inert but mechanically very stable polymer have been modified by a layer-by-layer deposition of chitosan and hyaluronic acid (HA) with the intent to make it more biocompatible. Hence, collagen fiber and cell density were significantly enhanced (Li et al., 2013). Further polymer coatings consist of simvastatin/collagen coating, as realized with polyethylene terephthalate (PET) used for anterior cruciate ligament (ACL) reconstruction in rabbits (Zhang et al., 2016a). There, the graft-bone healing was significantly enhanced in the group receiving the simvastatin/collagen/PET graft compared to a collagen/PET group (no simvastatin). This was attributed to the slow release of simvastatin, a bone promoting agent which stimulated angio- and osteogenesis. Though not primarily enhancing the strength of the reconstructed ACL, nevertheless the interface between the bone and the ligament was strengthened significantly (Zhang et al., 2016a).

New formulations of cues to act as supporting means in tendon healing are also achieved in the combination of core-electrospun meshes, where in the core of the fibers a lubricant like HA could be incorporated; in addition, the sheath could be spiked with silver (Ag) (Chen et al., 2015). Such a highly sophisticated implant material was fabricated by Chen and coworkers (using polycaprolactone (PCL) as the shell of the hollow fibers) and was intended to act as an antiadhesive as well as an anti-inflammatory agent in a rabbit flexor tendon model. Hence, a flexor *digitorum profundus* (FDP) tendon rupture model of the hind paw was used which is similar in the flexion mechanism to a human digit. Three weeks postoperation, the range of motion (ROM) was significantly higher in the HA–PCL–Ag implant compared to the control (no implant); however, also less sophisticated combinations such as HA–PCL or only PCL acted quite well when the ROM was under view. Nevertheless, when the adhesion formation was measured by the pull-out force needed to separate the regenerated tendon from the surrounding tissue, the HA–PCL–Ag system was clearly the best (lowest force needed) and was the only combination that was statistically significantly lower than the control (Chen et al., 2015). Although biomechanical properties of the extracted tendons did not show any improvement, core-electrospinning might offer a good option to incorporate also other bioactive molecules than HA, especially when sustained release kinetics of the corresponding molecules is intended.

The use of decellularized matrices to provide a basis of extracellular matrix (ECM) in tendon or ligament grafts has been investigated widely (Ning et al., 2012; Fox et al., 2013; Ozasa et al., 2013; Pan et al., 2014). However, the impact of the fabrication processes needed to achieve decellularization on the properties of the resulting graft, especially on its biomechanical characteristics, had only been investigated lately (Herbert et al., 2015). Time zero mechanical properties are extremely important to know because immediate early postoperative stability prevents gap formation and rerupture, one of the two mostly occurring problems after tendon or ligament repair (Elliot and Giesen, 2013). The mechanical properties should be sufficient, and even a fully load-bearing construct was suggested to be necessary for ACL reconstruction (Nau et al., 2015); it was reported that in a study where the opinion of 300 orthopaedic surgeons was assessed about the necessary properties of an ACL-graft, the majority mentioned this need for an immediate mechanical integrity (Nau et al., 2015; Rathbone et al., 2012). Therefore, Herbert and coworkers examined the decellularization processes of a flexor tendon more closely. They determined the effect of fat reduction by acetone, chloroform, or methanol and the sterilization by antibiotics and peracetic acid on the biomechanics of low strained porcine flexor tendons (Herbert et al., 2015). It was found that a combination of acetone and antibiotics performed best: although classical material properties were not affected (ultimate stress, elastic modulus), the differences were found in the early phase of loading—with the acetone-antibiotics group exhibiting the best transition strains and transition stresses (Herbert et al., 2015). The comparison of different bioprocessing steps leading to the desired product may be judged as an emerging trend and as literature on this topic is currently still rare, it is a field where improvements are needed.

## 12.3 New cellular approaches

As shown in different reports in the past, the use of cells to enhance biocompatibility and ECM production in artificial grafts is still under development. New trends consist in the fact that cells are modified before being seeded and applied at the corresponding reconstruction site. For example, adipose-derived stem cells (ASCs) were infected with Runx-2 adenovirus in order to promote and support the tendon-to-bone healing in ACL reconstruction in the rabbit model. Moreover, a slow releasing fibrin glue matrix was used in order to provide the Runx-2-ASCs in a sustained manner. Especially at earlier time points of the series of postoperative biomechanical analysis, the load until failure was significantly higher when Runx-2-ASCs were supplied compared to the fibrin glue alone treatment or compared to the application of the cells at once; however, at time point 52 weeks postsurgery, all three groups turned out to have the same biomechanics (Zhang et al., 2016b).

Although the choice of cell type used for tendon tissue repair usually falls upon stem cells with their inherent capability to differentiate into tenocytes among others, also adult fibroblasts/fibrocytes and tenocytes (specialized fibrocytes) are being applied in tendon tissue engineering. However, with regard to tendon tissue engineering, a thorough comparison of cell types including also muscle-derived cells (MDCs) has only been reported recently. Chen and coworkers looked for the optimal cell source in terms of biomechanics when a cord-shaped polyglycolic acid (PGA) scaffold had been chosen as a tendon scaffold material (Chen et al., 2016). First, cell-seeded grafts (murine tenocytes, fibroblasts, or MDCs) were seeded and cultivated for 2 weeks in vitro. Then, such prepared constructs were implanted subcutaneously in the back of nude mice and were sutured to *muscularis fascia* of the dorsum in order to be exposed there to mechanical loading evoked by muscle movement during 12 weeks in vivo. Not only were the MDCs found to be faster in proliferation in vitro, but also did they lead to a higher ultimate stress, elastic modulus, and collagen fiber diameter in vivo, when they were compared to the tenocyte-seeded PGA graft or the fibroblast analogue (Fig. 12.2) (Chen et al., 2016). Although critical readers may argue that the surrounding tissue at the implant site did not reflect the "real" situation encountered by a tendon graft in these experiments, it is nevertheless noteworthy to emphasize that the two most important biomechanical material characteristics were significantly higher with $p < 0.01$ when MDCs were used compared to fibroblasts or tenocytes.

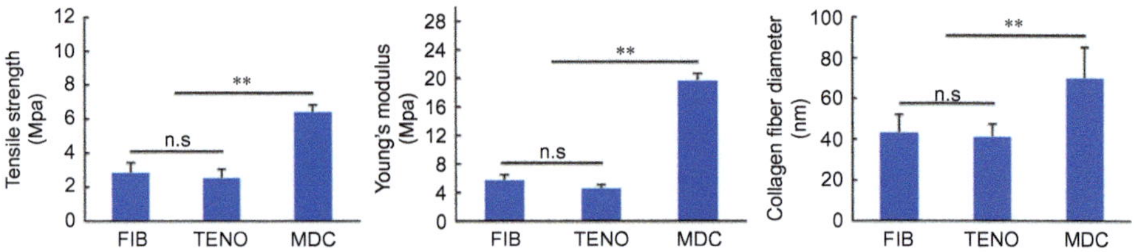


**Fig. 12.2** Quantitative analyses of engineered tendon on tensile strength (left), Young's modulus (center), and collagen fiber diameter (right), after 12 weeks in vivo (nude mouse model, scaffold subcutaneously sutured to *muscularis fascia* of the dorsum). **Significant difference with $p < 0.01$. *FIB*, fibroblast; *TENO*, tenocyte; *MDCs*, muscle-derived cells; *n.s.*, no significant difference between two groups.
From Chen, B., Ding, J., Zhang, W., Zhou, G., Cao, Y., Liu, W., Wang, B., 2016. Tissue engineering of tendons: a comparison of muscle-derived cells, tenocytes, and dermal fibroblasts as cell sources. Plast. Reconstr. Surg. 137, 536e–544e, with permission from the publisher Wolters Kluwer Health.

Another trend emerging is to use not only one cell type but cocultures or even a mixture of more cell types. For example, a convincing approach that led to an accelerated rehabilitation in a rat mid-substance Achilles tendon (AT) defect was based on a tendon hydrogel that was applied alone or with platelet rich plasma (PRP), with ASCs or with both, PRP+ASCs (Chiou et al., 2015). ASCs with bioluminescence positive luciferase and labeled with green fluorescent protein (ASC-luc2-eGFP) demonstrated a significantly increased total photon flux for hydrogel+PRP+ASCs compared to hydrogel+ASCs only, indicating the positive effect of the PRP. Moreover, biomechanical assessments at 8 weeks showed a significantly higher ultimate load in the hydrogel+PRP+ASCs group when compared to the control (saline). Hence, two easily accessible autologous sources, blood to extract PRP and fat for the isolation of ASCs, proved to be promising cellular approaches in this tendon hydrogel system (Chiou et al., 2015).

PRP has also been reported to be beneficial when applied with the support of a bovine ECM in a full transection ACL mini-pig model (Fleming et al., 2015). Importantly, the concentration of the applied PRP played a pivotal role in the bioenhancement; only the physiological concentration of PRP (and not the threefold nor the fivefold concentrations) led to a better outcome in terms of cartilage damage, meaning less cartilage damage. Also the stiffness and the ultimate load were highest when PRP was applied physiologically and not at higher concentrations (Fleming et al., 2015).

In contrast to the beneficial mixture of ASCs with PRP, the combination of ASCs with whole blood has been reported not to affect the biomechanical properties of repaired ACLs. An ECM-based matrix was saturated with whole blood, and either stem cells from adipose tissue (the retropatellar fat pad) or from peripheral blood were used for ACL reconstruction in a mini-pig model (Proffen et al., 2015). After 15 weeks, the groups including both, whole blood and stem cells, did not lead to any biomechanical enhancement when compared to the group receiving the ECM-based matrix with whole blood only (Proffen et al., 2015).

Furthermore, PRP has been shown to be beneficial when applied in combination with an electrospun collagen I, which was further polymerized with acid-solubilized collagen molecules at 4°C for 2 days to produce a 3D collagen gel (Oryan et al., 2015). The inactive pulverized PRP was then mixed with sterile phosphate-buffered saline and fully absorbed by the dry collagen gel. Thrombin acted as platelet activator. Such collagen-PRP-constructs were then applied in a 2 cm full AT defect in rabbits; in addition to this, collagen-PRP-group collagen gel without PRP (collagen implant only), no construct (shame-operation control), and native AT (normal) were tested. After 60 days, biomechanics revealed an improved outcome in the collagen-PRP-group compared to the other groups (Fig. 12.3). Not only structural properties (ultimate load and stiffness) were significantly higher than in the two other groups (collagen implant only or control), but also the material properties (ultimate stress and elastic modulus) were significantly enhanced, although they reached only around one-third of the native values at that time point (Oryan et al., 2015). Nevertheless, this approach seems promising and should be investigated further in a clinical trial.

Other cellular approaches including growth factors have been reported; however, the application of cells is not always successful even if the cell type is similar to tendon cells. For example, human foreskin fibroblasts (HFFs) were applied together with basic fibroblast growth factor (bFGF) on an electrospun PCL fiber mesh coated with collagen in order to act as a ACL-graft in nude rats (Leong et al., 2015b). For the fabrication of a 0.6 mm thick scaffold, four electrospun meshes were stacked. It was shown that the HFFs did not survive at 8 weeks postimplantation and, more importantly, at 16 weeks postsurgery the graft including bFGF only (no HFFs), led to a better biomechanical outcome compared to the cell-seeded grafts. In addition, the stiffness of the acellular graft was twice as high as that of the cell-seeded one and the peak load around 1.5 times higher. There was no beneficial effect of the HFFs; in contrast, the presence of those fibroblasts reduced biomechanics. It may be speculated that these fibroblasts supported scar formation rather than a true regenerative ligament healing (Galatz et al., 2015). A subsequent study by Leong and coworkers showed the positive effect of choosing a higher-molecular-weight PCL-polymer. Instead of the 80 kDa PCL used in the study mentioned before (Leong et al., 2015b), a 500 kDa PCL was applied; an electrospun mesh coated by collagen was likewise fabricated for rat ACL reconstruction (Leong et al., 2015a). Sixteen weeks post-operation, the high-molecular-weight PCL mesh resulted in better biomechanical properties of the neo-ligament compared to the lower molecular weight PCL, although by that time point the PCL had already degraded! It was found that the newly formed ligament had better aligned fibers, which resulted in a significantly higher ultimate load compared to the low molecular weight PCL ($25 \pm 4$N compared to $15 \pm 3$N, respectively) (Leong et al., 2015a).

The combination of cells, growth factor, and lubricant has been shown to improve digit function in an immobilized flexor tendon model in dogs (Zhao et al., 2016). In this study, bone-marrow-derived stromal cells (BMSCs) in combination with growth

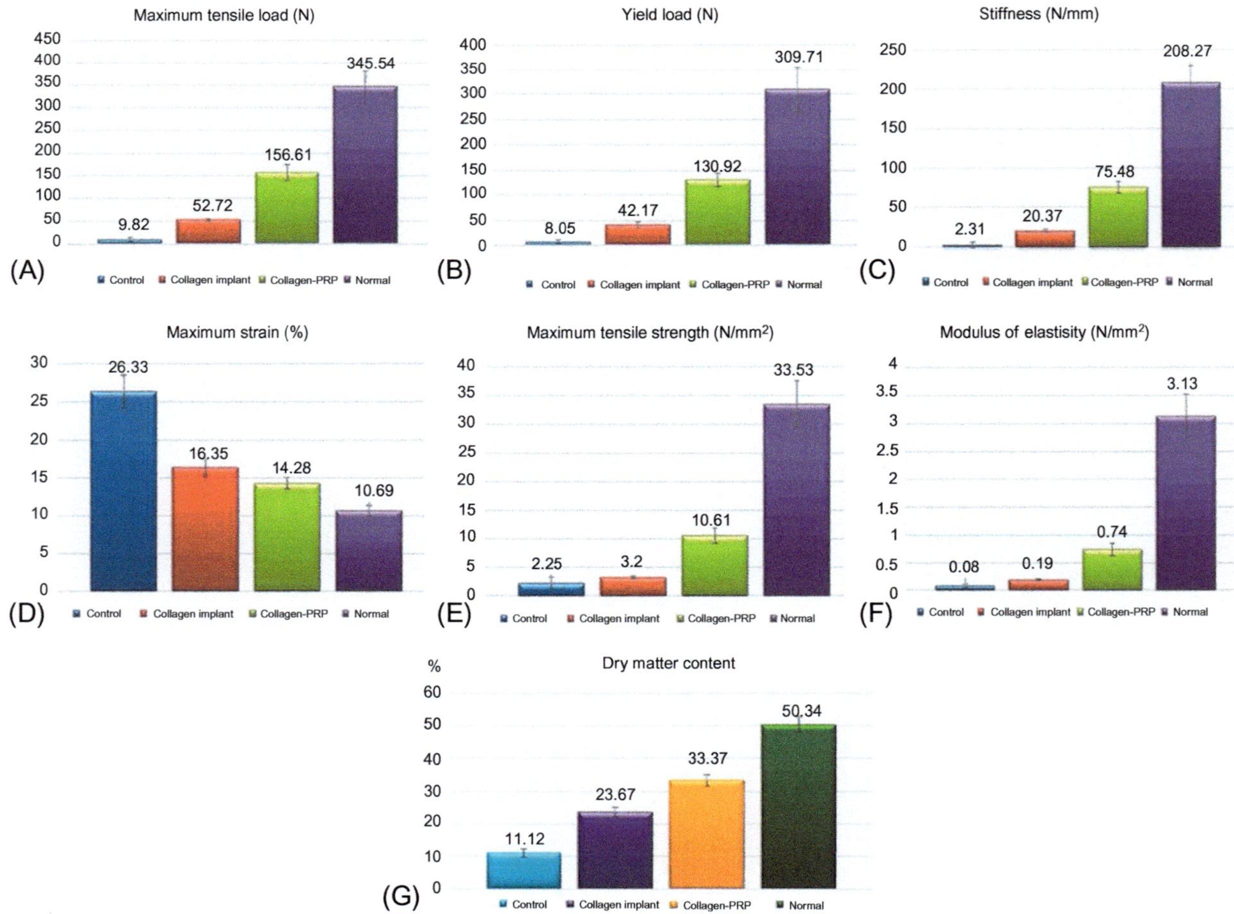


**Fig. 12.3** Biomechanical properties and dry matter content of the injured healing tendons at 60 days postinjury. Number of tendons in each group was 10. The results were expressed as mean ± standard deviation.

From Oryan, A., Moshiri, A., Meimandi-Parizi, A., Maffulli, N., 2015. Role of xenogenous bovine platelet gel embedded within collagen implant on tendon healing: an in vitro and in vivo study. Exp. Biol. Med. (Maywood), 240, 194–210, copyright © 2015 by Experimental Biology and Medicine, reprinted differentiation factor-5 on an interpositional graft treated with HA-lubricin were applied at the repair site of a fully transected FDP. Forty-two days postoperation the degree of adhesion was assessed semiquantitatively, and it was found that the treated tendons had less adhesions compared to the nontreated tendons. Although the failure load values were not significantly different compared to the nontreated tendons, the stiffness of the treated tendons was higher (Zhao et al., 2016). Interestingly, BMSCs labeled with DiI (a fluorescent lipophilic cationic indocarbocyanine dye) could still be tracked 42 days postoperation, implying their support in tendon regeneration also at later time points during healing.

Cells might also be transduced in order to reach more efficiency in wound healing. For example, rat mesenchymal stem cells (MSCs) were transduced with a lentivirus encoding short hairpin ribonucleic acid against TOB1 (Gao et al., 2016). TOB (=Transducer of ERbB2) is a negative regulator of the bone morphogenetic protein/Smad signaling pathway, and TOB1 deficiency should induce osteogenic activity. In the field of tendon regeneration, this functional activity is interesting when tendon-to-bone transitions are under view. Indeed, MSCs transduced in this way showed a higher proliferation in vitro and a better biomechanical outcome in vivo, as shown in a rat rotator cuff repair model. Furthermore, these beneficial findings led the authors to investigate the biochemical pathways of wound healing, resulting in the fact that TOB1 3′ untranslated region was found to be a direct target of miR-218. Overexpression of miR-218 enhanced the tendon-bone healing similarly, implying that expression of TOB1 might be regulated by miR-218 (Gao et al., 2016).

Another very promising transduction approach for ruptured human ACLs proposed a two-step procedure: first, stem cells were isolated from ligaments and $CD34^+$ stem cells were derived via fluorescence-activated cell sorting; second, the viral transduction, either to express VEGF (vascular endothelial growth factor) or its antagonist, FLT-1 (soluble fms-like tyrosine kinase-1). Why VEGF? After ACL reconstruction, a lack of vascularization in the regenerating ligament might lead to degeneration or micro ruptures; as VEGF is one of the most potent angiogenic growth factors, the transduction of ACL-derived stem cells to express VEGF was hypothesized to improve the ligament healing (Takayama et al., 2015). The transduced stem cells were used in an immunodeficient rat model for ACL reconstruction by an autologous flexor *digitorum longus* tendon graft that was wrapped with cell sheets secreting either VEGF—in moderate or high levels—or FLT-1. Load to failure measurements at 4 and 8 weeks postoperation revealed that the application of $CD34^+$ stem cells secreting a moderate level of VEGF led to the highest values, significantly higher than a high level, and also significantly higher than stem cells producing FLT-1 or cell-free grafts (Takayama et al., 2015). Such findings denote the pivotal importance of neoangiogenesis during ligament healing, because a sustained release of VEGF accelerates the healing process.

Clinical trials using cellular therapies for the repair of rotator cuff ruptures were reviewed (Ficklscherer et al., 2016). Therein, PRP application, tenocyte injection, tissue engineered scaffolds with tenocytes, as well as spongiosa beds with stem cells were discussed. Briefly, these cellular approaches have not been resulted in very convincing clinical outcomes so far, and the authors emphasized that they still should be optimized; furthermore, regulatory hurdles were and are quite high so that these strategies might be realized and get common only in a few years from then (Ficklscherer et al., 2016).

## 12.4 Special animal models

### 12.4.1 MRL/MpJ

In order to establish biological benchmarks for tissue engineered tendon constructs, one option is to characterize superior animal models as, for example, a special mouse strain that exhibits remarkable wound healing compared to wild type. Central patellar tendon (PT) injuries have been studied in the Murphy Roths Large (MRL) mouse model because the MRL/MpJ murine strain had been identified as a potential model of successful mammalian healing following injury in general (Lalley et al., 2015). The question why the superior healing that was also found for the PT occurred in this strain was assessed by gene expression at 3, 7, and 14 days postsurgery and by biomechanics 2, 5, and 8 weeks postoperation. While gene expression pattern was similar for both MRL/MpJ strain and wild-type mice, biomechanics were enhanced for the MRL/MpJ strain at 5 and 8 weeks. Moreover, the immune response was found to be lower in the MRL/MpJ strain compared to wild type. In addition, expression of the cyclin-dependent kinase inhibitor, p21, was not different between strains. Lalley and coworkers used this information to get insight into the mechanisms in the tendon healing process in this superior mouse strain. Furthermore, they announced to characterize the complete healing transcriptome to identify mechanisms driving the superior cellular response in the MRL/MpJ strain in order to provide better biological cues (Lalley et al., 2015).

### 12.4.2 Decorin or biglycan deficiency

In order to study small leucine-rich proteoglycans and their specific contribution to healthy tendons, a mouse model lacking decorin or biglycan was used (Gordon et al., 2015). Decorin-null C57BL/6 mice had been developed earlier (Danielson et al., 1997) like biglycan-null mice (Xu et al., 1998); nevertheless, the occurring trend is to use such established models to assess a better understanding in tendon healing and to apply and validate computational models to the experimental findings. The ATs of such mice were investigated structurally and mechanically. As expected, decorin- or biglycan-deficient mice exhibited interior mechanical and structural properties; while failure loads were significantly lower in the biglycan-deficient mice compared to the wild-type, decorin-deficient mice showed a trend toward lower failure loads ($2 \times$ lower failure loads), especially in old mice. As for the dynamic modulus, decorin- and biglycan-deficient mice had significantly lower values, which could be accurately predicted by a computational model (Gordon et al., 2015).

## 12.5 Mobilization

Convincing by the simplicity of the approach and the beneficial outcome was the report on a treatment of a collagenase injury rabbit AT model by a high and well-regulated postoperative mobilization (Imai et al., 2015). It was shown that the such induced Achilles tendinopathy had a better collagen fiber alignment (histologically assessed) and a higher fraction of collagen III at the final point of the experiment (38 days) when one side was actively moved up and down at various time points between days 21 and 38 postinduction, while the contralateral side was not moved and acted as a control. In addition, biomechanical analysis of viscoelastic properties revealed that ATs undergoing an active motion program had lower storage and loss modulus—indicating a better recovery of the biomechanical function compared to the nonactivated group (Imai et al., 2015).

## 12.6 Summary

In this chapter, emerging trends in tendon and ligament repair and tissue engineering are discussed. While many recent publications deal with combinations of proofed beneficially acting agents and cues, also single components are still investigated. However, as a general trend observed, researchers in this field tend to combine cellular therapy with additional drugs, antibacterial agents, or antiadhesives. Very often these active substances are incorporated in highly sophisticated matrices like core-electrospun meshes, fabricated by hollow fibers including bioactive agents in the core. As such, the obvious trend is to generate multiple good effects in the healing process by the synergism of different components.

Another still highly investigated approach is to use allografts and decellularize them before usage. Here, the trend is to optimize the decellularization step, in other words to tune the structural and mechanical properties by different solvents or sterilization processes to achieve the best starting conditions before implantation. Especially the mechanical strength is considered in this regard, as the load-bearing capacity at time zero before implantation has been found to be one of the most important prerequisites for successful functionality after implantation.

## Bibliography


Akdemir, O., Lineaweaver, W.C., Cavusoglu, T., Binboga, E., Uyanikgil, Y., Zhang, F., Pekedis, M., Yagci, T., 2015. Effect of taurine on rat Achilles tendon healing. Connect. Tissue Res. 56, 300–306.

Buchmann, S., Sandmann, G.H., Walz, L., Reichel, T., Beitzel, K., Wexel, G., Tian, W., Battmann, A., Vogt, S., Winter, G., Imhoff, A.B., 2015. Growth factor release by vesicular phospholipid gels: in-vitro results and application for rotator cuff repair in a rat model. BMC Musculoskelet. Disord. 16, 82.

Chen, C.H., Chen, S.H., Shalumon, K.T., Chen, J.P., 2015. Dual functional core-sheath electrospun hyaluronic acid/polycaprolactone nanofibrous membranes embedded with silver nanoparticles for prevention of peritendinous adhesion. Acta Biomater. 26, 225–235.

Chen, B., Ding, J., Zhang, W., Zhou, G., Cao, Y., Liu, W., Wang, B., 2016. Tissue engineering of tendons: a comparison of muscle-derived cells, tenocytes, and dermal fibroblasts as cell sources. Plast. Reconstr. Surg. 137, 536e–544e.

Chiou, G.J., Crowe, C., Mcgoldrick, R., Hui, K., Pham, H., Chang, J., 2015. Optimization of an injectable tendon hydrogel: the effects of platelet-rich plasma and adipose-derived stem cells on tendon healing in vivo. Tissue Eng. A 21, 1579–1586.

Danielson, K.G., Baribault, H., Holmes, D.F., Graham, H., Kadler, K.E., Iozzo, R.V., 1997. Targeted disruption of decorin leads to abnormal collagen fibril morphology and skin fragility. J. Cell Biol. 136, 729–743.

Elliot, D., Giesen, T., 2013. Primary flexor tendon surgery: the search for a perfect result. Hand Clin. 29, 191–206.

Ficklscherer, A., Pietschmann, M.F., Bendiks, M., Rossbach, B.P., Muller, P.E., 2016. Clinical management of rotator cuff tears: current concepts in cell-based therapy strategies. Orthopade 45, 143–148.

Fleming, B.C., Proffen, B.L., Vavken, P., Shalvoy, M.R., Machan, J.T., Murray, M.M., 2015. Increased platelet concentration does not improve functional graft healing in bio-enhanced ACL reconstruction. Knee Surg. Sports Traumatol. Arthrosc. 23, 1161–1170.

Flury, M., 2016. Patch augmentation of the rotator cuff: a reasonable choice or a waste of money? Orthopade 45, 136–142.

Fox, P.M., Farnebo, S., Lindsey, D., Chang, J., Schmitt, T., Chang, J., 2013. Decellularized human tendon-bone grafts for composite flexor tendon reconstruction: a cadaveric model of initial mechanical properties. J. Hand Surg. 38, 2323–2328.

Galatz, L.M., Gerstenfeld, L., Heber-Katz, E., Rodeo, S.A., 2015. Tendon regeneration and scar formation: the concept of scarless healing. J. Orthop. Res. 33, 823–831.

Gao, Y., Zhang, Y., Lu, Y., Wang, Y., Kou, X., Lou, Y., Kang, Y., 2016. TOB1 deficiency enhances the effect of bone marrow-derived mesenchymal stem cells on tendon-bone healing in a rat rotator cuff repair model. Cell. Physiol. Biochem. 38, 319–329.

Gordon, J.A., Freedman, B.R., Zuskov, A., Iozzo, R.V., Birk, D.E., Soslowsky, L.J., 2015. Achilles tendons from decorin- and biglycan-null mouse models have inferior mechanical and structural properties predicted by an image-based empirical damage model. J. Biomech. 48, 2110–2115.

Herbert, A., Jones, G.L., Ingham, E., Fisher, J., 2015. A biomechanical characterisation of acellular porcine super flexor tendons for use in anterior cruciate ligament replacement: investigation into the effects of fat reduction and bioburden reduction bioprocesses. J. Biomech. 48, 22–29.

Imai, K., Ikoma, K., Chen, Q., Zhao, C., An, K.N., Gay, R.E., 2015. Biomechanical and histological effects of augmented soft tissue mobilization therapy on Achilles tendinopathy in a rabbit model. J. Manipulative Physiol. Ther. 38, 112–118.

Lalley, A.L., Dyment, N.A., Kazemi, N., Kenter, K., Gooch, C., Rowe, D.W., Butler, D.L., Shearn, J.T., 2015. Improved biomechanical and biological outcomes in the MRL/MpJ murine strain following a full-length patellar tendon injury. J. Orthop. Res. 33, 1693–1703.

Leong, N.L., Kabir, N., Arshi, A., Nazemi, A., Jiang, J., Wu, B.M., Petrigliano, F.A., McAllister, D.R., 2015a. Use of ultra-high molecular weight polycaprolactone scaffolds for ACL reconstruction. J. Orthop. Res. 34 (5), 828–835.

Leong, N.L., Kabir, N., Arshi, A., Nazemi, A., Wu, B., Petrigliano, F.A., McAllister, D.R., 2015b. Evaluation of polycaprolactone scaffold with basic fibroblast growth factor and fibroblasts in an athymic rat model for anterior cruciate ligament reconstruction. Tissue. Eng. Part A 21, 1859–1868.

Li, H., Jiang, J., Ge, Y., Xu, J., Zhang, P., Zhong, W., Chen, S., 2013. Layer-by-layer hyaluronic acid-chitosan coating promoted new collagen ingrowth into a poly(ethylene terephthalate) artificial ligament in a rabbit medical collateral ligament (MCL) reconstruction model. J. Biomater. Sci. Polym. Ed. 24, 431–446.

Nau, T., Redl, H., Teuschl, A.H., 2015. Comment on: in vivo evaluation of electrospun polycaprolactone graft for anterior cruciate ligament engineering. Tissue Eng Part A. 2015;21 (7–8):1228–1236. Tissue Eng. Part A 21, 2775.

Ning, L.J., Zhang, Y., Chen, X.H., Luo, J.C., Li, X.Q., Yang, Z.M., Qin, T.W., 2012. Preparation and characterization of decellularized tendon slices for tendon tissue engineering. J. Biomed. Mater. Res. Part A 100A, 1448–1456.

Oryan, A., Moshiri, A., Meimandi-Parizi, A., Maffulli, N., 2015. Role of xenogenous bovine platelet gel embedded within collagen implant on tendon healing: an in vitro and in vivo study. Exp. Biol. Med. (Maywood) 240, 194–210.

Ozasa, Y., Amadio, P.C., Thoreson, A.R., An, K.-N., Zhao, C., 2013. The effect of surface modification on gliding ability of decellularized flexor tendon in a canine model in vitro. J. Hand. Surg. Am. 38, 1698–1704.

Pan, J., Liu, G.-M., Ning, L.-J., Zhang, Y., Luo, J.-C., Huang, F.-G., Qin, T.-W., 2014. Rotator cuff repair using a decellularized tendon slices graft: an in vivo study in a rabbit model. Knee Surg. Sports Traumatol. Arthrosc. 23 (5), 1524–1535.

Proffen, B.L., Vavken, P., Haslauer, C.M., Fleming, B.C., Harris, C.E., Machan, J.T., Murray, M.M., 2015. Addition of autologous mesenchymal stem cells to whole blood for bioenhanced ACL repair has no benefit in the porcine model. Am. J. Sports Med. 43, 320–330.

Qu, J., Thoreson, A.R., An, K.N., Amadio, P.C., Zhao, C., 2015. What is the best candidate allograft for ACL reconstruction? An in vitro mechanical and histologic study in a canine model. J. Biomech. 48, 1811–1816.

Rathbone, S., Maffulli, N., Cartmell, S.H., 2012. Most British surgeons would consider using a tissue-engineered anterior cruciate ligament: a questionnaire study. Stem Cells Int. 2012, 303724.

Takayama, K., Kawakami, Y., Mifune, Y., Matsumoto, T., Tang, Y., Cummins, J.H., Greco, N., Kuroda, R., Kurosaka, M., Wang, B., Fu, F.H., Huard, J., 2015. The effect of blocking angiogenesis on anterior cruciate ligament healing following stem cell transplantation. Biomaterials 60, 9–19.

Xu, T., Bianco, P., Fisher, L.W., Longenecker, G., Smith, E., Goldstein, S., Bonadio, J., Boskey, A., Heegaard, A.M., Sommer, B., Satomura, K., Dominguez, P., Zhao, C., Kulkarni, A.B., Robey, P.G., Young, M.F., 1998. Targeted disruption of the biglycan gene leads to an osteoporosis-like phenotype in mice. Nat. Genet. 20, 78–82.

Zhang, P., Han, F., Li, Y., Chen, J., Chen, T., Zhi, Y., Jiang, J., Lin, C., Chen, S., Zhao, P., 2016a. Local delivery of controlled-release simvastatin to improve the biocompatibility of polyethylene terephthalate artificial ligaments for reconstruction of the anterior cruciate ligament. Int. J. Nanomedicine 11, 465–478.

Zhang, X., Ma, Y., Fu, X., Liu, Q., Shao, Z., Dai, L., Pi, Y., Hu, X., Zhang, J., Duan, X., Chen, W., Chen, P., Zhou, C., Ao, Y., 2016b. Runx2-modified adipose-derived stem cells promote tendon graft integration in anterior cruciate ligament reconstruction. Sci. Rep. 6, 19073.

Zhao, C., Ozasa, Y., Shimura, H., Reisdorf, R.L., Thoreson, A.R., Jay, G., Moran, S.L., An, K.N., Amadio, P.C., 2016. Effects of lubricant and autologous bone marrow stromal cell augmentation on immobilized flexor tendon repairs. J. Orthop. Res. 34, 154–160.

# Index

Note: Page numbers followed by *f* indicate figures, and *t* indicate tables.

## U